T0135296

Studies in Big Data

Volume 84

Series Editor

Janusz Kacprzyk, Polish Academy of Sciences, Warsaw, Poland

The series "Studies in Big Data" (SBD) publishes new developments and advances in the various areas of Big Data- quickly and with a high quality. The intent is to cover the theory, research, development, and applications of Big Data, as embedded in the fields of engineering, computer science, physics, economics and life sciences. The books of the series refer to the analysis and understanding of large, complex, and/or distributed data sets generated from recent digital sources coming from sensors or other physical instruments as well as simulations, crowd sourcing, social networks or other internet transactions, such as emails or video click streams and other. The series contains monographs, lecture notes and edited volumes in Big Data spanning the areas of computational intelligence including neural networks, evolutionary computation, soft computing, fuzzy systems, as well as artificial intelligence, data mining, modern statistics and Operations research, as well as self-organizing systems. Of particular value to both the contributors and the readership are the short publication timeframe and the world-wide distribution, which enable both wide and rapid dissemination of research output.

The books of this series are reviewed in a single blind peer review process.

Indexed by zbMATH.

All books published in the series are submitted for consideration in Web of Science.

More information about this series at http://www.springer.com/series/11970

Todor Tagarev · Krassimir T. Atanassov ·
Vyacheslav Kharchenko · Janusz Kacprzyk
Editors

Digital Transformation, Cyber Security and Resilience of Modern Societies

 Springer

Editors
Todor Tagarev
Institute of Information and Communication
Technologies
Bulgarian Academy of Sciences
Sofia, Bulgaria

Vyacheslav Kharchenko
Kharkiv Aviation Institute
National Aerospace University
Kharkiv, Ukraine

Krassimir T. Atanassov
Bioinformatics and Mathematical
Modelling, Institute of Biophysics and
Biomedical Engineering
Bulgarian Academy of Sciences
Sofia, Bulgaria

Janusz Kacprzyk
Systems Research Institute
Polish Academy of Sciences
Warsaw, Poland

ISSN 2197-6503   ISSN 2197-6511   (electronic)
Studies in Big Data
ISBN 978-3-030-65724-6   ISBN 978-3-030-65722-2   (eBook)
https://doi.org/10.1007/978-3-030-65722-2

This Springer imprint is published by the registered company Springer Nature Switzerland AG
The registered company address is: Gewerbestrasse 11, 6330 Cham, Switzerland

Preface

The rapid development and massive incorporation of advanced technologies transform industries, services, conflict, government, health care, leisure and social interaction. In the strive for competitive positioning, developers and users often underestimate safety and security considerations, which in turn provides ample opportunities for exploitation by malicious actors.

The safety and security of modern societies can be enhanced by proper management and implementation of information technologies, monitoring cyberspace, early identification of threats and continuous situational awareness, a good balance of preventive, protective and reactive measures, and through enhanced resilience.

To reflect on these challenges and speed up the development and the implementation of innovative solutions, leading researchers from the Institute of Information and Communication Technologies of the Bulgarian Academy of Sciences (http://iict.bas.bg/), the Bulgarian Defence Institute (https://di.mod.bg/) and the European Software Institute-Central Eastern Europe (ESI-CEE, https://esicenter.bg/) decided to launch a series of scientific conferences.

The series of international conferences "Digital Transformation, Cyber Security and Resilience" (DIGILIENCE) brings together scientists, practitioners and policymakers with the aim to establish the state of the art and future demands in the provision of security and resilience of processes, services and systems that are heavily reliant on information technologies. Of particular interest are studies that examine systems in their interdependencies or place their operation in a human or wider policy contexts, as well as evidence- and data-based studies and presentations of the respective data sets.

The first in the series of DIGILIENCE conferences took place in Sofia from 2 to 4 October 2019. It brought together over 100 participants from 16 countries, ENISA and the European Defence Agency, delivering 55 presentations. Ten of the peer-reviewed papers were presented by young scientists and Ph.D. students.

In the opening session, senior policymakers presented the leadership perspective on the security challenges and research priorities.

The bulk of the peer-reviewed papers was organised in the following sessions:

- Effective, Efficient and Cyber Resilient Organizations and Operations;
- Information Sharing and Situational Awareness;
- Emerging Methods for Cyber Security and Resilience: AI, Blockchain, Fuzzy Sets;
- Policies and Solutions for Industry and Critical Infrastructure Protection;
- Cyber, Hybrid Influence and the Role of Social Networks;
- Human-centric Cyber Security and Resilience.

The final session served to exchange project experience and explore how research-based innovation is contributing, or could contribute further, to the cyber security and the resilience of our economy, the public administration, security and defence forces, the research infrastructure and the society at large.

After the conference, authors of selected papers were invited to provide revised and amended versions for the post-conference publication. The result is presented in this volume, including 32 papers organised in six parts.

The first part looks into how modern advanced IT can make modern organisations effective, efficient and resilient to effects from cyberspace. Individual papers address the challenges to command and control (C2) in multi-domain operations, the use of cloud technologies, the management of relevant standards, and in providing resilience of the research infrastructure. In their paper, Drs. Shalamanov and Penchev study the issue of collaboration in cyber security research networks [1]. Reflecting on a comprehensive study of the requirements to the governance of networked organisations [2], the authors propose a methodology of organisational design of such networks. Another paper, authored by a team from ESI-CEE, presents the novel concept of a serverless computing environment with application to cyber security education and training [3].

The second part looks into the requirements, challenges and solutions allowing the exchange of information on cyber security, leading then to proper situational awareness. Of key concern are the interoperability [4] and the adherence to respective norms. Two papers focus on the exchange of information on the situation in the maritime domain while adhering to international norms, including norms for the protection of privacy. Two other papers look at present models and standards for information exchange and options for their implementation to provide awareness of the status of critical infrastructures. The final two papers in the part examine specific cases of malware, raising thus the awareness of cyber threats and possible ways of identification and protection against such threats.

The third part is dedicated to the cyber security of critical infrastructures and industrial systems [5, 6]. The main challenge in the examination stems from the interdependencies and the potential cascading effects of a cyberattack on one sector throughout other sectors of critical infrastructure. Since a growing number of sectors and domains depend on the proper functioning of cyberspace and are interdependent, the first paper suggests a framework of mapping analysis scenarios allowing then to track the exploration of cyber security of critical infrastructures and services. A contribution by a team from Coventry University examines ways of testing automotive components for cyber security [7]. The remaining three papers suggest solutions, employing respectively a generalised net model [8], suggesting a knowledge management model to manage standards to meet security requirements and implementing an integrated security management into Industry 4.0 enterprise management [9].

The fourth part presents emerging cyber security technologies and solutions in the provision of cyber security, including blockchain technologies, genetic algorithms, generalised net models and advanced image recognition techniques.

The fifth part is dedicated to the examination of the human factor in providing cyber security and resilience, including presentation of theoretical advances [10] and analyses of insider threats, cognitive aspects, requirements to cyber ranges and the effectiveness of advanced learning methods and techniques.

The final part is dedicated to the analysis of hybrid influence, i.e. the influence by a malicious actor using a variety of tools and attack venues. Such influence has been used throughout history [11], while the globalisation and the recent explosion in the implementation of information and communication technologies made the problem even more salient [12]. The section starts with a paper comparing and outlining trends in the evolution of frameworks for analysis of hybrid influence and its impact [13]. The remaining papers look into the ways social networks are used for hybrid influence and suggest some remedies. A paper examining the hybrid influence on the national security system wraps up the section, as well as this volume.

Building on the momentum of the success of the first conference, the organisers announced the consequent steps in the exploration of the digital transformation and its security implications. The second conference, DIGILIENCE 2020, will be hosted by the Bulgarian Naval Academy, located in the city of Varna on the Black Sea coast. The third conference, DIGILIENCE 2021, will take place in the medieval capital of Bulgaria, Veliko Tarnovo, again at the end of September and the first days of October 2021.

We believe that the DIGILIENCE conference series will strengthen the cyber security knowledge and will contribute to building the network of like-minded professionals in the pursuit of a common goal—the prosperity and security of modern, democratic societies.

The editorial article of this volume presents the rationale for the DIGILIENCE series of conferences aiming to establish the state of the art and future demands in the provision of cyber security and resilience of processes, services and systems that rely heavily on information technologies. Thirty-two papers examine ways of providing effectiveness, efficiency and resilience of organisations and operations, solutions for information sharing and cyber situational awareness, cyber security of critical infrastructures and industrial systems, emerging cyber security technologies and solutions, the central role of the human factor in the provision of cyber security and resilience, ways of exercising hybrid influence and the role of social networks.

Sofia, Bulgaria   Todor Tagarev
Sofia, Bulgaria   Krassimir T. Atanassov
Kharkiv, Ukraine   Vyacheslav Kharchenko
Warsaw, Poland   Janusz Kacprzyk

References

1. Shalamanov, V., Penchev, G.: Methodology for organizational design of cyber research networks (2020 in this volume)
2. Tagarev, T.: Towards the design of a collaborative cybersecurity networked organisation: identification and prioritisation of governance needs and objectives. Future Internet **12**(4), 62 (2020) https://doi.org/10.3390/fi12040062
3. Papazov, Y., Sharkov, G., Koykov, G., Todorova, C.: Managing cyber-education environments with serverless computing (2020 in this volume)
4. Rantos, K., Spyros, A., Papanikolaou, A., Kritsas, A., Ilioudis, C., Katos, V.: Interoperability challenges in the cybersecurity information sharing ecosystem. Computers **9**(1), 18 (2020) https://doi.org/10.3390/computers9010018
5. Sklyar, V.: Cyber Security of Safety-Critical Infrastructures: A Case Study for Nuclear Facilities. Inf. Secur. Int. J. **28**(1), 98–107 (2012). http://dx.doi.org/10.11610/isij.2808
6. Ackerman, P.: Industrial cybersecurity: efficiently secure critical infrastructure systems. Packt Publishing, Birmingham, UK (2017)
7. Mahmood, S., Nguyen, H.N., Shaikh, S.A.: Automotive cybersecurity testing: survey of testbeds and methods (2020 in this volume)
8. Zoteva, D., Vassilev, P., Todorova, L., Atanassov, K., Doukovska, L., Tzanov, V.: Generalized net model of cyber-control of the firm's dumpers and crushers (2020 in this volume)
9. Dotsenko, S., Illiashenko, O., Kamenskyi, S., Kharchenko, V.: Embedding an integrated security management system into industry 4.0 enterprise management: cybernetic approach (2020 in this volume)
10. Happa, J.: Cyber resilience using self-discrepancy theory (2020 in this volume)
11. Abrams, S.: Beyond propaganda: Soviet active measures in Putin's Russia. Connections Q. J. **15**(1), 5–31 (2016) https://doi.org/10.11610/Connections.15.1.01
12. Yanakiev, Y. (ed.): Interagency and international cooperation in countering hybrid threats. Inf. Secur. Int. J. **39** (2018). https://doi.org/10.11610/isij.v39
13. Tagarev, T.: Understanding hybrid influence: emerging analysis frameworks (2020 in this volume)

Contents

Cybersecurity of Critical Infrastructures and Industrial Systems

Emerging Cybersecurity Technologies and Solutions

Human-Centric Cyber Security and Resilience

Hybrid Influence and the Role of Social Networks

Effective, Efficient and Cyber Resilient Organizations and Operations

An Insight into Multi-domain Command and Control Systems: Issues and Challenges

Salvador Llopis Sanchez, Joachim Klerx, Vicente González Pedrós, Klaus Mak, and Hans Christian Pilles

Abstract The modern battlefield requires to develop innovative solutions on holistic command and control systems. A digital transformation urges to revisit procedures and techniques to ensure a real-time synchronization of effects in operations across domains. The way joint forces operate depends on how the operational environment evolves in multiple domains, at the same time creating synergies and dependencies among them. Information is traditionally shared among different command levels using domain-specific systems. A paradigm shift emerges in order to achieve a unique view to support decision-making and integrating capabilities and enablers in a multi-domain concept of maneuver. This article analyses the state-of-the-art and the implications of current and future command and control systems, where the pace of technology advancement is changing dramatically both the commander's decision process and his perception about the situation.

Keywords Command and control · Command post · Multi-domain operations · Hybrid warfare · Intelligent agents · Information superiority

S. Llopis Sanchez (✉)
European Defence Agency, Brussels, Belgium
e-mail: info@eda.europa.eu

J. Klerx
AIT—Austrian Institute of Technology GmbH, Vienna, Austria
e-mail: joachim.klerx@ait.ac.at

V. González Pedrós
ISDEFE—Ingeniería de Sistemas para la Defensa de España, Madrid, Spain
e-mail: general@isdefe.es

K. Mak · H. C. Pilles
National Defence Academy, Vienna, Austria
e-mail: klaus.mak@bmlv.gv.at

H. C. Pilles
e-mail: hans.pilles@bmlv.gv.at

3

1 The Need for Reformulating the Concept of Command and Control

New technologies are expected to increase the speed of conflicts dramatically. The military strategy is confronted with the pervasive interconnectedness of sensors and other sources of information. The "internet of the defence things"—name adopted from the civilian concept of internet of things but tailored to the military context—is becoming a phenomenon which will bring radical changes to the digitalization of the battlefield [1]. Solutions are yet to be tested on how the abundance of information is going to leverage situation knowledge—perhaps by the implementation of big data technologies among others. This fast transformation will also affect the commander's situation awareness and the way the information is processed. Certainly, it is not all about technology but implies that the military thinking must take advantage of new means and resources in a rapid battle rhythm.

In this context, the existence of a global information grid (GIG) is key. It comprises a group of networks to connect ground, maritime, air, space and cyberspace assets, so that they can communicate among each other in a joint operation. The joint network must ensure a "secure register" to identify whether an information asset is trustful or not. Non-reliable assets may introduce "noise" in the situation understanding leading to potential wrong interpretations of the situation in a joint operational picture. One of the main challenges affecting the "internet of the defence things" is that the devices can be lost, reverse engineered and brought back to service by an opponent. To avoid that these devices undermine our own vision of the tactical landscape, a "secure register" is needed. In this analysis of information management in operations not only maintaining a traceability is key but to ensure the confidentiality, integrity and availability principles of the information handled by military communication and information systems (CIS). Interoperable secure sensors connected to the GIG, intelligent devices fed with artificial intelligence (AI) algorithms (capable of identifying which meaningful information should be exchanged according with mission conditions), the availability of secure clouds to store information and AI-supported tools to achieve tactical and strategic effectiveness are emerging trends in a full-digitalized scenario. In the realm of future enabling technologies, quantum computing is expected to be a key factor to reduce the complexity of the information processing at the battlefield. Quantum computing may enhance critical areas such as encryption, simulation and decision-making. However, does this conceptual approach on multi-domain operations (MDO[1]) prepare for the whole variety of future threats? How the next generation of command and control systems should deal with hybrid threats?

Achieving an information superiority is gaining predominance among theorists including the use of advanced techniques to get a much faster decision cycle which

---

[1]MDO is a conceptual approach towards achieving a fast and dynamic orchestration of military activities across all the five domains of operations (land, sea, air, space and cyberspace). Automation and enabling technologies play a fundamental role in the future data-rich battlefield with direct implication on how operations are planned and conducted.

could lead to a strategic or an operational advantage. A holistic or comprehensive approach to the new scenarios [2, 3] introduces non-military aspects in the operational environment and the need to implement information exchange practices with civilian actors. Critical infrastructures e.g. energy, play a fundamental role in support of operations. All the above considerations will influence, in one way or the other, any design of future C2 systems [4–6].

2 Results of a Global Screening of Existing and Future C2 Systems

In this section, the authors present an overview on the future trends in C2 systems by taking stock of the state-of-the-art and by applying a methodological mix of analytical processes in various steps described in Fig. 1.

A desktop research generated the baseline of existing C2 systems. Scientific publications about C2 systems are rather focused on theoretical concepts than on operational solutions making difficult to derive a future landscape. A further refinement was made to map existing C2 systems and cluster them into military domains providing a global overview. In a third step, the structured and mapped dataset was used to extract the core features of existing C2 systems considering the information needs at specific levels of decision and responsibility. The resulted C2 systems were customized to their contextual background.

The global overview of existing C2 systems, as reflected in Fig. 2, show that the operational domains such as space and cyberspace are lacking a diversity of solutions—most likely due to their relatively recent inception. One assumption about

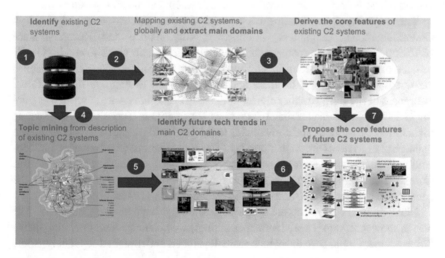

Fig. 1 Overview of the methodological approach. *Source* Own compilation

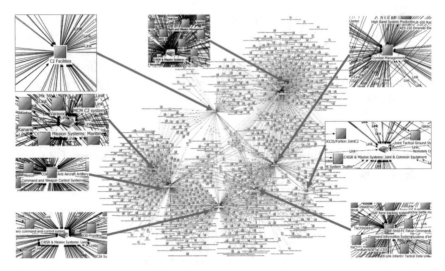

Fig. 2 Overview of existing operational C2 systems, 2019. *Source* Own compilation, multiple sources

the lack of cyber defence systems is that cyber could be integrated as a cross-cutting function to the other domains or it can be attributed to the fact that cyberspace is a new operational domain and there are not complete developments on how to address C2 in cyberspace. This is particularly true when addressing solutions to achieve a cyber situational awareness capability and the need to shift from a physical to a virtual domain perspective. The same reasoning may be applicable to the space domain.

An increased number of C2 systems are related to the physical domains—air, land and maritime—with a more prevalence of C2 systems in the air domain.

The majority of the existing C2 systems are referred to Command, Control, Communications, Computers, Intelligence, Surveillance and Reconnaissance (C4ISR[2]) [7, 8] mission management systems in the air force, followed by C4ISR mission management systems in land and maritime respectively. Most of the systems are not interconnected with others and are developed to improve the situational awareness of the staff. Another cluster of existing C2 solutions are described as joint for C4ISR mission management including C2 facilities. These solutions include higher level C2 systems and components to integrate and support the C4ISR mission management systems in other domains. The most important sub-cluster includes combat management and information systems, as well as specific control systems. Combat management systems are developed to provide an overview about the situation in an operational theatre on different command levels.

Multi-domain C2 systems are at this stage a concept beyond joint C2 systems. The results show that there is a predominance of domain-specific systems operating in five different operational domains which were developed and designed separately. Some

---

[2]Cyber is added to the new C5ISR acronym.

developments are proprietary versus commercial-off-the-shelf. A simple system integration endeavour will most likely stall trying to put all this information together in a situation understanding. Finally, there is a small but specific cluster of C2 systems providing support to joint space operations centres. They have a specialized purpose and are not generally integrated in other combat management systems or C4ISR mission management systems. In general, C2 systems do not address completely the management of intelligent autonomous devices or "internet of defence things" [9, 10]. Hybrid warfare C2 systems are not even a topic in scientific discussions. Yet hybrid is expected to be one of the most important challenges in future conflicts.

The issue is not to have a one-size-fits-all solution but a comprehensive and complete C2 system well adapted to the threat landscape [5]. To meet these challenges is essential to deliver a truly interactive real-time common battlespace picture to get acquainted with full mission capabilities. Surveillance assets can provide tailored situational awareness for each military domain. Joint C4ISR mission management systems could be interpreted as a "system of systems" and could be considered as the starting point towards a multi-domain C2 system. For an agile combat support, a revolution in data management is needed. It is 'all about data' and its availability for decision making. Data science will become paramount to achieve mission goals. Standard architectures and data protocols should drive the design of future C2 systems' developments.

Modeling, simulation and training are an essential part of a generic C2 system. They provide the opportunity to rapidly evaluate any proposed courses of action and ensure a "wargaming" platform where to exercise tactical and strategic planning.

3 A Further Look into Future C2 Systems

The structure and capabilities of future C2 systems will depend on the evolution of a data-rich digitalised battlefield. It is far from being understood the way to conduct MDO and how build a core C2 functionality to support them. For tactical or operational purposes, an optimal C2 system should provide all relevant information as early as possible to whoever needs it. This implies, that future C2 systems will work across all military domains C2, taking whatever relevant information is available to be provided to the end-user of one of the sub-systems. This abstract definition of a multi-domain C2 needs to be operationalized.

Each end-user of the system has different information needs. Too much information can be as harmful as to less information. A flexible common information space, corresponding to the nature of the conflict and customized to the information needs of each single user is essential. Systems must change from the "need to know" to the "need to share" principle. But what are the criteria for "relevant information"?

Decisions can only be made by testing, training and improving processes, in a first step by humans and in a second step by intelligent agents. The second step may enhance data processing speeds. A possible solution to overcome an information overload in C2 systems consists of using intelligent software agents [11–13] to aid

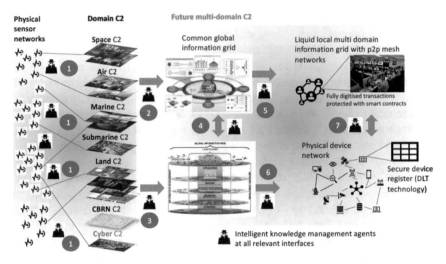

Fig. 3 Information flows of a future multi-domain C2 system. *Source* Own compilation with open-source information

information exchange and planning at different levels of abstraction and capable of making decisions by 'general consensus' at all interfaces. Figure 3 visualizes the functionality of these intelligent software agents at different interfaces.

C2 systems in the future will be more interoperable leading to a full integration into a multi-domain command and control. But what does it mean? The jointness of the battlefield coupled with the emergence of unmanned systems, intelligence sensors and platforms including the internet of things [14] applied to military obliges to rethink about the isolation of domain-specific C2 systems. The synchronization of events and capabilities is no longer a subject for one individual domain or combat system.

AI-supported operation in a data-rich landscape requires standardized information flows. Information sharing is handled by intelligent agents, developed as self-learning agents for data exchange and data analytics. The first type of data agent (number 1) is necessary at sensor level to decide, whether a "pattern" is relevant enough to be transmitted to a domain- specific C2 system. From there two agents decide, whether the information should go to the next step and be transmitted to the global information grid (number 2) or to the local information grid at the battlefield (number 3). Both grids (local and global) establish a data transfer regime (number 4). Information is retrieved from the digital infrastructure for its exploitation e.g. de-centralized command posts linked in a mesh network (number 5). The local grid monitors the network of "internet of defence things" and establish a secure register with a distributed ledger technology (number 6). A data agent (number 7) facilitates the interactions between the physical device network and the mesh network. It is not solved how to implement such an information and knowledge management function and how could the underlying network support quick data transfers and searches from big volumes of data repositories (data lakes) to reach the end-user across multiple layers.

4 Conclusions

The "fog of war" can be interpreted in different ways in the context of C2 systems. There could be fog when a military commander lacks information about a mission. If we referred to weather conditions, fog is a constraint to operations because it makes difficult to operate and observe. A fog may create confusion or uncertainty. The term can also be applied to a situation where there is a misleading or contradictory information hampering decision making.

The military is often confronted with radical changes in the tactics. Along the history there are examples of fundamental transformative changes. For instance, the employment of artillery or the air support to ground forces among many others. Information superiority enables rapid decisions, the abundance of data from different sensors and the pervasive communication of any device will change the "tempo" between action and reaction.

C2 is a science in itself, closer than ever to a data science. Multiple layers of simultaneous standoff in all domains coupled with the emergence of enabling technologies push forward a reformulation of the concept on C2. The same technologies are expected to increase the speed of conflicts dramatically and drive a fundamental change in the character of tomorrow's conflicts. It must be differentiated the command exercised by an authority to the control and how both—command and control are interrelated in a C2 system facilitating the planning and conduct of operations. The system displays the required information at the right time with tools and services used by a military staff in the analysis of the course of actions. The situation of the assigned forces, the logistic planning or the communications with subordinate units are visualised. The increase of data volumes coming from a myriad of sensors and intelligent devices in the mission may cause a disruption of future activities if it is not properly managed. The advent of disruptive technologies like AI may introduce an additional layer of complexity in the system. The OODA loop (observe, orient, decide and act) could be challenged because decisions will be much faster. The future may lead to a fundamental change about which level of automation could be allowed (mandate) and which tasks can be considered as routine and delegated to machines.

In the current operational planning process, all the mission analysis is reported to a single authority at three levels of command: strategic, operational and tactical [15, 16]. What if (because of the rapid technological pace and disruptiveness) the single authority—military commander—and the three command levels are subject to a transformation, Is this a revolution in human or cognitive affairs? Are the essentials of the hierarchical chain on command no longer valid in the future? Answers to these questions is what MDO is trying to elucidate benefiting the modernisation of the art of planning and conducting operations.

The analytical results and expert considerations described in the article show that many challenges are to be addressed in the design of future C2 systems embracing a multi-domain C2 concept [17]. Some of the challenges are related to data and knowledge management but also on the speed and unity of actions. Only with an "integrated all-domain C2 concept" is possible to collect sensor data from different domains and integrate them into a holistic view of the situation. C2 systems are inherently dependent on CIS. MDO will require an unprecedented network interconnectivity including radio spectrum needs. CIS are expected to facilitate mobility and support flexible network configurations in order to keep pace with the given mission. One of the most complex aspects for the future C2 systems is to keep a balance between an unprecedented interconnectivity and the information sharing constraints to cross different security boundaries. In terms of sustainment of operations, the focus would be on optimising software developments and select those innovations that allow an efficient troubleshoot, diagnose and fix of software problems. MDO will constitute a significant paradigm shift in the years to come.

Disclaimer The contents reported in the paper reflect the opinions of the authors and do not necessarily reflect the opinions of the respective agency/institutions. It does not represent the opinions or policies of the European Defence Agency or the European Union and is designed to provide an independent position.

References

1. Kott, A.: Ground Warfare in 2050: How It Might Look. US Army Research Laboratory, Office of the Director, ARL, ARL-TN-0901, Aug 2018
2. Klerx, J.: An intelligent screening agent to scan the internet for weak signals of emerging policy issues (ISA). In: Proceedings of 8th International Conference on Politics and Information Systems, Technologies and Applications: PISTA 2010 in the context of the 4th International Multi-Conference on Society, Cybernetics and Informatics: IMSCI 2010, June 29–July 2, Orlando (2010)
3. Klerx, J.: A political agent for multi-agent simulation of spatial planning policy. In: Carrasquero, J., Welsch, F., Oropeza, A. (eds.), Proceedings PISTA 2004, Volume II: Informatics, Government, Ethical Voting and Political Parties, July 21–25, Orlando, pp. 73–78 (2004)
4. Exploring command and control in the information age. Seminar Command and Control Centre of Excellence (C2COE), 4–6 March 2014, Tallinn, Estonia
5. Joint concept note (2/17): future of command and control. Ministry of Defence, DoD UK (2017)
6. New command and control challenges. Seminar, Command and Control Centre of Excellence (C2COE), 19–21 March 2013, Bratislava, Slovakia
7. Hoeben, B.A., Kainikara, S.: 5th Generation Air C2 and ISR: Exploring new concepts for Air Command & Control and Intelligence, Surveillance & Reconnaissance related to F-35 employment in the RAAF and RNLAF. Royal Netherlands Air Force, International Fellowship Paper (2017)
8. Madden, D.: Advanced imagery analysis supports GEOINT success. Overwatch Geospatial Solutions, Sterling, VA (2013). Available from: https://eijournal.com/print/articles/advanced-imagery-analysis-supports-geoint-success

9. Russell, S., Abdelzaher T.: The internet of battlefield things: the next generation of command, control, communications and intelligence (C3I) decision-making. In: MILCOM 2018—2018 IEEE Military Communications Conference (MILCOM), 29–31 Oct, Los Angeles, CA, USA (2018)
10. Varghese, V., Desai, S.S., Nene, M.J.: Decision making in the battlefield-of-things. Wireless Pers. Commun. **106**, 423–438 (2019). https://doi.org/10.1007/s11277-019-06170-y
11. De Lucia, M.J., Newcomb, A., Kott, A.: Features and operation of an autonomous agent for cyber defense. U.S. Army Research Laboratory, Maryland, USA, 301-394-0798 (2019). Preprint version of the paper published as: De Lucia, M., Newcomb, A., Kott, A.: Features and operations of an autonomous agent for cyber defense. CSIAC J. **7**(1), 6–13 (2019)
12. Kott, A.: Intelligent autonomous agents are key to cyber defense of the future army networks. Cyber Defense Rev. **3**(3) (Fall 2018), 57–70 (2018). Published by Army Cyber Institute. Available from: https://www.jstor.org/stable/10.2307/26554997
13. Theron, P., Kott, A.: Towards an active, autonomous and intelligent, cyber defense of military systems: the NATO AICA, Reference Architecture. In: 2018 International Conference on Military Communications and Information Systems (ICMCIS), 22–23 May, Warsaw, Poland (2018)
14. Kott, A., Alberts, D.S.: How do you command an army of intelligent things? Computer **50**(12), 96–100 (2017). https://doi.org/10.1109/MC.2017.4451205
15. Clemente, F., Gray, S.: The Future of the Command Post: Part 1. A NATO Command and Control Centre of Excellence Study, NATO C2COE (2018)
16. Van der Veer, J.: The Future of the Command Post: Part 2. A NATO Command and Control Centre of Excellence Study, NATO C2COE (2018)
17. Gouré, D.: A new joint doctrine for an era of multi-domain operations. Lexington Institute, Real Clear Defence, May 24, 2019. Available from: https://www.realcleardefense.com/articles/2019/05/24/a_new_joint_doctrine_for_an_era_of_multi-domain_operations_114450.html

Cloud Technologies for Building a System of Data Centres for Defence and Security

Rosen Iliev and Kristina Ignatova

Abstract This chapter provides analysis of cloud technologies as a current trend in the development of IT infrastructure, to focus on their main characteristics, their security levels and the increased requirements they must meet when used in defence and security. The main standards and requirements that modern data centres have to meet to ensure a high level of availability of the IT services provided are considered. Specific requirements have been formulated for building a sustainable system of modern data centres for defence and security needs, and attention has been paid to data protection when using cloud technologies. A solution is proposed for implementing cloud technologies, along with an approach for building an integrated system of data centre for the needs of defence and security in organizing collaborative work between officials within the organization.

Keywords Cloud technologies · Data centers · Defense · Cloud computing · Security

1 Introduction

Cloud technologies have entered very fast into all spheres of the modern world. By using them, the hardware is reduced and the reliability of the information services is increased, the development of information infrastructures of different organizations is optimized. A major advantage of cloud computing is the ability to allocate costs and resources between users, as well as access to information services for them regardless of where they are located. The important thing to achieve this is by providing available network connectivity.

R. Iliev · K. Ignatova (✉)
Bulgarian Defence Institute, Sofia, Bulgaria
e-mail: k.ignatova@di.mod.bg
URL: https://www.mod.bg/bg/EXT/InstitutOtbrana/index.htm

R. Iliev
e-mail: r.iliev@di.mod.bg
URL: https://www.mod.bg/bg/EXT/InstitutOtbrana/index.htm

© The Author(s), under exclusive license to Springer Nature Switzerland AG 2021
T. Tagarev et al. (eds.), *Digital Transformation, Cyber Security and Resilience of Modern Societies*, Studies in Big Data 84,
https://doi.org/10.1007/978-3-030-65722-2_2

Cloud computing is a technology which provide computing services to users, and these services (software and information) are typically in the form of WEB. It also provides access to the hardware and system resources of the data centres that offer these services. In this way of organization and operating computer systems, the provided computer resources (processor time and computer memory) can be optimally distributed and dynamically enhanced through virtualization technologies.

Cloud computing includes collaboration, agility, scaling and availability and provides opportunities for cost savings through optimization and efficient resource management. This is also the solution for outsourced software, platforms and infrastructure. This technology allows ubiquitous access to cloud computing configurable resources, such as networks, servers, storage arrays, applications and data centre services.

According to their visibility clouds are divided into private, public, hybrid and community [5].

Each of them has different security features and levels, which determine their applicability.

The public cloud is an IT infrastructure, platform, or service that is publicly available on the Internet and maybe free of charge or active against payment. This cloud is deployed in the field of CRM, communications, offices, and so on. In the field of defence and security, it is appropriate to support information systems for work with outside clients and organizations. An example of such type of cloud is a Gmail platform.

The Private Cloud is a cloud infrastructure that combines the IT services of a company or organization and is not accessible to external organizations and individuals. It is managed by a private (internal) data centre organization and thus the company assures the integrity and security of its data. The private cloud has a higher price and security level than the public cloud. It is protected by a firewall and can only be accessed through an internal secure network. Processes, services, and information are managed within the organization itself, so there are no additional safeguards, legal requirements or network constraints in the cloud that exist in public cloud structures. Cloud service providers and customers build optimized and controlled infrastructure with increased security by eliminating network access for external users.

Private cloud is preferred option for building internal information infrastructure in organizations related to defence and security. An example of a private cloud is the IT infrastructure of a bank.

The Hybrid Cloud is a combination between private cloud and public cloud, and aims to reduce the cost of performing different functions within the same organization by increasing the flexibility of the infrastructure, going beyond the corporate physical data centres. An example of a cloud organization is a bank that manages data and IT systems in-house, but uses a public cloud backup to store the backup encrypted copy of the data. It is precisely this type of cloud that would be very suitable for use in the defence and security sphere, as it offers higher security for the data it stores.

The Community cloud is an infrastructure shared by several organizations that form a community that shares close interests such as: security, terms of use, compatibility requirements and more. This cloud type offers a higher degree of privacy,

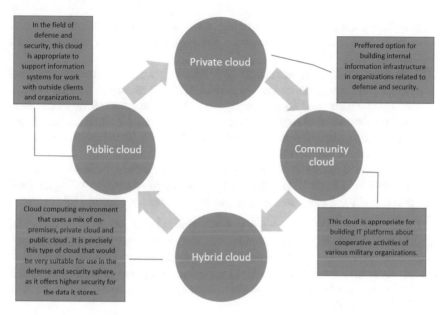

In the field of defense and security, this cloud is appropriate to support information systems for work with outside clients and organizations.

Private cloud

Preffered option for building internal information infrastructure in organizations related to defense and security.

Public cloud

Community cloud

Cloud computing environment that uses a mix of on-premises, private cloud and public cloud . It is precisely this type of cloud that would be very suitable for use in the defense and security sphere, as it offers higher security for the data it stores.

Hybrid cloud

This cloud is appropriate for building IT platforms about cooperative activities of various military organizations.

Fig. 1 Cloud types according their feasibility in defence and security

security, and compatibility policy. It is appropriate for building IT platforms for cooperative activities of various military organizations. An example of such a cloud is the Google Gov Cloud project.

Figure 1 shows the four main types of clouds according to their visibility, focusing on their most important action in terms of their use in defence and security.

One of the first and most known cloud provider companies are Amazon (Amazon Web Services), Google (Google AppEngine), VMWare, Microsoft (Microsoft Azure), Apple (Apple iCloud), IBM, Citrix Systems, Oracle in Bulgaria—Cloud.bg and others [10].

The types of cloud services are: Software-as-a-Service (SaaS), Platform-as-a-Service (PaaS) and Infrastructure-as-a-a service (IaaS) [17]:

Software as a Service (SaaS)—provide easy access to various applications and services through Internet connection. It is not necessary to install a special software for this. This saves money and pays only for what they are used for. The software package is automatically upgraded so the end user to be always up to date. One of the biggest advantages of SaaS is that the user has access to their work and services from anywhere in the world where an internet connection is available. Examples of SaaS are Gmail from Google, Microsoft Office, and Cornerstone and more.

Platform as a Service (PaaS)—is a set of services that provide an environment for developing, implementing, managing and integrating applications in the cloud [9]. The service is aimed at software developers, enabling the development of software solutions that can be extended. PaaS providers have many web-based tools for reducing development time and reducing costs. There are many examples of PaaS today, such as the Google Engine App, Amazon EC2, and Microsoft's Azure platform.

Infrastructure as a Service (IaaS)—Provides computing and storage infrastructure in a centralized, transparent environment [7]. Infrastructure includes storage, servers, bandwidth and network equipment, which includes software that monitors infrastructure usage and allows the user to pay only for what they use. Some of the most popular examples of IaaS include Go Grid's ServePath and Amazon's Elastic Compute Cloud (EC2).

Cloud computing is closely related to the term **virtualization**. It allows abstraction and isolation of lower level functionality. This enables portability of higher-level features and the sharing and/or aggregation of physical resources [23].

Cloud computers are connected to virtualization as many aspects of the computer are virtualized including software, memory, storage, data and networks. It consolidates the servers. It also allows more users to be supported for hardware applications, starting faster [1]. Cloud core is such an application, computing technologies make it possible to have key features for multifaceted work, massive scalability, fast elasticity and measured cloud computing.

2   Data Centres Based on Cloud Technologies. Standards

Modern data centres are engineering and technical facilities designed to provide information resources with a high level of security and affordability. This places higher demands on the performance and reliability of their systems for processing and storing information, the communications environment, power supply, air conditioning, fire protection, security and other support systems [4].

With the development of cloud technologies and the need to consolidate services, modern data centres are becoming increasingly important and emerging as cores for processing, storing and providing information. New standards are introduced for their construction, which determine the level of availability (reliability) of the centre, its cabling, the types of spaces (premises), the permissible operating environment conditions, as well as the requirements for disasters and fire. These standards are developed by various institutes, associations and companies, such as Uptime Institute Tier Certifications, TIA 942, EU Code of Conduct in Data Centres, ISO/IEC 24764, BDS EN 50173-5, DCA, ASHRAE Standard, NFPA Standard, and more.

The Uptime Institute Tier Certifications [14] defines four levels of accreditation related to the design, construction and stability of the data centre (Table 1). Service availability levels (centre operation) are determined from the allowable interruption of the data centre for one year. For example, at tier 1, the allowable break is 28 h per year, and at the highest level (tier 4), only 26 min per year [15]. Depending on the requirements and availability needs, it is determined at which level the data centre is built.

Table 1 Data centre levels

Level	Availability requirements
Tier 1	Ensure 99.671% availability
Tier 2	Ensure 99.741% availability
Tier 3	Ensure 99.982% availability
Tier 4	Ensure 99.995% availability

The Telecommunications Industries Association (TIA) [19] has developed a data centre standard called TIA 942. It defines the factors leading to the creation of a sustainable data centre in terms of architecture, electricity, mechanics, telecommunication components.

The Bulgarian State Standard (ENS) EN 50173-5 is based on the European standard EN 50173-5 and prescribes general wiring requirements for the provision of IT services as well as the connection of large quantities of equipment within the limited space in the data centres.

The International Organization for Standardization (ISO) also offers a standard for cable systems in data centres. ISO/IEC 24764 [20] includes requirements for wiring systems, wires and hardware devices.

The European Code of Conduct in Data Centres [11] addresses the environmental and economic impacts of energy consumption in centres.

3 Requirements for Modern Data Centres

Based on the standards listed above, modern data centres must to correspond to different requirements, such as location selection, for example, not to be located in danger areas with potential for natural disasters but to be close located to several energy sources near fire department and the police. It is also necessary to avoid places that limit the ability to supply more bulky facilities, to allow for the future expansion of the data centre, etc.

There are special criteria that must be met by data centre premises. For example: no external windows that are a security risk; floor/wall material to be antistatic; the lighting must be at least 500 lx in horizontal line and 200 lx in vertical line; the minimum freight distribution per floor should be 7.2 kPa, the recommended minimum being 12 kPa (according to TIA-942 standard). The size of the premises must depend on the specific requirements for the type of equipment, with the possibility of future expansion, etc.

There are also many requirements for the physical environment: maintaining room temperature need to be from 20 to 25 °C and humidity from 40 to 55% (according to ASHRAE Thermal Guidelines for Data Processing Environments); it is advisable to use a double floor for better cooling control. The cables used must meet the standards ANSI/TIA/EIA-568-C.2 and ANSI/TIA/EIA-568-C.3, etc.].

Last but not least, the building needs to have a system for protection against electromagnetic interference, using shielding through the steel structure of the building or by building a Faraday cage of sensitive equipment. The premises are also required to be provided with a physical security system (perimeter security system, intruder alarm system, access control system, CCTV system).

General requirements may be imposed on technical devices such as: to be resistant to changes and short-term downtime and interruptions to mains voltage, electrostatic discharges, interference from electromagnetic fields, etc. according to BDS EN 50130-4:2011, etc. The power supply must be basic, redundant and uninterrupted with grounding and lightning protection; protection against static electricity, as well as with a control and monitoring system. Data centres are also required to have systems for sustainability such as ventilation and air conditioning, a water detection system, fire alarms, fire extinguishers and more.

Data centres must be equipped with uninterruptible power supplies to ensure that the equipment operates in the event of a power failure for a minimum of 30 min, to have a remote monitoring and control port and replacing the batteries should be carried through work.

The data centre information and communication environment must also meet various requirements for cabinets used, server configurations, data management and storage, cloud computing and virtualization platform, backup, data protection systems and more.

Most modern data centres are stationary or they are built in a specific area, without the ability to change their location. They can be also mobile (removable, container type, etc.) data centres, which are usually located in a container and can be moved from one place to another, which is especially useful for defence needs, for example in the deployment of military formations. In this case, the entire mobile data infrastructure or communication and support equipment must be integrated within the space of a standard transportable container. The requirements for this type of data centre are similar to those for stationary use, but there are additional requirements related to the frequently changing operating conditions of IT equipment.

The mobile data centres are subject to quite a few specific requirements, some of them provide the possibility of easy transportation to any place by rail, road or sea; rapid deployment through standard technology blocks (by All-In-One concept). They must also allow easy installation and dismantling of reusable equipment; provide scalability, system modularity and variability (more accessible types); mechanical resistance to damage and abuse (protection against vandalism). In addition, portable centres must be able to operate in a wide range of outdoor climatic conditions—indoor temperature—from 18 °C to a maximum of 28 °C and humidity not exceeding 55%. They need to be able to provide service, access control and monitoring as well as to have backup energy equipment. Last but not least, mobile data centres need to have a smoke capture, fire extinguishing and condensation management capabilities. Requirements include bypass maintenance and emergency shutdown, and the average cabinet power available needs to be tailored to the designed server support equipment that provides the necessary IT services.

Each mobile data centre container must be a functional standalone unit and be capable of multiplication or expansion by attaching other containers to it.

Good practices for building modern data centres are outlined in [2, 8, 12, 18]. They include the design of data centre and the architecture of the support systems and its overall structure. However, for the needs of defence and security, the integration of the individual centres into a data centre system needs to be done in such a way as to provide the necessary redundancy at the level of "cloud infrastructure." To achieve this the following approach is proposed:

Identification of the specific services subject to virtualization in the cloud computing of the defence and security authorities (MoD, Ministry of Interior, etc.) that are used by the respective users and the determination of the single point of failure and their need for redundancy;

Summarizing data on the average number of concurrent users who use certain services;

Gather data on the load of hardware resources during work and the amount of information for storing each service;

Determining the capacity of the hardware resources needed to ensure the performance of a given service in the virtualized cloud computing environment, booking the work of the same;

The size of data centres for defence and security needs depends on the volume, degree of availability and availability of the services offered, and also the number of users of these resources.

In the past, the connection of the individual centres was only at the level of data exchange, while today they are integrated at the level of resources and services. Such integration provides the Cloud Infrastructure described above, which allows multiple data centres to work together to provide the necessary information to users, regardless the location of the data in the cloud. When designing and building such data centre system, the individual centres need to be the same, with a single hardware and software platform allowing increased compatibility, interchangeability, easier training of administrative, operational staff and etc.

When building up modern data centres, the following three important requirements must be met:

The continuity of information resources and access to them;

High consolidation, scalability and flexibility, based on cloud technologies and virtualization;

Reliable protection and storage of information.

Behind the fulfilment of these conditions, there are many communication-information and support systems that provide the necessary communication, climate, physical and information protection, redundancy and overall—high reliability.

4   Implementation of Cloud Technologies in Defence and Security

By developing of technologies worldwide, the modernization of the communication infrastructure is underway and, in many places, transmission is being replaced by optical, which provides higher capacity and reliability of the information transmitted. The use of modern information systems and software products has imposed high demands on hardware resources such as fast action, capacity, energy consumption, management and more.

The possibilities for building a system of data centres for defence needs are related to solving a number of issues regarding: the number of individual centres, their size, capacity, integration between them, the platforms used, the level of service provision (Tier 1–4 according to TIA-942 standard), information security, management and more. After their construction, the existing information systems and software products should be consolidated and placed (if is possible) on the virtualized platform of the data centre system. Obligatory conditions for cloud infrastructure include the provision of storage space and processing resources for the platform. Officials should be provided with single access points for a large number of services with the same software tools they used, regardless of where they were on the network at any given time. The information must be protected of unauthorized access or interference with systems, programs and services should be kept to a minimum.

During the integration of data centres, the necessary redundancy at cloud infrastructure level must be ensured. The services in cloud computing that need to be virtualized have to be define in the Ministry of Defence's and also need to be identified the degree of criticality and the need for redundancy determined. This is done by evaluating the data in terms of how many users are simultaneously using certain services, how important they are, what the load of hardware resources is, and also what is the required amount of data to be stored (for each service).

The design and construction of the individual data centres must be of the same type, with a single hardware and software platform, allowing for better interoperability, interchangeability when needed, easier administrative and operational staff training. When building cloud infrastructure, the performance of existing systems must not be disrupted. Migration from the old to the new platform needs to be smooth, with the least possible refusal of end-user services and retention of the data at its disposal. Training of staff involved in the administration of the system and services should be provided.

Implementing cloud-based solutions to defence and security will optimize the cost of building high- performance information centres, provide easy access to a wide range of services, enable integrated IT solutions, and make more effective use of information resources. One proposed solution for organizing collaborative work between officials based on the use of cloud technologies in data centres is through:

> Creating highly organized virtual platforms to provide cloud resources of the three types—"Software as a Service" (SaaS), "Infrastructure as a Service" (IaaS), and "Platform as a Service" (PaaS);

Creating online user applications available through a web browser to meet the computing needs of users while data and software are stored on servers from the centralized infrastructure;

Reserve critical information in outreach Data Recovery Centres;

Building IC-environments to integrate multiple services into a single point of access by the user, in accordance with the imposed security policies;

Establishment of *Groupware systems* to ensure the high integration of physically remote users involved in collaborative processes and to provide a face-to-face exchange of information between them;

Providing users with integrated services such as knowledge sharing, group calendars and schedules, document processing in a group environment, organizing workflows, etc.

Integrating information and communication resources into a single entity and building *unified communications* to integrate audio, video and data, as well as user services such as *instant messaging*, presence information telephony (including IP telephony), video & audio conferencing, data sharing, call control, voice recognition, integrated voice mail, e-mail, SMS and Fax), etc.

Building intelligent management systems based on Soft-Collision and Decision Support Systems by the Managing Authorities. These systems are particularly applicable in the presence of inaccurate and incomplete information on the problems solved, in the necessity to make quick decisions in the absence of sufficiently trained specialists, as they successfully "mimic" the processes of reasoning and decision- making by man;

Optimizing the energy saving infrastructure (Green Energy);

Provide enough space for data storage and processor resource in the cloud, as well as the ability to create copies of the documents produced on the local computer;

Providing "continuity" at the time and place of activity of a user on the web, and this "mobility" of access to documents and services must provide high reliability and security;

Organizing effective information protection and minimizing the probability of unauthorized access or interference with systems, programs and services.

For defence and security purposes, building data centres by applying cloud technologies is particularly important. The cloud computing will improve the availability and accessibility of services (from anywhere on the network after authorization), making them more reliable and faster (due to large computing capabilities) and providing more redundancy to information resources (low "denial of service"). Cloud computing, distributed over interconnected data centres, will enable self-service (automatic redistribution of resources to achieve optimal use), large capacity (disk space), and flexible administration, monitoring and management of services. It will help ensure faster disaster recovery, as the information is stored in more than one data centre in the cloud.

In recent years, a prototype of a virtualized server platform has been developed at the Defence Institute at the Ministry of Defence to provide cloud computing such as

PaaS and SaaS platforms. A platform for deploying multiple virtual servers with built-in specialized software (E-mail server, DV server, WES server, GIS server, servers for Domain Controllers, Groupware, etc.). The information services provided by this cloud infrastructure are organized in a common information WEB environment to work by creating a public information portal for the exchange of information and the ability to use e-mail, document sharing), instant messaging, electronic signature, organization of information and document flow between officials and others. In-depth analysis of virtualization systems in cloud architecture design is done in [6] and [3] is presented A model for defining the software system state.

5   Benefits of Applying Cloud Technologies for Defence and Security

Cloud technologies are exposed to continuous threats and attacks that may adversely affect organizational data (missions, functions, image or reputation), organizational assets, individuals or other organizations. Their protection is subject to in-depth research to detect and eliminate threats [21, 22].

Defence and security data centres are also the subject of malicious attacks using both known and unknown vulnerabilities to compromise the privacy, integrity, or availability of the information being processed, stored, or provided.

When building security and defence data centres, attention needs to be paid to the risks associated with project management, investment, legal responsibility, safety, security of information, etc. Risk management must be a cyclical process, consisting of a set of coordinated activities for supervision and control of individual risks. This process is aimed at enhancing strategic and tactical security and involves the implementation of a risk mitigation strategy and the use of control techniques and procedures for continuous lifecycle monitoring of IT systems and data centres as a whole [16].

Building a cloud-based data centre system at the Ministry of Defence will unite the hardware information resources and allow flexibility and scalability in their use. This will increase the quantity and quality of services because virtualization involves fewer physical devices (servers, disk arrays, etc.). Cloud computing will improve the availability and accessibility of services (from anywhere on the network after authorization), making them more reliable, faster (due to the large computing capacity) and providing more redundancy of information resources (low "opt-out"). The cloud environment will allow for self-service (automatic reallocation of resources to optimize their use), large capacity (disk space) for use by users, as well as easy and flexible administration, monitoring and management of resources. This will allow personalization of services (as required by individual groups or individual users) and will improve control. Since the cloud stores information in more than one data centre, faster disaster recovery will be guaranteed.

6   Conclusion

Cloud computing and virtualization are the most modern areas at the moment in which many resources and know-how are invested. It has emerged as the main platform for enabling the relative independence of software solutions from hardware, upgrading and multiplying information resources, centralized management and decentralized use of services, integration of various security solutions.

For example, the European Union will fund a € 15.7 million project, called the Vision Cloud (Virtualized Storage Services Foundation for Future Internet), to explore new cloud computing technologies, including data mobility and access control [13].

Building a system of modern defence and security data centres will improve not only the integration of information resources and services, but will also increase the reliability and reduce the denial of service.

Modern virtualization and cloud infrastructure technologies create an effective storage, accessibility, and computing environment to meet the ever-growing challenges of today's information world. Building a sustainable system of modern data centres is a serious solution not only for the needs of defence and security but also outside them.

References

1. Armbrust, M.: Above the Clouds: A Berkeley View of Cloud Computing. Reliable Adaptive Distributed Systems Laboratory, UC Berkeley (2009)
2. Bell, M.: Use best practices to design data centre facilities. https://www.it.northwestern.edu/bin/docs/DesignBestPractices_127434.pdf. Accessed on 15 Aug 2019
3. Bozhilova, M.: A model for defining the software system state. In: Proceedings of Sixth International Scientific Conference "Hemus-2012," pp. II-51–II-56 (2012)
4. Cisco Data Center: https://www.cisco.com/c/en_uk/solutions/data-centre-virtualization/what-is-a-data-centre.html. Accessed on 9 May 2019
5. Cloud Technologies: https://www.icn.bg/bg/blog/novini-ot-icn-bg/oblachni-tehnologichni-modeli-za-predo. Accessed on 9 May 2019
6. Genchev, A.D.: Analysis of virtualization systems in cloud architecture design. In: Proceedings of Conference "Military Technologies and Systems," MT&S-2013, ISSN 2367–5942, pp. II-25–II-40, Sofia, Defence Institute
7. Creeger, M.: Cloud computing: an overview. Queue 7(5) (2009). https://doi.org/10.1145/1551644.1554608
8. Greenberg, S., Mills, E., Tschudi, B., Rumsey, P.: Best practices for data centres: lessons learned from benchmarking 22 data centers. In: 2006 ACEEE Summer Study on Energy Efficiency in Building, pp. 3-76–3-87 (2006)
9. IBM: IBM Cloud Computing: PaaS—United States. https://www.ibm.com/cloud-computing/us/en/paas.html
10. https://www.softwaretestinghelp.com/cloud-computing-service-providers/. Accessed on 29 Nov 2019
11. Acton, M., Bertoldi, P., Booth, J., Newcombe, L., Rouyer, A., Tozer, L.: Best Practice Guidelines for the EU Code of Conduct on Data Centre Energy Efficiency, Version 9.1.0, JRC Technical

Reports, European Commission (2018). https://publications.jrc.ec.europa.eu/repository/bitstr eam/JRC110666/kjna29103enn.pdf. Accessed on 30 Aug 2019

12. https://www.colocationamerica.com/blog/data-centre-design-best-practices. Accessed on 15 Aug 2019
13. https://cordis.europa.eu/project/rcn/95928/factsheet/en. Accessed on 20 June 2019
14. https://uptimeinstitute.com. Accessed on 20 Aug 2019
15. https://networkalliance.com/datacentre-part-2-whats-a-tier. Accessed on 20 Aug 2019
16. https://ws680.nist.gov/publication/get_pdf.cfm?pub_id=919234. Accessed on 9 May 2019
17. Carlin, S., Curran, K.: Cloud computing technologies. Int. J. Cloud Comput. Serv. Sci. **1**(2), 59–65 (2012)
18. https://www.techxact.com/blog-what-are-top-ten-data-centre-best-practices. Accessed on 15 Aug 2019
19. https://www.tiaonline.org/what-we-do/standards. Accessed on 10 May 2019
20. https://www.iso.org/home.html. Accessed on 10 May 2019
21. Mahlianov, D., Stoianov, N.: The security of the Internet of things. In: Proceedings of Eighth International Scientific Conference "Hemus-2016", ISSN 1312-2916, pp. III-217–III-225
22. Tselkov V., Stoianov, N.: A Formal model of cryptographic systems for protection of informa- tion. In: Computer Systems and Networks, Bulgarian Cryptography Days—BulCrypt 2012, Proceedings, Sofia, pp. 15–29 (2012). ISBN 978-954-2946-22-9
23. Vouk, M.A.: Cloud computing—issues, research and implementations. In: Information Technology Interfaces, ITI 2008. 30th International Conference, pp. 31–40 (2008)

Methodology for Organisational Design of Cyber Research Networks

Velizar Shalamanov and Georgi Penchev

Abstract This chapter presents the approach, developed in the framework of the ECHO project, for the design of collaborative networked organisation focused on the cybersecurity domain. The research is focused on identifying the steps to support optimal decision making in defining processes and structures needed for the network governance and management. The approach links Enterprise Architecture model, COBIT framework and network analysis aiming to define a standard and comprehensive framework for analysis of needs and objectives, propose alternatives and select the most suitable one to be further designed for resilient governance and management of the networked organisation.

Keywords Cybersecurity · Process design · Enterprise architecture · Governance · Management · Network analysis · Modelling and simulation

1 Introduction

During the past decade, the cybersecurity landscape has evolved significantly with the enhanced use of IT in all social and economic activities. The growing concerns about cybersecurity cannot be considered and resolved only from technological side. Current and future issues in this field are and will be complex, influenced by many factors and will require complex solutions. Such solutions and adaption to fast changing security posture are not within the reach of a single organisation. The solution can be found in establishment of network type of organisation with effectively managed process and pooled resources among network participants.

V. Shalamanov
Institute of Information and Communication Technologies, Bulgarian Academy of Sciences (IICT, BAS), Sofia, Bulgaria
e-mail: shalamanov@acad.bg

G. Penchev (✉)
University of National and World Economy (UNWE, Sofia), Sofia, Bulgaria
e-mail: gpenchev@unwe.bg

© The Author(s), under exclusive license to Springer Nature Switzerland AG 2021
T. Tagarev et al. (eds.), *Digital Transformation, Cyber Security and Resilience of Modern Societies*, Studies in Big Data 84,
https://doi.org/10.1007/978-3-030-65722-2_3

The establishment and change management of network organisations in Cybersecurity is the purpose of many national and international initiatives. NATO started its Digital Endeavour [1] in order to provide a "digital enterprise" of network type with effective IT architecture and services. In consideration of Digital Single Market security, in EU were started four pilot projects with final goal to establish European Cybersecurity Competence Network [2].

On the other hand, the network (and respectively network organisations) are complex structures and their complexity rises with each additional participant or additional rule. There are many theoretical approaches for analysis, design, implementation of organisation as a single-unit. Applicable to network (or alliances) studies is significantly smaller in number.

The aim of the paper is to explore the possible approaches for building a new network organisation in cybersecurity. The research is focused on selection of activities, processes and structures needed for the network governance and management, aligned with the draft regulation of the EU for establishing the European Cybersecurity Industrial, Technology and Research Competence Centre and the Network of National Coordination Centres.

In order to prove possible links between the most common approaches for IT related analysis and management, the study considers Enterprise Architecture approach, COBIT framework and net-work analysis. In addition, for planning of the organisational development and strategic planning of the activities of the organisation well known methods as Balanced Score Cards, ADKAR (Awareness, Desire, Knowledge, Ability, Reinforcement) and CMMI [3] are integrated.

The application in practice of these approaches for change management initiatives of network organisation will be considered, as well as the current status and possible future development of cybersecurity network on EU level (under Regulation on establishing the European Cybersecurity Industrial, Technology and Research Competence Centre and the Network of National Coordination Centres [4]) will be analysed.

The complexity of the approaches and initiatives requires to define the assumptions and a framework of the research in more details with an effort to select the full set of concepts and adequate methodology for the design of the collaborative networked cybersecurity organisation.

2   Study Assumptions and Framework

There are many possible views and approaches to describe network organisations, their purpose and activities. As an example, from the economics point of view, the network organisation can be considered as an alliance. Alliance can be targeted to increase market share or to optimize supply chain through sharing customer information or production processes. Network can also be establishing just for information and data sharing.

The network organisation can bring benefits from the network effect. The network effect takes place, when adding user of service or product increases utility of the service or the product to other network users—new and existing [5]. Adding new users to the network will give new opportunities to connect or to use services or products, thus increasing value for all network participants.

The relationships of enterprises within the network can be of very different types—from very loose informational type to highly integrated connections of joint ventures organisations or can be centralised within one or more centres—so the span of options is large enough from established information sharing system to sharing of the assets.

Decision to establish or to participate to a network is an independent decision for each network unit and can have different motives, but in any case there is a need for joining process, based on certain type of accreditation or certification.

The paper aims not to study the benefits, the variety and the complexity of network organisations. The research is focused on organisations related to cybersecurity. The main research question is: How network organisation can be designed and improved with a focus on specific dimension of the network in cybersecurity—national or multi-national, sectoral or multisectoral and specific case of coordination in a horizontal domain as defence and space?

In this regard, options for design, governance, management and improvement of network organisation are central to the research. This mean to find solid methodological ground for modelling and for assessment of network organisations, selection of the most suitable alternative, strategic planning of its operation and required change management efforts to position the organisation, initially, for success.

There are many frameworks for modelling of organisations as a single unit which can control all factors of their internal environment. On the other hand, the network organisations do not have full control over all participants and their decisions. Each organisation dedicates part of its (or specific) services, processes or resources to the network, except the network centres (if there are any). This poses many questions to the modelling of network organisations related the governance and decision making. In addition, the network through its representative bodies or a joint platform, provides new services, requiring network level governance and management.

In order to find possible link between single entity and network organisational studies the paper explore relevant parts of Enterprise Architecture Modelling and Business Processes Analysis frameworks; standard for IT organisation governance and modelling—COBIT 2019 and applicable network analysis methods.

3   Enterprise Architecture

The idea and the concept of Enterprise Architecture (EA) was introduced in 1987 by John A. Zachman in his article "A Framework for Information Systems Architecture" [6]. The article suggested conceptualization of enterprises' architecture from multiple perspectives—objectives and scope, operational architecture model, system model and technical model. The conceptualisation can be done by different architectural descriptions of data and functions. The paper also proposes a framework as a matrix with 30 cell with documentation suggested for the specifics of each cell. The documentation describes data by analysing entity-relations models or process models.

The EA has changed in parallel with development of business practices since 1980s. In 2012 Ahlemann set following definition of EA: "a management practice that establishes, maintains and uses a coherent set of guidelines, architecture principles and governance regimes that provide direction for and practical help with the design and the development of an enterprise's architecture to achieve its vision and strategy" [7]. First government agency adopting EA is the US Department of Defense (Department of Defense Architecture Framework (DoDAF)). The architecture approach become popular in IT and management related concepts in national and international administration. The UK Ministry of Defence developed its framework (MoDAF) in support of defence planning and change management activities. NATO adopted its architectural framework (NAF) and on EA basis US administration developed a comprehensive Federal Enterprise Architecture. The EA general frameworks were developed and introduced—one of the most popular—TOGAF was developed by the Open Group. Architectural models with specific focuses were also developed—as an example—SABSA, focusing on IT security.

The enterprise architecture is a complex framework and EA analysis, modelling and management can be very detailed. Nevertheless, there is a common understanding that human beings have cognitive limitation to analyse and use EA complex models. Therefore, the architecture granularity has to be controlled and only the most important aspects have to be chosen and have to be managed. Currently organisations are often use rough and subjective estimations [8] within EA analysis.

There are no commonly accepted definition of EA. Up until 2008, according to Schoencherr [9], 49 different concept were given by 49 authors and in 2015 a research on EA development stated that all EA concept until that date are incomplete [10]. Searching compatibility with other concepts that are discussed in the following sections of the paper, the TOGAF view to EA will be used. The most cited and widely accepted TOGAF framework consider the EA as a system formed by four subsystems (or layers): business, data, application, and infrastructure. The main concern when developing layers is the behaviour and goals of the enterprise stakeholders.

In general, the EA can be considered from two main prospective—structure and process point of view. Having in mind, that within a network organisation participating units dedicate just a part of their resources and functions to the network we will consider the process view to the architecture as a primary one. The task is to establish an approach for identification of these processes within single network unit that are related to the network organisation. Finding these relations will give opportunity to model and analyse the network as governance and management model. The IT related and cybersecurity related point of views are also important for the analysis.

4 COBIT and Process View

One of the most comprehensive framework for analysing enterprise governance and management models is the Control Objectives for Information and Related Technologies (COBIT) framework, developed by Information Systems Audit and Control Association (ISACA) [11]. COBIT (as most of the industry standards) is a good practices based framework which is oriented toward good governance of IT related activities within an organisation and the framework is in development since 1996. It comprises of five main principles and many interconnected sub-frameworks for implementing the principles and for managing IT processes. Only two aspects of these frameworks will be discussed here. First, the relationships with EA and, second, the governance and management objectives and models.

The general scheme of the EA based concept for selection and development of processes dedicated to the network for a single organisation or for the network organisation as whole, could be combined with the COBIT framework, as it is shown on Fig. 1.

Main steps for selecting and developing network components can be identified as follows:

- identify business mission, vision strategy and objectives;
- define business attributes and develop related components in order to achieve the goals;
- define controls for the goals;
- blueprint of program to design and implement controls;
- describe and plan components life-cycle implementation in terms of logical, physical and component architectural views.

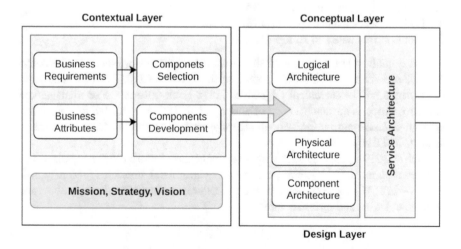

Fig. 1 Integrating COBIT and EA in design

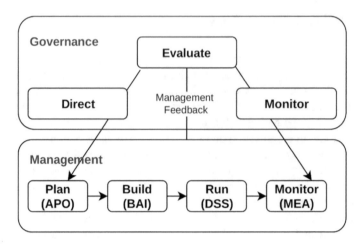

Fig. 2 COBIT 2019 Management objectives

Steps can be further analysed in order to create detailed plans for implementation and monitoring. The important implication here is that the processes aspect of organisation can be mapped to its architectural views and can be studied and planned with standard and compatible management frameworks.[1]

The COBIT framework divides the processes into governance and management areas. The two areas contain a total of 40 objectives in 5 domains, organized as follows:

1. Evaluate, Direct and Monitor (EDM)
2. Align, Plan and Organize (APO)
3. Build, Acquire and Implement (BAI)
4. Deliver, Service and Support (DSS)
5. Monitor, Evaluate and Assess (MEA).

The domain relationships with the governance and management of the enterprise IT are given on Fig. 2 (The source of the figure is COBIT 2019 Framework: Governance and Management Objectives [11]). Each governance or management objective supports the achievement of alignment goals that are related to main enterprise goals. The main enterprise goals are mapped through specific "cascade" to align to the COBIT objectives.

COBIT 2019 defines the following components in order to build the organisation's governance model and system: processes, organisational structures (and responsibilities—in RACI matrix), policies and procedures, information flows, culture and behaviour, skills, and infrastructure.[2]

---

[1] The idea for mapping EA views and COBIT framework is given according to Ghaznavi-Zadeh [12].
[2] In COBIT 5 the seven components were called "enablers".

Components of the governance model can be either generic or modification of the generic component. Generic components are described in the COBIT Core Model and can be applied in principle to any situation with appropriate customization.

Focusing on processes components, each governance and management objective includes several process practices. Each process has one or more activities. Example metrics accompanies each process practice, to measure the achievement of the practice and its contribution to the achievement of the overall objective [11].

The organisation implementing COBIT principles can choose (and modify according to the generic standard) process component in order to implement initiative for their monitoring and improvement. Thus, applying practical guide and measures toward implementation of organisational change, while staying within good practices standard.

The process component in COBIT also has its implementation part—monitoring and improvement of performance, which can be done by defining and controlling for processes' maturity. Here COBIT evaluation is based on use of Capability Maturity Model Integration (CMMI) levels.

Participation to the network organisation suggests an organisational change and it can be modelled, implemented and improved using partial selection of components dedicated to the network. Figure 3 depicts possible scheme of such kind of selection.

The network organisation requirements as goals, decision-making rules, processes, competences and products sharing rules can be used as main objectives for COBIT based design of network dedicated components.

When the components are selected the type and of relationships between organisation (networks nodes) can also be studied from the network point of view—e.g. whether these relationships are tight or loose, how the decisions are passed, accepted and approved, etc.

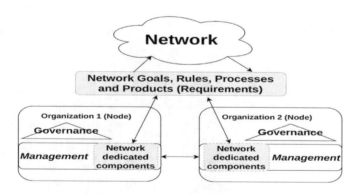

Fig. 3 Network and dedicated components

5 Network Organisation Aspects

The graph theory can be used to describe network in general as a set of $x_1, x_2, \ldots, x_n$, nodes (vertices) and edges (arcs) connected to them. Graphs are suitable to describe network organisations and for network analysis application to EA. There is a fast growing field of research using network analysis with EA analysis [11].

Homogeneity of the network (the graph) is an important network structure parameter. Networks can be uni- or multi-modular. Different nodes of the in multi-modular network can be considered as different layers in EA analysis or as different type of organisation in network organisations.

In order to find how different layers of nodes are connected to each other in heterogeneity networks, clustering algorithms can be applied, thus finding communities of similar nodes, as well as the most important nodes within communities and within network as whole. The measure of modularity represents the randomness to groups (sub-network) connections. The results from application of popular Louvain modularity clustering method [13] is presented on Fig. 4.

Mapping of EA to complex networks was suggested by Santana et al. [14]. The main idea is to find three types of connections within network of business organisations as follows:

- Business unit to Business unit in business EA view;
- Business process to Business process, also in business view;
- and Business object to Business object in EA information view.

Studying of these three networks can be done by applying appropriate network measures [15]. Thus, the most important connection and nodes, central for the network, can be identified. It can be argued that the identified three networks related

Fig. 4 Communities within the graph

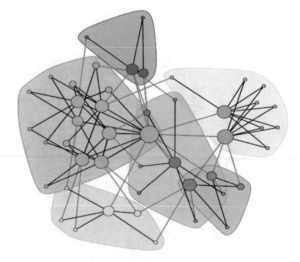

to units, processes and information can be used for simulations and optimization of the network and its structure.

As an example, the measure of Betweenness centrality can show the nodes and processes that are important intermediary channels of information. The Eigenvector can represent process that are connected with other significant (well-connected, with high level of Betweenness centrality) business processes. These well-connected nodes can be considered as structural nodes for the network.

6 Methodology for Design and Development Planning of the Cyber Research Networks

Initial design of the governance model is based on a general framework, derived from the study of good practices. It is selected to be flexible enough so to be aligned to the strategy and change management plan proposed. Strategic plan is looking 5–10+ years ahead with ambitious goals aligned with the mission of the network. In order to position the network to be able to achieve these goals, initial institutional building and change management plan is to be developed to achieve a status of the network, most suitable for implementing of the strategic plan. Change management effort is of 3–5 years.

The proposed approach for establishment of ECHO [16] network governance and management model has following phases and inputs:

Case studies on existing networks' structures and models;

1. Mission, vision and strategy analysis and definition of requirements and criteria for the model;
2. Generation, analysis and selection of alternatives;
3. Modelling, optimization, verification and approval;
4. Strategic and transition planning;
5. Implementation and improvement, based on transition plans and on monitoring and evaluation;
6. Extending the network.

Main tools which will be used during the third phase is the Analytic Hierarchy Process method. This method will be applied to alternatives which have been identified during the case studies analysis phase. The criteria for selection are developed during the mission, vision and strategy analysis phase.

The modelling phase will use as input the results from previous stages with detailed process identification trough COBIT components design and analysis. Optimization will be focused on achieving best process model for the organisation goals.

Optimisation can be done by modelling for several possible modifications of selected during phase three alternative. Thus, the network modelling and analysis can provide insides about the network characteristics (number of nodes, connections, clusters, etc.) as it is discussed in the previous sections.

The proposed framework and architecture has clear properties of a network organisations, especially in of national and EU level cooperation. The vision, the mission and the strategy are oriented toward processes change and improvement. As the EU also plans to strengthen its Cyber capabilities one of the key fields that have to be changed is the Cyber research, training and competences.

EU has previous experience in establishing complex IT networks (especially in scientific research) and the EU networks are more diversified than NATO Enterprise. The EU networks are more heterogeneous and comprises from many different actors ranging from SME and NGO to big national agencies and governments. This heterogeneity is caused probably by the main target of European Commission: "to strengthen the EU's cybersecurity capacity and tackle future cybersecurity challenges for a safer European Digital Single Market" [2]. The increase of the competitiveness of EU cybersecurity enterprises is another important field of this initiative.

So, when the general design of the cyber research network is finished there are still 3 more steps:

- identify and assess the current situation;
- implement strategic planning methodology to define Strategic initiatives to achieve the goals for the next 10–15 years;
- identify the change initiatives for implementing change management program in the first 3–5 years in order to position the network for a successful implementation of the strategic plan.

Methodology is established around COBIT, but also includes all other reviewed above concepts and framework as EA, CMMI, Network analysis, ADKAR, ITIL and for the selection of the most suitable alternative–AHP (Analytical Hierarchical Process) method. Implementation of the COBIT principles imposes following requirements:

- Engage all the stakeholders (during the ECHO project definition of the stakeholder categories and engagement strategies are developed).
- Cover the whole process from stakeholder engagement through service provision and organisational development to stakeholder satisfaction—the initial here focus have to be on following three processes:

  - partnership development (new partner acceptance);
  - service provision and Service Level Agreement (SLA) development;
  - internal organisational development, strategic planning and change management.

- Holistic approach which has to cover not just IT aspects, but all elements of the development and operation of ECHO Network.
- Integrated Framework, developed with COBIT principles at the core, but also including (and compliant with) following frameworks, approaches, methods and standards:

  - Enterprise Architect—for organisational structures design;
  - Business Processing Modelling Notation (BPMN)—for processes modelling;

- Project Management in Controlled Environment (PRINCE II)—for project management;
- IT Infrastructure Library (ITIL)—for service management;
- Optimal Responsibility Assignment Matrixes (RACI) to develop the organisational procedures and documentation;
- BSc and Strategic maps to do strategic planning;
- ADKAR to plan and implement change;
- Performance management system to link goals and objectives from the Strategic plan and Change management (transition) plan to individual performance through cascading of performance indicators along the organisational structure;
- Capability Maturity Model Integration (CMMI)—for auditing of the maturity level and development assessment and improvement plans;
- Customer Relationship Management (CRM)—for definition of stakeholders engagement and satisfaction program;
- Delineation between the management and governance.

7 Governance Model Framework and Research Activities

Development and implementation of the Governance model for the ECHO network is one of the key goals of the Project. The Working Package 3 (WP3) "ECHO Governance model" is dedicated to achievement of this goal, as an instrument to transform the large community of partners in effective, efficient, adaptable and resilient network for cyber research and training, as well as for providing a set of network services.

The governance framework in general is presented at Fig. 5. On the figure key stakeholders are identified in relation with policy, funding environment. The main elements of the physical infrastructure connecting the entries are also presented. This infrastructure is the Federated Cyber Ranges (FCR) with an Early Warning System (EWS), which will further strengthen the community of partners, as whole.

The ECHO Governance model is envisioned as an instrument to bring together all ECHO partners and to attract new partners in order to build a network with central hub for development of a portfolio of services in following several areas:

- Multi Sector Analysis Framework (MSAF);
- FCR;
- EWS;
- Cybersecurity skills framework with training and certification programme;
- Technology roadmap development;
- Certification schemes development.

The focus falls in following primary (vertical) sectors:

- Energy;
- Maritime transportation;
- Health care.

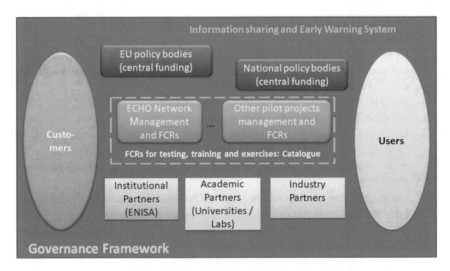

Fig. 5 Governance framework (management needs and objectives—business model)

The task includes also the exploration of the following sectors as some secondary relations:

- Defence;
- Space.

Figure 6 presents the agreed process for providing the deliverable D3.2 Governance alternatives of ECHO's WP3.

The short list of four alternatives were selected as an input. The selection of the was based on previous analysis done within the deliverable D3.1 Governance needs and

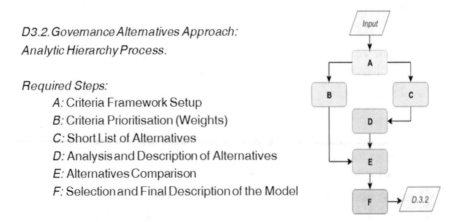

Fig. 6 The process of development, evaluation and selection of the most suitable alternative for the ECHO Governance model

objectives, which was dedicated to study of the existing networks, interviews with potential stakeholders and customers, as well as to literature and legal documents analysis. The resulting identification of alternatives was done by classification of the existing networks along following dimensions related to the Governance model:

- Centralization—Decentralization;
- Public—Commercial funding;
- Full direct representation—Limited representation by elected representatives of the stakeholders;
- Majority voting rule—Consensus.

The methodologic framework for the design of the governance model for ECHO network organisation is presented on Figs. 6 and 7. These two figures highlight the process of the development of specific project deliverables as follows: D3.2 Governance alternatives (e.g. evaluation and selection of alternatives); D3.3 Governance model description (further detailed design and description of the most selected alternative in D3.2); and D3.4. Governance model implementation plan (establishing the developed governance model).

The aim for using of this framework is to rationally identify the goal, criteria and to develop guidance, in order to integrate all business areas of ECHO community in one network organisation. Based on this vision, the selected most suitable alternative will be further developed from top to bottom in details as description of individual processes, RACI matrix, organisational structure with legal charter and Standing Operating Procedures (SoPs).

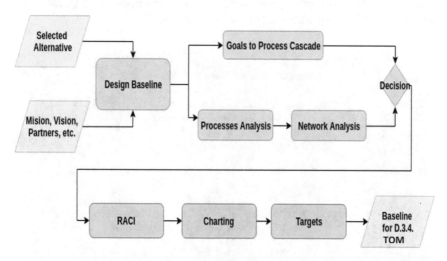

Fig. 7 Process and organisational design of the ECHO Governance model, based on the selected most suitable alternative (The abbreviation TOM means Target Operating Model)

8   The Conceptual Governance Model

The general governance model (framework) as integration of the governance models of the different business areas, tested through the demonstration and dissemination phase of the project is presented on Fig. 8.

The general governance model (framework) as integration of the governance models of the different business areas, tested through the demonstration and dissemination phase of the project is presented on Fig. 8.

There are two areas of integration.

First, integration on the *governance and management level* is aimed at the development and testing of the framework governance structures such as Board of Directors (BoD), Strategic Advisory Committee (SAC), Finance Committee (FC), Audit Committee (AC) and various Working groups.

The first level envelopes also the framework management positions such as Chief Executive Officer (CEO), Chief Operation Officer (COO), Chief Financial Officer (CFO), Chief Technology Officer (CTO), Chief Partnership Officer (CPO), Chief Support Officer (CSO), Human Resource Officer (HR), Legal Adviser (LA), Internal audit (IA) to form an Executive Management Board.

Second level is related to *demonstration and integration of the services* in different business areas, that initially are developed under their own specific governance and management arrangements, but trough implementation process are to be integrated under ECHO governance an management. The implementation and integration process finishes with creation and publication of the network Catalogue of Services.

Based on the above conceptual model, the key processes and aspects of the Governance model were identified as presented on Fig. 9.

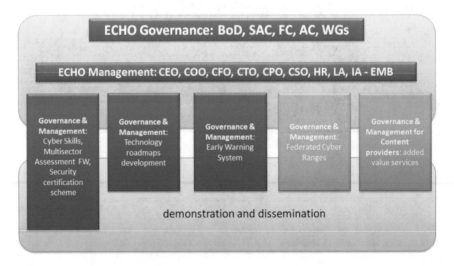

Fig. 8 Governance conceptual model

Fig. 9 Governance model key processes

General architecture of the Governance model was developed under this study to steer the generation of alternatives as it is depicted on Fig. 10.

It is important to stress that ultimately European Commission and EU Military Staff (EUMS) are over the top on the Fig. 10. The EUMS has its important role for defence domain and European Space Agency (ESA) play sole for the Space domain. Other important ECHO stakeholders in the institutional framework of the ECHO Governance are the EU Network and Information Security Agency (ENISA) as a cybersecurity agency and the European Cybersecurity Organisation (ECSO) in industry-EC public–private partnership area.

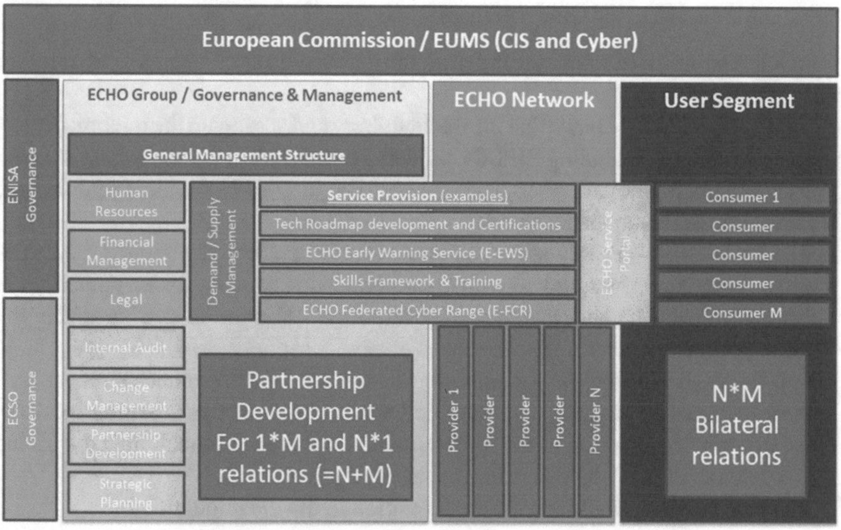

Fig. 10 Architecture of the ECHO Governance model

Further on, the Governance and Management of ECHO Group (as a central hub for the ECHO network) are developed along the service lines in the Figure's centre with key management processes on the left and relations with the consumers (or users) on the right. Obviously the ECHO Network is federating different partners through the ECHO Service Portal to deliver services to the customers through the governance and management framework established. This is simplifying the relations between partners and customers in service provision and streamlines research and development of the new services.

9   Transition and Implementation

There are two final Methodology's steps that have to be explained.

First step is the preparation of the transition plan (deliverable D3.4) from the current operating model (COM), which is identified through initial audit of the partnership network and of the Project's governance and management, to the target operating model (TOM), defined in D3.3. deliverable.

The transition has to be supported by establishing the process of change with effective management of the implementation of the model, as well as by maintaining of the parallel effort of enlarging the partners network.

The second (and last) step is to organize the internal auditing process of TOM implementation under the transition plan. The audit should to assess the maturity, to provide correction plans on yearly basis and to prepare the final report on the implementation of the ECHO Governance.

Unique practical experience was gained through the project implementation within the following steps and activities:

Step 1: 2 months—Organizing the work in a team (team structure, kick off, minutes and tasking).

Step 2: 2 months—Developing the methodology and setting up the toolbox (Enterprise Architect, including BPMN modelling and RACI matrix organisational design, Expert Choice and tools to support CRM, CMMI Audit, BSc, Strategic Maps, Performance management).

Step 3: 1 month—Presentation and approval of the Methodology for the development of alternatives, linked with the initial results of study on best practices in different aspects of business model, governance, information sharing within the network organisations.

Step 4: 1 month—Defining options and clusters for the Governance model, based on best practices analysis and identified needs and objectives. Setting the Goal and criteria for selection of the most suitable Governance model, linked with the study on needs and objectives for the network governance.

Step 5: 1 month—Ranking criteria in relation to the Goal identified.

Step 6: 4–5 months—Developing of 4–5 alternatives by different teams to describe effective, efficient, adaptable and resilient governance options for the collaborative networked organisation. Presenting the descriptions of the alternatives, their assessment, selection the most suitable alternative and its sensitivity analysis.

Step 7: 3–4 months—Process design of the selected alternative and its optimisation.

Step 8: 3–4 months—Organisational design of the selected alternative (architecture, RACI structure, ToR). As part of the development of SoPs potential piloting of the e-Platform to support the governance and management.

Step 9: 4–6 months—Developing of the transition (change management) plan, having in mind the long term goals of the Strategic plan for the organisation and required steps to achieve TOM in the first 2–3 years.

The e-Platform for governance and management of the cybersecurity research network is required in order to support the implementation of the designed model. An adequate Enterprise Business Application (EBA) in the context of the ECHO project is agreed to be the initial SharePoint environment for the project management, used as a baseline to develop a prototype of a system, supporting the main processes with a feasible level of automation.

Critical for success is to agree on Key Performance Indicators (KPI) for the process of Governance model development and implementation. The following KPI can be used as an example:

- number of alternatives developed (min 3–max 5);
- number of key processes defined (min 2–max 4);
- consensus on ToR for ECHO Group and Network;
- consensus on Transition plan for ECHO Network along draft EU Regulation on cybersecurity network to be implemented (during the period 2021–2027).

Finally, when defining the governance and management model for the ECHO Group and ECHO Network it will be done around the following lines:

- Mission, Vision and Strategy for the Network, including a Business model with its KPI;
- Processes and organisational design, including information sharing and EWS[3] to clarify issues as:

  - Customer base segmentation;
  - Partnership network and contractors;
  - Demand management and partnership development;
  - Program, project, portfolio management;
  - Service management;
  - Capacity management;
  - Compliance framework;

---

[3] Key processes in this field are (1) partnership development (new partner acceptance), (2) service provision and SLA development, (3) internal organizational development by strategic planning and change management.

- – Organisational KPI.

- Human resources management, including recruitment, education and training, career development, performance management;
- Technology and innovation management with special focus on FCR (linked with the role of CTO on the management level). These activities require answers of following questions:

  – Technology and innovation level of ambition;
  – Technology taxonomy scope;
  – Service Catalogue development and management;
  – Funding and decision making in support of innovation;
  – Partnership with industry, especially SME to foster innovation;
  – Governance of demonstrations and exercises for joint innovation with customers and users;
  – Technology KPIs;
  – Innovation KPIs.

- For Financial management (link with the role of CFO on the management level) following issues should be considered:

  – Initial and sustainable funding model;
  – Customer funding model;
  – Customer rates;
  – Operational fund and break-even point;
  – Financial audit;
  – Financial KPIs.

- KPIs for the ECHO network (Integration of all aspects above in the integrated ECHO Network Dashboard of KPIs);
- ToR for the Governance body and relations with the ToR for the Management body;
- There are also following additional external elements which have not to be missed from management sight:

  – External relations with the EU stakeholders;
  – External relations with the Defence Sector—EUMS, European Defence Agency and NATO;
  – External relations with the Space Sector—ESA.

10   Expectations and a Possible Solution in the Context of the EU Draft Regulation

Four projects—CONCORDIA, ECHO, SPARTA and CyberSec4Europe, funded by Horizon 2020 have to provide operating pilot example for a European Cybersecurity Competence Network and to develop a common European Cybersecurity

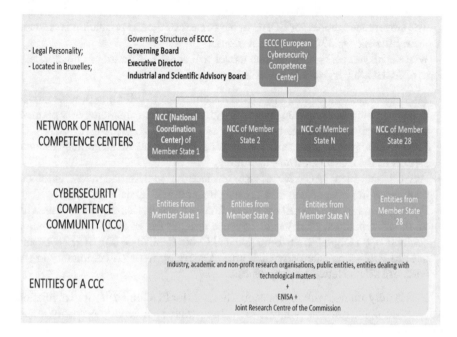

Fig. 11 Institutional framework for the EU network on cybersecurity, according to the draft regulation

Research and Innovation Roadmap (EC 2019). Institutional framework of the draft EU regulation is presented on Fig. 11.

In this context the work on development of a governance model for ECHO project will focus on accomplishment of following tasks:

- Prototyping the governance model for European Cybersecurity Industrial, Technology and Research Competence Centre in parallel to ECHO Group and Network Governance;
- Prototyping of the governance of the National Coordination Centre in parallel to ECHO Group and Network Governance;
- Prototyping of the governance of the Network of National Coordination Centres in parallel to ECHO Group and Network Governance;
- Integrating of the ECHO Group and ECHO Network in the European Cybersecurity Industrial, Technology and Research Competence Centre and the Network of National Coordination Centres construct if the project leadership decides to have ECHO Group and ECHO Network after the completion of the project;
- Investigate funding mechanisms of Digital Europe (cybersecurity and trust), Horizon Europe, European Defence Fund, European Space Agency and others, including funds from member states or from Joint Endeavour/Action funding as part of the governance models described above.

The Governance model for ECHO in the context of the EU regulation is oriented to answer the *WHO, WHAT, HOW questions.*

WHO is responsible—Governance model will consist of Board of Directors in a form of ECHO Supervisory Board (ESB), supported by:

- Multisector Innovation and Engagement Committee (MSIEC) to provide advice to ESB on innovation and engagement with stakeholders, including new partners joining the network, customers and others as ENISA, ECSO, EUMS, EDA, NATO;
- Finance Committee (FinCom) for advising ESB on all financial and resource aspects of the ECHO Network operations;
- Audit Committee (AC) for advice on internal or external audit missions, continuous improvement and risk management in ECHO network;
- For engagement with the EU cyber community and customers, as well as with all the stakeholders, ESB has to organise through the ECHO Group an annual conference on Digital Transformation and Cyber Resilience (DIGILIENCE) and have to publish an International Journal to develop research community in the areas of ECHO Network competences.

ESB is fully aligned with the concept of the BoD for ECC under the draft regulation and MSIEC could provide members to the Industrial and Scientific Advisory Board to the BoD of ECC.

ESB comprises of representatives from all partners—directly or through selected National representative with National Partnership Coordination Councils (NPCC) established for every member state from the partners. This will motivate partners form the member states to develop national cyber communities and to establish national partnership coordination function, aligned with the NCC, (as it is constructed in the draft regulation).

NPCCs could serve as advisory bodies to the ECC, after nomination by the member states to ECC.

The operational management of the ECHO Network is to be provided by ECHO Group with key management functions and organisational structure with an e-Platform to support the key management processes.

The key management functions in ECHO Group are as follows:

- Strategic planning and change management;
- Business planning and execution;
- Financial management;
- Innovation management and development of the Service Catalogue;
- Partnership management and Customer relationship management;
- Human resource development, education and training (and talent management);
- General services and infrastructure management.

The head General Manager or CEO of ECHO Group will be supported by COO, CFO, CTO, CPO, LA, HR, General Services Manager (GSM) and other experts as required and decided by ESB.

WHAT responsible persons are doing—ESB is the key decision making body with authority and responsibility for the following activities:

- Providing Strategic direction and guidance to the ECHO Group and Network;
- Approving annual goals and objectives for the ECHO Group on behalf of all the participants;
- Approving the Strategic plan every 5 years;
- Approving annual Business Plan and Financial Plan (with 3 year horizon) and respective annual reports;
- Approving the new partners to join the ECHO Network and structure of the ECHO Group;
- Approving the Service Catalogue and service rates and customer rates for the partners in the ECHO Network (under the customer funding regulatory framework);
- Selection of the Head of ESB secretariat and Head of ECHO Group as well as the Chairman of the Board of Directors.

The key services along FCR and EWS have to make ECHO unique, because the Project developed a good model for information sharing and trust in support to ECHO Governance and management in multisector and multinational environment with opportunity of FCR tests and exercises to feed EWS with warning about real life risks and even to share hints in real time.

Other unique features of ECHO Network are the technology roadmap development services, education and training services and multisector analysis services, certification scheme services, as well as consulting services in the area of governance and management.

Development of these services will require specific decisions to be taken on ESB level, supported by its substructure in order to maintain effective, efficient, adaptable and resilient collaborative network organisation.

HOW the things are done—ESB will meet 3–4 times annually for formal sessions to take decisions. ESB is supported by the secretariat and advised by the respective committees as decided by the Board. For the operation of ESB and its substructure, as well, the Secretariat and ECHO Group and Network operate an e-Platform, developed to support key processes and information management requirements.

As an example, MSIEC could be leading body to organise the annual conference with the stakeholders and to maintain a program committee for the conference, that is editorial board for the Journal of the ECHO Group, as well. The same program committee could serve as an innovation team for MSIEC, running innovation competition along the annual conference.

Further details and detailed answers on related governance questions *WHO*, *WHAT*, *HOW* will be developed within the Charter of ECHO Group and Network, process design with respective RACI matrix and organisational structure, including key SoPs and Job Descriptions.

11 Conclusions

The goal to find links between architectural and process aspects of organisational building was accomplished by studying of Enterprise Architecture approach and COBIT framework. The research is focused on IT related activities of organisation and this is the reason to select COBIT. The research is extended to possible application of these two approaches in design and implementation of network organisations. The overview of approaches' applications in network environment provides evidence for usability of network analysis combined with EA and COBIT.

Combining this theoretical base with the practical methodology to identify needs and objectives, establish goal and criteria for selection of the model among the set of developed alternatives is demonstrated through the implementation of the methodology to design the governance model for a collaborative network organisation for ECHO project under Horizon 2020.

Summarizing, the endeavour for competence enlargement and improvement through connecting existing organisation and pooling resources within a stable network is possible, but it is a complex task which can be and has to be based on solid analysis, sound understanding and agreement on goals and rules by all participants.

All activities for the establishment of the European network should be executed in networking environment, where the EU Centre is supported by National Coordination Centres, which on their turn coordinate the Community of enterprises related to cybersecurity on national level. In terms of the current paper analysis, the network will be of multi-modular type with three distinctive type of nodes—central node, regional or functional structural nodes—national or functional area centres and participants' nodes.

Considering this set-up of the Network following issues can be identified:

- How information for community capabilities within one national centre will reach other national centres?
- Are the capabilities and capacity within one national community comparable with other national communities?
- Who will manage compliance of community enterprises with EU Centre quality requirements?
- How capability gaps will be identified and how capacity for supply of services within the Network will be assessed for planning and contract management purposes?

Even if one of the above issues is given only in responsibility to the EU Centre it will require huge work force and will produce complicated and slow processes. Using the full range of theoretical analysis of such complex organisational object will produce also very complex analysis, far beyond the human cognitive limitation.

One possible solution is to create a practical accreditation processes, based on solid methodological approach, so the results of structured procedure will be useful for further analysis of the network organisation complexity. It is also preferable to give this procedure in responsibility of participating organisation, not to the centres, thus alleviating centre's work and structure.

Acknowledgements This work was supported by the ECHO project which has received funding from the European Union's Horizon 2020 research and innovation programme under the grant agreement no. 830943.

The work was also supported by the by Bulgarian National Research Program on ICT in Science, Education and Security.

References

1. Doyon, S., Hazebroek, S., Kalff, V., Katsampas, V., Kuehne, S., Szczesniak, P.: Project Report on NATO Digital Endeavour. NEDPD-10 (2019)
2. EC: Four EU pilot projects launched to prepare the European Cybersecurity Competence Network. Text. European Commission, Digital Single Market. February 26 (2019)
3. CMMI Institute: CMMI V2.0 (2019). https://cmmiinstitute.com/cmmi. Accessed September 5
4. EC: Council Regulation (EC) 2018/0328 (COD) of 12 September 2018 on proposal for establishing the European Cybersecurity Industrial, Technology and Research Competence Centre and the Network of National Coordination Centres COM(2018) 630 final. European Commission (2018)
5. Shapiro, C., Varian, H.R.: Information Rules: A Strategic Guide to the Network Economy. Harvard Business Review Press, Boston, MA (1998)
6. Zachman, J.A.: A framework for information systems architecture. IBM Syst. J. **38**, 454–470 (1999). https://doi.org/10.1147/sj.382.0454
7. Ahlemann, F., Stettiner, E., Messerschmidt, M., Legner, C. (eds.): Strategic Enterprise Architecture Management: Challenges, Best Practices, and Future Developments. Management for Professionals. Springer-Verlag, Berlin Heidelberg (2012)
8. Schmidt, R., Möhring, R., Härting, R.-C., Reichstein, C., Zimmermann, A., Luceri, S.: Benefits of enterprise architecture management—insights from European Experts. In: Ralyté, J., España, S., Pastor, O. (eds.), *The Practice of Enterprise Modeling*, pp. 223–236, 235. Springer International Publishing, Cham (2015). https://doi.org/10.1007/978-3-319-25897-3_15
9. Schoenherr, T.: Logistics and supply chain management applications within a global context: an overview. J. Bus. Logist. **30**, 1–25 (2009). https://doi.org/10.1002/j.2158-1592.2009.tb00109.x
10. Kotusev, S., Singh, M., Storey, I.: Consolidating enterprise architecture management research. In: 2015 48th Hawaii International Conference on System Sciences, pp. 4069–4078. IEEE, HI, USA (2015). https://doi.org/10.1109/HICSS.2015.489
11. ISACA: COBIT 2019 Framework Governance and Management Objectives (2019)
12. Ghaznavi-Zadeh, R.: Enterprise Security Architecture—A Top-down Approach (2017)
13. Blondel, V.D., Guillaume, J.-L., Lambiotte, R., Lefebvre, E.: Fast unfolding of communities in large networks. J. Stat. Mech: Theory Exp. **2008**, P10008 (2008). https://doi.org/10.1088/1742-5468/2008/10/P10008
14. Santana, A., Kreimeyer, M., Clo, P., Fischbach, K., de Moura, H.: An Empirical Investigation of Enterprise Architecture Analysis Based on Network Measures and Expert Knowledge: A Case from the Automotive Industry: 11 (2016)
15. Scott, J.: Social Network Analysis: A Handbook. Sage Publications, Inc., Thousand Oaks, CA, USA (1991)
16. ECHO: European network of Cybersecurity centres and competence Hub for innovation and Operations (2019)

Managing Cyber-Education Environments with Serverless Computing

Yavor Papazov, George Sharkov, Georgi Koykov, and Christina Todorova

Abstract This chapter reports on the experience of the authors in their work in applying an innovative computation paradigm—"serverless" computing—for the management of cyber education environments through the Course Manager platform, developed by the authors' team. The serverless paradigm, also referred to as Function-as-a-Service (FaaS), helps the developers abstract or automate almost all infrastructure and operation overhead, allowing for what is often touted as "infinite scalability" applications, which can be a good fit for the rapidly increasing demand for practical cyber environments. The chapter provides an in-depth overview of the architecture and frontend of a cyber-education environment management framework, designed to work in a serverless environment and analyses the lessons learnt from using this framework for the provision of cyber trainings to students and IT professionals for more than a year.

Keywords Cyber education · Cybersecurity education · Cyber range · Serverless · FaaS

Y. Papazov · G. Sharkov (✉) · G. Koykov · C. Todorova
European Software Institute—Center Eastern Europe, Sofia, Bulgaria
e-mail: gesha@esicenter.bg

Y. Papazov
e-mail: yavorpap@gmail.com

G. Koykov
e-mail: georgi.koykov@esicenter.bg

C. Todorova
e-mail: tina@esicenter.bg

G. Sharkov
University of Plovdiv "Paisii Hilendarski", Plovdiv, Bulgaria

© The Author(s), under exclusive license to Springer Nature Switzerland AG 2021
T. Tagarev et al. (eds.), *Digital Transformation, Cyber Security and Resilience of Modern Societies*, Studies in Big Data 84,
https://doi.org/10.1007/978-3-030-65722-2_4

1 Background

Cybersecurity, as a field, is a very much multi-disciplinary and case-specific domain, thirstily in need of a strong, practice-based and simulation-heavy immersive education. Hence, the ability to carry out a successful, hands-on technical cybersecurity training, clings on to the availability of an effective cyber-education learning, simulation and exercise environment that allows for its users to practice. However, such a system should also be dependable, resilient, agile, easily maintainable and manageable, as it needs to be able to support a wide range of IT infrastructures or functions.

In the past, to develop, manage and maintain such an environment, would have been possible and practical only for large, industry-driven initiatives, supported by a strong financial backbone. Nevertheless, nowadays, the rapid development of serverless technologies, is shifting the paradigm of application development, allowing for organizations of all sizes, maturity levels and sectors to be able to undertake such projects at a dramatically lower cost, empowering them to change the focus of resource distribution from operational overhead, to the actual product value [1].

Serverless platforms and cloud computing in general have become increasingly ubiquitous [2] throughout the recent years, accelerating innovation, promising cost-effective capabilities for deploying scalable microservices and delivering superior digital experience at a faster speed to market [3]. "Serverless" is somewhat of a marketing term, referring to the Function-as-a-Service (FaaS) products, initiated by Amazon Web Services (AWS)[1] in 2014 and now offered by most of the major cloud service providers, such as Azure Functions,[2] IBM/Apache Open Whisk,[3] Google Cloud Functions,[4] Oracle Cloud Fn[5] and others. This paper will, however, put its focus on AWS, as this is the platform used and featured to create the serverless solution used for the management of cyber-education environments, which will be presented below.

Regardless of the platform under consideration, however, at the heart of the term "serverless" is the potential to program the cloud in an auto-scaling, pay-as-you go manner [4]. Serverless platforms allow for pieces of code to be deployed in the cloud with minimal to no operational concerns regarding the infrastructure they are running on. Therefore, the application development in itself, becomes decoupled from the infrastructure maintenance, monitoring, resource management and fault-tolerance, enabling the developers to focus on the actual application development. As a result, the application development process focuses highly on abstractions, such as functions, events and queries, and build applications that infrastructure operators could map to concrete resources and supporting services [3].

---

[1] https://aws.amazon.com/.

[2] https://azure.microsoft.com/en-us/.

[3] https://www.ibm.com/cloud/functions.

[4] https://cloud.google.com/functions/.

[5] https://fnproject.io/.

In this paper, we discuss our experience with building a cybersecurity education environment based on the AWS serverless solution—Lambda. In Sect. 2 of this paper, we discuss our choice to employ serverless computing for cybersecurity education environments management. In Sect. 3, we provide an overview of the design of the developed environment, which we divide into three core sections, namely 3.1 Architecture, 3.2 Deployment and 3.3 Frontend. In Sect. 4, we discuss the implemented framework for the underlying education environments, and in Sect. 5, we present our conclusions so far, our future plans and development opportunities for this project.

2   Serverless Computing for Cybersecurity Education Environments Management

According to the Wipro's State of Cybersecurity Report for 2018 [5] the usage of cloud services has been steadily increasing with an indicated an augmented preference for the adoption of Platform as a Service (PaaS) and Software as a Service (SaaS), at 40% and 55% respectively, according to the primary research survey. Regardless, there is still a high usage (48%) of Infrastructure as a Service, which can be attributed to the fact that it is not easy to migrate traditionally-built monolithic applications to a newer model. Last but not least, they report that Function-as-a-Service (or serverless computing) is slowly gaining wider acceptance with 17% of the respondents of Wipro's initial survey having already adopted FaaS.

Regardless of it steadily gaining momentum, serverless computing is still not widely recognized as a solution for educational purposes, such as for building and managing cybersecurity education environments. According to Wipro, among the challenges to employ FaaS could be the lack of subject-related competency or awareness about serverless technology, as much as the difficulty in migrating traditionally-built monolithic applications to a serverless infrastructure. However, especially for cybersecurity education, serverless technologies bare an enormous potential, especially when it comes to academic institutions [6].

Employing serverless technologies for instructional purposes and design of educational experience will shift the development focus to the functionality of the applications, enabling a more sophisticated educational orchestration, resulting in more realistic exercises and demonstrations and improved practical participant involvement, along with the betterment of the logic and the functional requirements of the system, all as a result of not having to allocate as much resources to the management of infrastructures, the scalability and the capacity of the environment. Furthermore, a great advantage to employing FaaS for cybersecurity environments is the ability to quickly implement changes, updates, bug-fixes and improvements.

By the same token, the application of the serverless paradigm, ensures that all necessary resources are being dynamically allocated, meaning that resources are being dedicated when there is a trigger for a function, allowing for a value offering essentially equivalent to cyber education environment-as-a-service, or in a nutshell, deployment of infrastructure based on the demand. Not having to maintain and pay for idle resources as well as running resources with low upfront investment and limited human and computing resources means that organizations, with limited employees or with scarcer cybersecurity trainings schedule, could pay for the used infrastructure based on the duration of the execution of the functions that are being triggered and used.

Due to those reasons, we decided to use the serverless paradigm to create the "Course Manager"—an AWS Lambda powered application that we developed and have been using for the past year to orchestrate cybersecurity course environments for our cybersecurity education program.

Among the core functionalities of the Course Manager we needed was to create VM instances (EC2 instances), hosting the exercises for the different courses we carry out. As our organization hosts public cybersecurity courses based on a schedule, we needed to be able to use the resources when we need them—or in other words, when we have a course, and not having to pay for them when idle.

Furthermore, another issue we needed addressed as a small organization, was the human investment needed to support some of the necessary logistical processes around the courses we carry out. Among those processes is the need to send out follow-up e-mails to all participants from a given course, with the link to the exercise platform for the respective course, as well as unique credentials to access it, along with the slides from the course and other relevant information.

Along the same lines, we needed the Course Manager to be able to manage the collection of feedback from the course participants and manage the information collected.

Last but not least, we needed to be able to schedule the data collection (namely the performance of the participants with the exercises) and also importantly, to schedule the termination of the EC2 instances launched for the course. The execution of all of that is explained in details below, in Sects. 3 and 4 of this paper.

3   Methodology

3.1   Architecture

The goals of the Course Manager architecture are to: (1) enable a highly-scalable and dynamic capacity of the system, as well as (2) to automate error-prone manual tasks around managing a training course. The system is designed to scale up to the AWS service limits and in particular the horizontal limits for EC2 instances. The execution of this design limits the usage of AWS to high-scaling services, such as Lambda, DynamoDB, S3 and (only horizontally) EC2.

The Course Manager solution employs a highly decoupled and mostly stateless architecture, somewhat inherent to properly designed Serverless applications. Some of that decoupling is necessary to accommodate the lack of native persistent storage mechanisms within AWS Lambda, while in other instances it is necessary to allow flexibility and agility within the frontend component, as well as within the cyber education environments that are the main deliverables of the Course Manager.

The core abstraction behind the Course Manager is, unsurprisingly, the Course. The name "Course" itself is somewhat of a misnomer, as it implies a training must take place within the duration of the "Course", which is not necessary from an architectural (or implementation, for that matter) point of view. A more apt name (as already used) would be Cyber Education Environment. In practice, Courses are usually associated with actual trainings. Due to this, it is generally assumed that Courses have a limited number of users (in practice less than 25), negating the need for vertical scaling. In case a Course needs a significantly higher number of users, it is up to its deployment to ensure service level, either through a larger EC2 instance, or through a distributed system.

The main assumptions with regards to Courses that need to be honoured to allow integration within the Course Manager are:

- **Self-containment**

  The Courses need to be entirely self-contained units that do not depend on any third-party code, infrastructure or outside services after their setup period. Some justification for this decision can be found in Sect. 4. This requirement is critical for allowing infinite scalability, as coordinating potentially thousands of interdependent services and their updates would present an almost insurmountable overhead on operations. By requiring each individual education environment to be independent of others, the system is enabling trivial horizontal scaling. Additionally, only a very limited form of vertical scaling for the environments themselves is necessary, as trainings typically include a limited number of participants and can be satisfied with a relatively 'large' VM (EC2) instance.

- **Self-deployment**

  While this is not a requirement, imposed by the architecture of the Course Manager, it is currently expected that, Courses can continue running their own setup procedure without the need for further interference (but with the possibility of continuous monitoring) from the Course Manager.

  The team has considered and validated the possibility of using various implementation paradigms, including (e.g.) Course Manager running Terraform, Ansible, and others to deploy the Courses.

- **Limited Runtime**

  One of the core functionalities of the Course Manager is the ability to 'clean up' any infrastructure that was set up during the deployment of a Course. This implies a discrete and clearly identified endpoint of the education environment, after which it is to be disposed of. While the runtime period may be significant in length (e.g. several months), the requirement for a specified termination date is currently critical. Although not presently implemented, the system can support open-ended courses (the termination of which is not scheduled at all, rather than just scheduled years in the future) and the delay of course termination (i.e. "snoozing" on termination) with relatively little modification. The business case for such

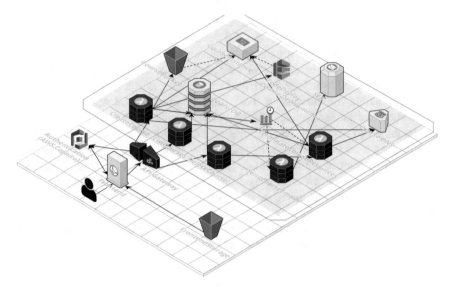

Fig. 1 Simplified course manager architecture (Created with cloudcraft.io)

functionality is yet to present itself, however. The limited runtime also presents an elegant, yet partial, solution to the update issue—any vulnerability in the execution environment is immediately phased out on newly created instances, while, even if left vulnerable, old instances have limited lifetime and vulnerability impact due to user segmentation. This can be, however, mostly mitigated by activating automatic security updates for the respective package manager/update manager for the instance.

The architecture of the Course Manager system is presented in Fig. 1. Almost all of the services used are provided by AWS. Below is a rundown of those services and their role within the system:

- Lambda—the main computational platform for the Course Manager itself. Enables the usage of the FaaS paradigm.
- DynamoDB—the core storage of the project. DynamoDB is the preferred and recommended persistence mechanism for Serverless (FaaS) projects, due to its scalability and availability focused design from the start. Contains all the metadata for all Course instances. Any data not contained within DynamoDB can be considered ephemeral.
- S3—used as file storage mechanism. Most AWS services can accept an S3 reference in place of an actual file. S3 buckets are also practically infinitely vertically scalable (as well as having the limitation of 1000 buckets per AWS region) [7]. Within the Course Manager, file storage is necessary mostly for static or initialization artefacts, that are not modified later during the runtime.

- EC2—the core computational power for the Course instances themselves. While having limited vertical scaling, EC2 provides very high horizontal scaling, which can allow for a higher number of Courses in parallel.
- Cognito – this service is Amazon's managed Web identity and authentication provider. This service is used for the Course Manager Dashboard component (discussed in details in 3.3) only, as individual environments employ their own authentication/authorization mechanism(s).
- SES—used for delivering training materials to end users.
- Route53—used for registering domains/subdomains for Courses.
- API Gateway—used for exposing secure access to the provided Lambda functions for use from the frontend.
- CloudWatch Events—used for scheduling of events, in particular sending Course materials and terminating the Course resources. While it may appear that AWS SQS (Simple Queue Service) is a better fit for this purpose, SQS only supports scheduled events within the next 15 min, which is woefully inadequate for the purposes of trainings, spanning multiple days or weeks.
- Code Deploy—used for bootstrapping the provision process of the Courses themselves, as well as for providing visibility into any defects of the process.

While the current version of the Course Manager is strongly dependent on the facilities, provided by AWS in terms of IaaS, PaaS and SaaS offerings, the architecture is developed with some degree of cloud-agnosticism—AWS-specific functionalities have been isolated and alternative approaches have been identified for services that are exclusive to AWS and don't have an equivalent. The codebase contains wrapper abstractions over the raw AWS APIs utilized, which allows for an easier portability to alternative providers.

The implementation of the Backend (Course Manager core) is discussed below in 3.2, while the implementation of the Frontend (Course Manager Dashboard) is detailed in 3.3.

3.2 Deployment

At its base, the Course Manager is a set of APIs, accessible upon authentication through an API Gateway. Those APIs are proxies that call their respective Lambda functions. Below, we present a brief, high-level overview of the motivation and purpose of all the Lambda functions involved:

- **CreateCourse**
  The "workhorse" of the Course Manager. Creates the cyber education environments and schedules follow-up activities (e.g. termination, sending course materials). Discussed in greater detail below. Not exposed through API Gateway due to the Lambda-HTTP execution model mismatch (see below).
- **CreateCourseAsync**

An asynchronous wrapper over CreateCourse that returns quickly and does not wait for the entire deployment process. This Lambda function is necessary, as API Gateway does not allow longer timeouts for Lambda integrations. This is also reasonable in terms of HTTP, as the protocol (as well as the RESTful paradigm) eschews long-running requests in favour of an early return with incomplete data.

- **ListCourses**
Provides the complete data behind the Courses in JSON, the native format of DynamoDB. This can also be augmented with filtration and sorting parameters to relief some of the burden of the query.
- **SendCourseEmails**
Sends the course materials via email with a link to the education environment, as well as (e.g.) presentation slides. Not exposed directly through API Gateway, as this is an internal Function.
- **TerminateCourseResources**
This Function essentially undoes what CreateCourse does for Courses, going over all the allocated resources and cleaning them up. The notable exception is the file storage of the Course and its logs, which are retained by default for the purposes of better visibility and accountability. Not exposed directly through API Gateway, as this is an internal Function.
- **ExtractCourseSolutions**
Performs a 'pull' type check in the Course instance to extract the usernames of the participants who have completed all exercises (in case the Course consists of 'exercises'). Used to issue completion certifications for participants. Not exposed directly through API Gateway, as this is an internal Function.
- **DeleteCourse**
Destroys Course resources. While the overlap with TerminateCourseResources is obvious, the former is exposed through API Gateway and supports optional flags for deleting resources that are usually retained.
- **PurgeCourse**
Erases a Course entirely, including records in DynamoDB for its existence. Can only be applied on an already terminated course.

While usually Serverless projects are implemented in terms of microservices architecture, allowing for a greater degree of independence of each microservice (or Function), all of the Functions, defined in Course Manager, share a single codebase, written in JavaScript/Node.js. This approach leads to certain versioning hurdles, in particular the necessity to support multiple versions of the codebase in parallel for different Functions during more significant refactor efforts. Nevertheless, the approach described allows for the reduction of the common codebase—the entirety of the Backend of the system is less than 3.7 thousand lines of code (including a Gulp-based build system and unit and automated end-to-end tests). Another (aspect of the same) benefit is the higher density of test coverage, due to code reuse.

A rough sketch of the functionality, provided by the CreateCourse Function, is provided below. Each step is accompanied by an appropriate update of the Status field in DynamoDB.

1. Create a Course object in the DB;
2. Acquire access token for the SCM (a Git provider);
3. Clone the appropriate (optional branch or tag) version of the source code for the respective EC2 instance;
4. Create the required (configurable) EC2 instance;
5. Register the FQDN for the Course to point at the EC2 instance;
6. (Optional) Create a participant survey and/or an extraction key;
7. Customize the cloned repository with necessary variables—FQDN, user data, survey URL, extraction key;
8. Run CodeDeploy on the EC2 instance;
9. Schedule termination and (optional) extraction for the instance;
10. (Optional) Schedule sending materials;
11. Mark completion of creation.

3.3 Frontend

For the Frontend component of the Course Manager, the implementation team came upon the decision to employ a minimalistic design approach in order to focus on the intuitive usability of the user interface. Furthermore, the minimal design of the Course Manager API allows for a significant level of freedom in terms of the implementation of the Course Manager Dashboard, i.e. the project Frontend component. To complete the scalability of the Course Manager Backend, the Frontend takes a scalability-first approach. To achieve this, the entire Frontend is static, served through AWS CloudFront, a content delivery service designed with scalability in mind [8] (Fig. 2).

During the lifetime of the Course Manager platform, the team has developed two Frontends: a simple Bootstrap-based static HTML/JS/CSS site, and a more modern approach—using the Nuxt.js framework. The migration of the API calls to the Backend turned out to be trivial, empirically proving the loose degree of coupling in the architecture.

The latter Frontend utilized Nuxt for its ability to generate static HTML/JS/CSS for Vue.js-based user interfaces. Nuxt.js additionally allowed for a more rapid development due to retaining the ability to debug locally with instant code reload.

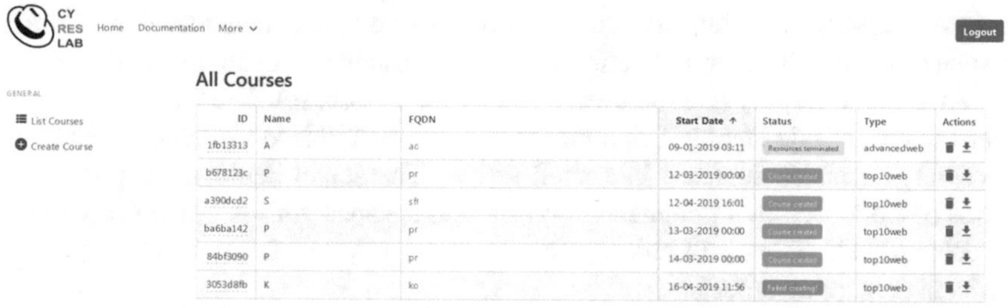

Fig. 2 Course manager dashboard screenshot

The flexibility of this architecture allows for a trivial update procedure—upon triggering the automated deployment procedure, a static version of the current codebase is built and uploaded to S3. After that, a CloudFront invalidation is triggered, to ensure CDN cache is updated. After the completion of this invalidation, the new version of the Frontend is deployed.

4   Framework

An integral part of the architecture of Course Manager is the contract between the Course Manager APIs and the Cyber Exercise Environments. This contract is referred to as the Framework for the purpose of this chapter.

The goal of this contract is to ensure the possibility for scaling, defined by AWS EC2 horizontal, rather than vertical service limits, as well as per-user group (Course) isolation and non-scheduled termination.

In general, the Course Manager takes a "laissez-faire" approach to the Course instances—following the initialization of the instance, the Manager is not involved in its runtime with two exceptions—when (optionally) collecting the list of users that have solved the challenges, and when the Manager destroys the instance.

At an abstract level—the Course Manager is responsible for starting the Course setup and providing services, common to all Courses—surveys, slides, termination scheduling, extraction, DNS, etc., while the Course itself needs to setup any internal resources and is expected not to depend on external resources (the "Self-Containment" requirement from 3.1) after its bootstrap phase is complete. A logical consequence of this is the necessity to support the "graceful destruction" of the Course EC2 instance, however this should be granted in most cases when no external resources are utilized. Beyond the logs of the Lambda execution and the EC2 instance itself (through CloudWatch Logs agent), the list of user solves is the only artefact preserved after the Course instance termination.

These requirements are deemed necessary to ensure no interdependence exists between different Course instances, removing entirely the overhead, imposed by the need for synchronization and its own limitations.

We will be referring to the source code (including provisioning) for the particular Course Type (predefined blueprints of Cyber Execution Environments) as the Codebase within the rest of this chapter. The bootstrap from the viewpoint of the Course instance consists of the initialization of the EC2 instance with an installation script for AWS Code Deploy (can be overridden by the Codebase). The Course Manager then creates a Code Deploy Deployment with the customized Codebase (including necessary variables declared in a shell script). The actual deployment process in terms of Code Deploy implementation (through appspec.yml) is left entirely to the Codebase itself. The result and logs are automatically synced by Code Deploy to assist in debugging defects.

The Course is then expected to become available in a "reasonable" time to serve client requests, usually (but not exclusively) through HTTP(S). Courses can and must export a variety of configuration options to the Course Manager, including source code repository URL, DNS prefixes, EC2 instance details, timeouts and supported languages.

5 Conclusion

Within the past few years, the multi-spectrum utilization of serverless computing has been gathering momentum and not without a reason. From a business standpoint, serverless computing serves as a unique opportunity to achieve better cost-efficiency, a quicker way to deliver new features and ways to abstract or automate away almost all infrastructure and operation overhead.

Course Manager was designed and developed by a small team to automate the business processes, related to the organization of cybersecurity trainings, as well as the deployment of the relevant infrastructure, providing practical exercises and demonstrations.

For the 18 months of exploitation, this system managed to significantly reduce delays and operator errors during deployments. Additionally, the system displayed capacity to provide scaling, necessary to serve the rapidly increasing demand for cybersecurity training and education that can be observed both locally and globally.

In this article, we provided a brief overview of the most common capabilities and applications of serverless computing. We further presented our experience with the application of the serverless computing paradigm to create and manage a cyber-education environment management framework, designed to work in a serverless environment.

Within the third section of this paper, we summarized the architecture, deployment and frontend of the cyber-education management system and within the fourth section, we analyzed the framework of the of the education environment, designed to work in a serverless environment and analyses the lessons learnt from using said framework in providing cyber trainings to students and IT professionals for more than a year.

This experience surfaces opportunities for the future improvement of this implementation of cybersecurity education environments especially in terms of resource utilization and automation design.

References

1. Amazon Web Services: Serverless: Changing the Face of Business Economics—A Venture Capital and Startup Perspective. Whitepaper (2018)
2. Pendse, G.: Cloud Computing: Industry Report and Investment Case. Report (2017). https://business.nasdaq.com/marketinsite/2017/Cloud-Computing-Industry-Report-and-Investment-Case.html

3. van Eyk, E., et al.: Serverless is more: From PaaS to present cloud computing. IEEE Internet Comput. **22**(05), 8–17 (2018). https://doi.org/10.1109/MIC.2018.053681358
4. Hellerstein, J., et al.: Serverless Computing: One Step Forward, Two Steps Back (2018)
5. Wipro: State of Cybersecurity Report 2018. Foresight for the global cybersecurity community. Report (2018). https://www.wipro.com/content/dam/nexus/en/service-lines/applications/latest-thinking/state-of-cybersecurity-report-2018.pdf
6. Vasileiou, I., Furnell, S.: Cybersecurity Education for Awareness and Compliance, Advances in Information Security, Privacy, and Ethics (1948–9730), pp. 16, 110. IGI Global (2019). ISBN 152257848X, 9781522578482
7. Amazon Web Services: Bucket Restrictions and Limitations (2019). Available at: https://docs.aws.amazon.com/AmazonS3/latest/dev/BucketRestrictions.html
8. Amazon Web Services: Hosting Static Websites on AWS. Prescriptive Guidance Whitepaper (2019). Available at: https://d1.awsstatic.com/whitepapers/Building%20Static%20Websites%20on%20AWS.pdf

Unified Innovative Platform for Administration and Automated Management of Internationally Recognized Standards

Toddor Velev and Nina Dobrinkova

Abstract More and more companies are realizing that the use of internationally recognized standards is a prerequisite for rising their productivity, increasing their competitiveness and establishing their place on the world market. The unified Platform for Automation and Management of Standards (PAMS) is a high-tech innovative product in the field of information technologies. PAMS is a modern, integrated information and communication system that models, digitizes, registers, manages, stores and controls work processes and related information and documentation, in accordance with a variety of internationally recognized standards. By modelling different sets of standards with the unified objects and processes, PAMS is a highly effective in monitoring, control and operational management and can assure implementation of the requirements of the standards and their practical utility.

Keywords Information technology · ICT · Platform · Integrated system · Document system · Standards · Modelling · Automation

1 Introduction

International Standard is a standard developed by International Standards Organizations (ISO, IEC). The main task of these organizations is to promote the development of standardization and related activities in order to facilitate international trade in goods and services and strengthening cooperation in the intellectual, scientific, technical and economic fields. International standards are developed by experts from

T. Velev (✉) · N. Dobrinkova
Institute of Information and Communication Technologies (IICT), Bulgarian Academy of Sciences, Acad. Georgi Bonchev Str., Bl. 2, 1113 Sofia, Bulgaria
e-mail: ceo@perfectbg.com
URL: http://www.iict.bas.bg/EN/index.html

N. Dobrinkova
e-mail: ninabox2002@gmail.com
URL: http://www.iict.bas.bg/EN/index.html

© The Author(s), under exclusive license to Springer Nature Switzerland AG 2021
T. Tagarev et al. (eds.), *Digital Transformation, Cyber Security and Resilience of Modern Societies*, Studies in Big Data 84,
https://doi.org/10.1007/978-3-030-65722-2_5

ISO/IEC technical committees. An international standard is considered accepted only if it is approved by 75% of the member organizations.

The documents are records that reflect in discrete points the development of every action, event, resource, process and intellectual production in the company. The amount of all conventional and specialized records, figuratively called documents, defines and determine the company, its history, capacity, quality and competitiveness.

Documents in themselves are not independent separate units. They are linked in a company-specific document network. Each document is an end product, behind which there is a process of different complexity and scope. Each process consists of activities and stages. In order to be achieved the end product, as in any production, it needs resources and materials. Materials, in turn, are most often other documents that have their own processes and life cycles. For the company, this leads to a multidimensional net of records which connect with direct, relational and correlational links all elements of entire working environment. Unlike the Internet, however, a company's document network has defined paths, links, and forms. For this reason, the company documents system is seen not as the document network, but as a **documentary matrix** of the company.

Every organization or company, regardless of its scale, performs its operations in an internal and external work environment. The internal environment is determined by the company's resources (human, tangible and intangible assets, etc.) and by the work processes in it, related to the subject of activity. The outside environment is the customers, suppliers, partners and institutions, with which the company operates [1]. The document matrix reflects every element of the organization's internal and external work environment and it is logical, that it provides the tools and the possibilities for maximum control and monitoring.

The company documents matrix is the basis for the management of any internationally recognized standard. The standards are set of work processes, procedures, and many forms of control. They define, regulate, manage, monitor, record and archive the internal and external environment of functioning through the company's document system.

The purpose of the Platform for administration and automated management of internationally recognized standards (PAMS) is to use the latest technology advances in order to digitize, index, and revive the organization's documents matrix and allow operational speed of management.

When documents are in paper form, they are figuratively speaking dead. It is difficult to use them by multiple users at the same time, it is difficult to trace the numerous connections, it is difficult to multiply decisions, and it is difficult to control the processes on their base. When the documents are digitized, the links between them are visualized and the matrix is dressed in appropriate functionality, a highly effective system for monitoring, controlling and managing all levels and resources in a trading company or other organizations is created.

The basic concept is to structure an open PLATFORM that enables modelling and efficient management of systems, defined by internationally recognized standards.

2 Research and Analysis (R&A)

The platform model is the result of a four-stage research process:

- Stage 1: Research and analysis of information classes related to the production of an innovative product. Classification and structuring of information;
- Stage 2: Research and analysis of target groups;
- Stage 3: Research and analysis of communication channels and technical capabilities;
- Stage 4: Reengineering of working processes in the user groups and audit organizations.

2.1 Stage 1 R&A of Information Classes

This first stage consist of five steps as follows:

Profound research and analysis of the information sources for internationally recognized standards in aspects—content, infrastructure, and superstructure

Researched information sources are grouped by type into three categories:

1. Sources and publications related to internationally recognized standards [2–4];
2. Internationally recognized standards;
3. Sources and publications related to the platforms of information and communication technologies [5–7].

Standard is "a document, established by consensus and approved by a recognized body, that provides, for common and repeated use, rules, guidelines or characteristics for activities or their results, aimed at the achievement of the optimum degree of order in a given context" [8].

Given the wide variety of standards and practically unlimited volume of information about them, research has focused on the most common families of internationally recognized standards as: ISO 9000, ISO 14000, ISO 20000, ISO 22000, ISO 28000, ISO 31000, ISO 50000, OHSAS, HACCP, EMAS, PAS 220-223, IFS LOGISTICS, IFS FOOD, BRC, TL 9000, SA 8000, etc. [8–10]. Special attention is paid to the group of standards for information security ISO 27000, ISO 15408, IEC-62443, NERC CSS, NIST CSF, and ANSI/ISA 62443.

These are standards that relate mainly to the organization and management of the main processes in a company, regardless of the sector of the economy in which it operates and to specific standards relating to its production activity. Has been made an analysis for each of researched standard in four aspects:

- Content of the standard—name, aim, policies, scope, requirements, terms, definitions, type and area of applicability.
- Infrastructure—procedures, processes, documents, controls, registers, terms, tools and functional algorithms.

- Superstructure (external environment)—correspondents, customers, suppliers, partners and institutions, connected standards, possible integration options, audits, etc.
- Resources—organizational structure and staff, assets, tools and materials, required for responsibility, planning, maintaining and control of standard mechanisms.

Defining evaluation criteria

Criteria have been developed to evaluate the information gathered from the re-search of information sources. Criteria are grouped into four categories as follows:

- Criteria reflecting user needs;
- Criteria related to the organization of the information system;
- Criteria related to system servicing;
- Characteristics of platform.

The evaluation criteria for designing the platform for administration and automated management of internationally recognized standards are related to the main objectives to be achieved when it is implemented.

- To meet the needs of the users—for administration, automation and management of internationally recognized standards
- To be as comfortable as possible
- To provide an optimal organizational structure for real-time communication and content management.

The primary purpose of the PAMS is to satisfy users' needs for interoperability and functionality. Therefore, the volume and structure of the content of the in-formation in the system must be responsive to user demand and acquisition.

Because Information Search = Consumer Behavior, Information Search is directly related to the decision-making process.

Researchers of consumer behavior argue that information search behavior is a prerequisite for making a decision. In order to evaluate the different alternatives, the decision maker must know the attributes of each of them. When such information is not available, it has to go through the information search process. In any case, the goal is to maximize the value of decision-making and to reduce cognitive efforts.

Selection of the common set of information identifiers, constituting the information content of the platform

On the basis of the research and analyses carried out, is generated a full set of information descriptors with their attributes, that can be included in the platform. Identifiers are either common for a majority of the standards, or specific for some of them, or even applicable only for element of a standard.

Analysis have shown that standards, despite their diversity, have many common elements. One of them, for example, is the PDCA (Plan −> Do −> Check −> Act)

Fig. 1 PDCA model

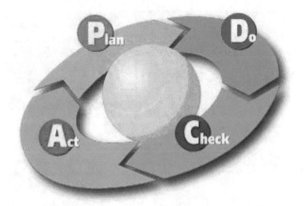

iterative four-step method, also known as the Deming circle. PDCA cycles are integral part of management and control algorithms, used for continuous improvement. PDCA model is used in ISO 9000, ISO 14000, ISO 20000, ISO 45000, ISO 27000, ISO 50000 and others. The method is used in different contexts: organization; needs, expectation and requirements of relevant parties; risk determining and assessing; process management and etc. In this sense, the descriptors of the method are used to describe a number of standards (Fig. 1).

On the basis of the developed evaluation criteria from the information array gathered during the research process, were selected the common set of information identifiers. The result is full set of descriptors with their definition and attributes, constituting the information content of the Platform for Automation and Management of Standards.

Structure of information identifiers (fields) in object-oriented information classes and subclasses, in accordance with their functional orientation in the description of the standards

The set of descriptors generated in the previous steps has been decomposed into logical information classes in accordance with their functional orientation in the description of the standards. Information classes are considered as program objects and sub-objects. They are described by their information attributes, which are called information identifiers. Their presentation is unified by using a specific technical format that defines the form of representation of the values of each at-tribute of the information objects. The technical format of the information classes and their attributes is described by the syntax of the unified modeling language (UML).

Information identifiers are structured in 5 levels information classes:

- Level 1—main class "Standard";
- Level 2—classes "Register", "Process", "Procedure";
- Level 3—classes "Document", "Document turnover";
- Level 4—classes "Node", "Application", "Activity";

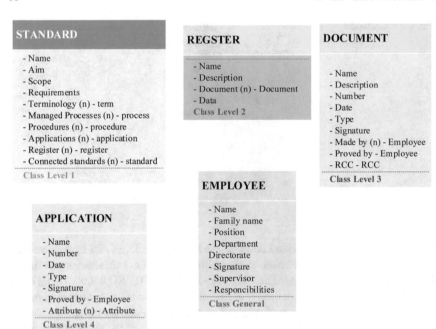

Fig. 2 Information classes

- and General level classes—"Term", "Form", "Attribute", "Employee", "RCC—Registration Control Card", "Correspondent" (Fig. 2).

Research and analysis of relationships and dependences between information objects and definition of a logical model

PAMC's mission is to integrate all working processes, associated with document flows into the client organization, model the company's document matrix, and to provide an integrated environment for digitization, storage, management and exchange. The Platform allows to keep track of the information flows, structured in-to separate types of document units and to provide reliable control.

The resulting logical model of the Platform for Automation and Management of Standards is defined after research and analysis of relationships and dependences between information objects. The model is shown in Fig. 3.

The resulting logical model is represented by class diagrams of the information classes and subclasses and their attributes and is described by the Unified Model-ling Language (UML) syntax.

The logical connections of the figure are obvious. The platform is built by adding N number of STANDARD core classes, depending on the number of standards the company uses. The base class contains all sub-classes with their determinants, needed for the particular modeled standard. In order to work the platform for a particular company, the data for the company (that is of the "correspondent" type) must be

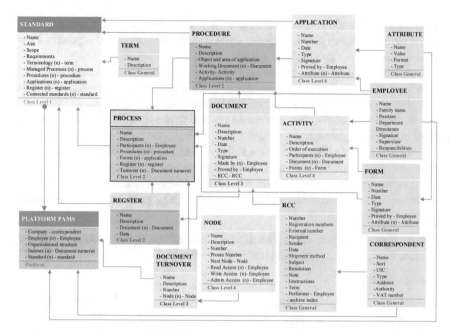

Fig. 3 Logical model of PAMS (PDCA model)

entered AND accordingly, the organizational structure and employees (employee class) should be defined.

All information is structured and classified through a well-maintained system of indices individually for each process and/or organizational structure. Each index is described with the information class "DOCUMENTS", thus providing flexibility of information flows, electronic modeling of activities and process dynamics.

2.2 Stage 2 R&A of Target Groups

The research aims to ensure the usability of the system, its competitiveness, innovative uniqueness and future adaptability in accordance with the trends of development of consumer groups. The specificity, information assurance and functional needs and requirements of the consumer groups and their business partners are analyzed, as well as the perspectives and trends for the development of the innovative product PAMS.

The following steps are completed.

Research and analysis of automation and administration processes

After analyzing the literature relating to internationally recognized standards and their operational management and exploring the practical aspects, the different phases of the automation and administration processes were defined, broken down by functional modules of the platform in the context of the behavior of the potential user groups in functional and hedonic aspects.

Consumer processes are examined and analyzed in the context of the platform's functional modules. The purpose of the analysis is to clarify what needs to be done to realize its functions. This is done by decomposing the functions of the system into their constituent elements and creating a logical model of the processes and data flows necessary for their implementation.

Prior to performing the logical conclusions is made at the physical level an analysis of the movements of the documents between the relevant processes and departments of the existing environment. The scheme of this movements allows to define the key tasks and to evaluate a number of aspects of the management and effectiveness of the system used.

Research and analysis of the functional needs of consumer groups, related to information assurance and electronic interaction

The results includes:

– Analysis of the functional needs and requirements of end-user groups differentiated according to the main world markets and their characteristics—economic development, specific traditions, level of use of technology and innovation, etc.
– Defining criteria and shaping target groups on a functional basis.
– Study of world literature, statistics and best practices in the field.

Potential customers of the product are private companies or public organizations that have certified or implement one or more internationally recognized standards. These are legal entities from all sectors of the economy engaged in various economic activities. Companies that have adopted internationally recognized standards are characterized by sustainability, quality of work processes and production, and a long-term vision for development.

R&A of the functional needs of audit organizations related to information assurance and electronic interaction with organizations introduced the standards

Study of the functional needs and requirements of business partners differentiated according to the main world practices and their characteristics—economic development, specific traditions, level of use of technology and innovation, have been made.

Certification is an objective verification performed by an independent competent certification body that seeks to determine whether a management system implemented in an organization meets the requirements of one or more international standards to which it has been developed. Compliance with the requirements of the international standard is verified by issuing a certificate of conformity.

Certifying organizations are a major business partner in the environment of the Unified Platform for Administration, Automation and Management of Internationally Recognized Standards. They carry out certification and control audits to verify compliance. The functional needs of the audit organizations related to information assurance and electronic interaction with the organizations that introduced the standard are directly derived from the certification process and its nature.

Research and analysis of trends in the development of consumer groups

Correspondingly have been made research and analysis of trends: in the development of consumer groups; in the development of internationally recognized standards; and trends in information and communication technologies. Sources are world literature, scientific publications and statistical indicators.

The evolution of document management is intrinsically linked to different generations of content, and for this reason the future of the application is mostly digital. Digital content evolves into two key aspects that directly affect the professional and consumer environments. The first concerns the volume of content generated. The size of the so-called digital universe is doubling every 18 months and this requires the interconnections between business departments of companies to be transformed. Policy development throughout the information management and security organization needs to be promoted and expanded. In the digital universe, actively created digital information is much less than the passively generated information that accumulates without the knowledge and consent of humans.

The huge amount of digital content, ease of use and social technologies are behind people's expectations. Today, software companies are required to offer a set of consumer senses that allow employees to handle corporate information in the way they handle personal information—easy access and communication with other employees. By bringing together the best of customer service models with proven enterprise content management capabilities the platform reach a much wider range of employees, creating new collaboration models between them.

Defining the basic functionality of the platform

Based on the research and analysis, the identification and description of the optimal functionality of the platform and the directions for its development in accordance with the identified user needs were made.

The functionality of the Unified Platform for administration and automated management of internationally recognized standards is conditionally divided into three main groups:

- Administrative functionality
- System functionality
- Operational functionality.

The **mission** of the platform is to integrate all working processes related to document flows into the client organization. It replaces many application software products working in this field (record keeping, document flow, archive, human resources, CRM, etc.) with a single information and communication and management system. The platform integrates information sections and information flows into a single information space and provides an integrated storage, management and sharing environment. This allows effective monitoring of the information flow, structured in different types of document units and reliable control.

The Platform modules build models related to managing different types of internationally recognized standards with rich tooling—typed processes; pre-developed scenario and scripts; a set of predefined templates; conventional registers developed; specialized registers; internal communication environment; group and individual video calls and conferences in real time; audit evaluation, control and support systems and other.

The platform provides the opportunity for further development and implementation of both new modules and elements to existing modules. The PAMS enables the flexible introduction of language versions of system interfaces based on the principle of "dictionary matching".

2.3 Stage 3 R&A of Communication Channels and Technical Capabilities.

The stage aims to analyze all possible communication channels for interaction with consumer groups and to identify possible innovative methods based on combinations of communication relationships and alternative automation methods relatable to consumer groups. The following analyzes were made:

- Research and analysis of the applicable communication capabilities and technical methods for interaction between users in the user group;
- Hardware research and analysis;
- Research and analysis of technological and operational software;
- Research and analysis of the technological architecture of the platform.

The widespread use of Internet technologies and the development of ICT in the world has led to the emergence of a network economy. It is also characterized by the formation of a critical mass of economic agents, the reorganization of forms of joint activity between enterprises, the modernization of infrastructure and the creation of networked institutional structures [11, 12].

Conceptually the unified Platform for administration and automated management of internationally recognized standards PAMS is an open platform with a multi-layer, heterogeneous, client–server, service-oriented architecture (SOA) based on component technology that enables the development of Business on Demand solutions, using of existing building blocks. Each adaptation of the system represents the necessary set of functions and sub-functions of the system modules, optimal for the client organization [13, 14].

The architecture of PAMS has the following major advantages:

- Independent functioning of the functional modules as autonomous systems;

- Combining the various modules in an 'on demand' solution that meets the client's needs and requirements. Pre-built functional modules can also combine partial functionalities without having to implement the entire module;
- Step-by-step implementation of the functional modules or completion of the built-in solution without interrupting the workflow of the entire system. The specific solution for each client company can be built or supplemented according to the management's vision and the proposed timelines, ensuring robust functionality.
- Multiplication of unified solutions for autonomous substructures.
- Possibility for adaptability and operational change—flexibility and resilience.

An internal web portal provides functionality of a single point of access to particular applications within the platform. It is the user's entry point to the PAMS resources. Additionally, the portal implements a range of information services, contact information and purely portal features as: news, forums, messaging and more. The portal integrates Single Sign-On technology to unify user profiles across the different system components and to reduce the complexity of administering the systems, available through the portal. Upon registration, the user is granted access to specific modules and functionalities, including the execution of single requests for reports from them.

The external portal is used by external users of the system—contractors and auditors. In order to comply with the requirements for communication and information security, the external portal is divided into a public part and an internal part, separated by a firewall. Electronic services as well as means of electronic communication is provided. In addition, reporting information services are implemented as well as portal functions such as forums, news and other functionalities of modern content management systems (CMS).

2.4   Stage 4 Reengineering of Working Processes in the User Groups and Audit Organizations

The main objective is to study workflows and to develop optimal work algorithms for integrating the innovative solution into the conventional business model of consumer organizations.

The following studies are parts of this stage:

- Exploring the points of contact and the possibilities for integrating traditional activities and innovative activities for servicing management systems of internationally recognized standards;
- Research and development of optimal procedures for servicing the work processes related to the operation of the innovation system;
- And researching and formulating a proposal for organizational assurance.

PAMS automates the procedures for the implementation of the entire workflow through predefined electronic routes for all types of documents and objects.

Routes are defined following the subordination of optimal organizational structures and teams. The documents automatically pass through different participants in the process, ensuring uniform objective standards and rules that meet the different types of activity for all participants. In this way communication between the different participants in the process is realized and quality order of execution is guaranteed.

The platform provides tracking and control of task activities subject to time constraints and processes, providing departments and end users with detailed information on the current status of each task, its steps, stakeholders and related responsibilities.

The implementation of a unified Platform for the administration, automation and management of internationally recognized standards requires additional staff competencies. In performing the activity, the existing organization and accordingly the job descriptions and profiles of the employees engaged in maintaining internationally recognized standards in the company are examined. Optimal organizational assurance is elaborated.

Optimal integration model

With new tools such as PAMS, much of the volume of activities that are conventionally performed to maintain standards in accordance with their different requirements are modeled and automated by the platform. There are new operations that need to be performed to service the system, and at the same time, manual operations that the platform automates have become unnecessary.

During the phase, the volume and scope of the duplicate activities were identified. An integration model has been developed that optimizes workflows in the activities of the user organization operating the platform. Variations of the platform's operating procedures are being developed to obtain an optimal model for simplifying workflow algorithms and meeting user group needs (Fig. 4).

The optimal integration model meet the needs of the customer groups and exploit the added value of the unified PAMS platform without hampering the workflows of employees of the company. If additional operations are to be carried out by the employees, it should be used as a lever for increasing the added value of the PAMS, which as a result should reduce the overall volume of operations and increase the efficiency and effectiveness of the teams and the company as a whole.

3 Conclusions

In the last few years, companies have shown increasing interest in the introduction of internationally recognized standards. The main motive for this is the recognized need to improve their management processes and to create opportunities for competitive advantages. In the highly demanding common market, customers prefer those who hold certificates that guarantee compliance with international standards.

Efforts are directed towards the development of procedures for the implementation of activities that interconnectedly lead to the improvement of processes ensuring high

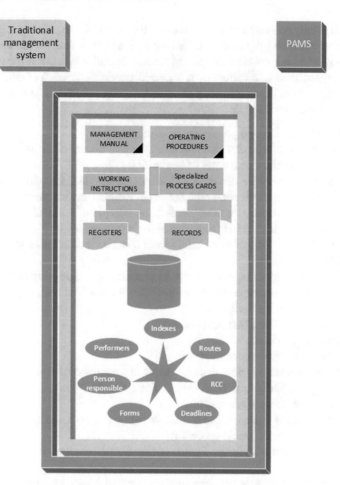

Fig. 4 Integration model

quality and security of products and services. Adopting this approach is about realizing a better economic alternative to investing preventively in refining the processes to ensure the outcome, rather than taking action after defects have been identified and the losses incurred are irreversible [12].

The results of implementation of unified platform for administration, automation and management of internationally recognized standards in an organization can be compared to the effect of introducing the production line into a manufacturing enterprise. It speeds up the transmission of data between different units, setting out the precise regulation of the actions at each stage of the data processing. As a consequence, it reduces the number of mistakes made, increases the specialization of the employees in carrying out the individual operations and increases the productivity of labor. The organization is able to carry out more business operations without increasing the number of its employees.

Traditionally, efficiency is assessed on the basis of the relationship between the performance and the cost of achieving it. It can therefore be increased in two ways—by reducing costs or increasing results. A unified platform for automation and management of internationally recognized standards makes it possible to implement both options simultaneously. The added value of PAMS and its functional advantages are expressed in:

- Ensure registration and processing of data;
- Summary of the basic data that determine the operational picture of the company;
- Ensure traceability and control of operations;
- Provision of up-to-date communication tools to enable fast and efficient connections. Better communication with the organizational environment;
- Provide the necessary prompt, up-to-date, complete, consistent and objective information at all levels of the organization's management;
- Ensure the protection of information from unauthorized access and in-formation security;
- Ensure uniformity of input and output, i.e. achieving single entry of data and their subsequent processing and their use in different functional units depending on the specificity of their activity;
- Accelerated decision-making process;
- Control of decisions;
- Monitoring of work processes;
- Comprehensive and correct documentation in accordance with the standards used;
- Tidiness, availability and consistency of company knowledge;
- Automated processes and audits;
- Digital and easily manageable records based on pre-prepared samples;
- Traceable and manageable processes;
- Indexed archives with the ability to quickly search and find the information you need;
- More productivity, better communication, improved team effectiveness and faster decision-making;
- Optimized reuse of information, easy shared access to documents and quick retrieval of a specific document;
- Centralized maintenance, archiving, administration and data recovery if needed;
- Reduction of paper documents;
- Reduction of non-productive working hours of employees;
- Speeding up information flows;
- Minimizing costs from losing intellectual capital;
- Minimal risk of customer dissatisfaction;
- Changing corporate culture.

The platform collects and digitizes unstructured data from the external environment of the company and, together with the internal information flows, organizes it into a structured, well-set information sphere.

References

1. Laudon, K.C., Laudon, J.P.: Management Information Systems: Managing the Digital Firm, 15th edn. Pearson Education, Harlow, UK (2018)
2. Marshall, G.H.S.: Evaluating Management Standards: Empirical Research into the Scottish Quality Management. University of Stirling (2006)
3. Jorgensen, T.H., Mellado, M.D., Remmen, A.: Integrated management systems—three different levels of integration. J. Cleaner Prod. **14**(8), 713–722 (2006)
4. Feigenbaum, A.V.: Total Quality Control: Achieving Productivity, Market Penetration, and Advantage in the Global Economy, 4th revised edn. McGraw-Hill Education (2015)
5. Kleppe, A., Warmer, J., Bast, W.: MDA Explained: The Model Driven Architecture Practice and Promise. Addison Wesley (2003)
6. Organization for the Advancement of Structured Information Standards (OASIS): Web Services Business Process Execution Language (WSBPEL) Technical Committee. https://www.oasisopen.org/
7. IT Service Management Standards: A Reference Model for Open Standards-Based ITSM Solutions, An IBM White Paper, April 2006. https://www-07.ibm.com/sg/governance/servicemanagement/downloads/itsmstandardsreferencemodel.pdf
8. International Organization for Standardization (ISO). https://www.iso.org/home.html
9. Bulgarian Institute for Standardization. https://www.bds-bg.org
10. National Institute of Standards and Technology: Information Technology Standards. Standards Incorporated by Reference (SIBR) Database
11. Resmini, A., Rosati, L.: Pervasive Information Architecture: Designing Cross-Channel User Experiences. Morgan Kaufmann, Burlington, MA (2011)
12. Patterson, D.A., Hennessy, J.L.: Computer Organization and Design; The Hardware/Software Interface, 5th edn. Elsevier, Amsterdam (2013)
13. Bass, L., Clements, P., Kazman, R.: Software Architecture in Practice. SEI Series in Software Engineering, 3rd edn. Addison-Wesley Professional, Boston, MA (2012)
14. Clements, P., Bachmann, F., Bass, L., Garlan, D.: Documenting Software Architectures: Views and Beyond, 2nd edn. Addison-Wesley Professional, Boston, MA (2010)

Digitalization and Cyber Resilience Model for the Bulgarian Academy of Sciences

Velizar Shalamanov, Silvia Matern, and Georgi Penchev

Abstract The paper is based on a three-year internal study on digitalization and cyber resilience of the academy of sciences in Bulgaria, and supported during the last months by the National Research Program for ICT in science and education and the ECHO project of H2020. A successful model for digital transformation is conceptualized, based on the best practices in IT capability development and service provision. It aimed to respond to several questions as how to manage projects, services, funding and, most importantly, people to develop, operate and protect ICT infrastructure and applications. The study explores options for these four areas and suggests the approach to select the most suitable, as well as how to integrate them through effective system for governance and management in the specific environment of the Bulgarian Academy of Sciences as a national academic institution for both fundamental and applied research and education.

Keywords Digitalization · Cyber resilience · Change management · Portfolio management · Service management · Personnel management · Innovation management

V. Shalamanov (✉) · S. Matern
Institute of Information and Communication Technologies, Bulgarian Academy of Sciences, Sofia, Bulgaria
e-mail: shalamanov@acad.bg

S. Matern
e-mail: silvia.totseva@gmail.com

G. Penchev
University of National and World Economy, Sofia, Bulgaria
e-mail: gpenchev@gmail.com

© The Author(s), under exclusive license to Springer Nature Switzerland AG 2021
T. Tagarev et al. (eds.), *Digital Transformation, Cyber Security and Resilience of Modern Societies*, Studies in Big Data 84,
https://doi.org/10.1007/978-3-030-65722-2_6

1   Introduction: Call for Digitalization of the Academic Institutions—The Case of the BAS

Academic institutions are typical knowledge-based organizations. In order to be used effectively for education as well as directly by industry their products need to be easily available. This could be achieved if the results are digitalized and offered in the network environment or in the cloud. In addition to planned research and education projects inside academic institutions there are unique instruments and scientific equipment at large that could be effectively utilized by other users, including industry if the access is easy and secure.

Investments in research infrastructure are very expensive and the best way to have a good Return on Investment (RoI) is to make this infrastructure largely available through customer funding for projects and it is even better to be provided as a structured set of services (Costed Customer Services Catalogue).

Digitalization of a large academic institution with many institutes and laboratories is a challenging task, because of the autonomy of the elements within the organization, their specifics of research environment and the characteristics of its organizational culture.

At the same time, as for every large organization there is a need to start with clear Mission, Vision and Strategy for digitalization in the context of the governance and management (business model). This will set up requirements to be able to address in change management effort [1]:

- Partnership/customer relationship management;
- Technology and innovation management;
- Research management;
- Personnel and competencies management, incl. education and training;
- Infrastructure and applications planning and development within projects, compliant with and project portfolio management;
- Service management (based on available infrastructure and software applications);
- External contract management;
- Funds management.

It has to be noted that the clarification of the roadmap of the digitalization has two levels. Firstly, for all the fields, covered by the academic institution, to include ICT research and education and secondly for the very specific field of digitalization of the academic institution, focusing on ICT research and development, education and training in support of digital transformation and cyber resilience of the academic organization itself.

This paper presents the experience gained in past 3 years in defining and study of the changes needed at the Bulgarian Academy of Sciences in order to develop and implement successful digitalization strategy. A successful digital transition at the Bulgarian Academy of Sciences will dramatically improve its contribution to the development of the country (as it was envisioned by its founders 150 years ago). At the same time, it will prove the ability to implement similar digitalization model for public administration, industry and society as a whole in the context of Digital Europe initiative[1] of the EU and NATO's Digital Endeavor,[2] where the country participates.

Specific aspects of the partnership and customer relationship management, technology and innovation management, research management, personnel and competencies management (especially education and training) and external contract management are under development within the concept of BAS digitalization, as separate, but interrelated research activities. This paper is focusing on projectized model for the development of ICT systems, transition to service provision—internal and external, funding models for ICT support. In this context some key aspects of selection and development of human resources and innovation and technology management are covered.

The general model for digitalization and cyber resilience of an academic organizations used in this study is based on the experience of the Digital Endeavor in NATO along the lines of establishment of NATO Communications and Information Agency (NCIA) as the primary service provision and capability development agency in NATO.

2 Concept of Digitalization and Cyber Resilience in BAS

Digital transformation is a transformation driven by the digitalization of processes with improved effectiveness, efficiency and cyber resilience. In essence, it is a transformation with all the challenges of this fundamental change and requires real strategic agility [2]. The model of transformation as a spiral involvement of different stakeholders is represented on Fig. 1.

Transformation in the process of digitalization requires specific Information Resource Management Organization (IRMO) within the BAS organizational structure with several key issues to be addressed as depicted on the mind map on Fig. 2.

The result of the transformation will be a transition to the new type of IT infrastructure and applications/cyber security arrangements along with improved processes at the BAS. Such a transition has key aspects to consider as shown on Fig. 3.

The Operational and Maintenance (O&M) costs for the old capability will increase with each new project and new requirement, as well as with amortization of the assets. We suggest linear (additive) trend of these costs, but if new research projects impose requirements near or over the capability capacity then the costs can grow rapidly [3].

---

[1]Digital Europe program: Funding digital transformation beyond 2020: https://www.europarl.eur opa.eu/thinktank/en/document.html?reference=EPRS_BRI(2018)628231.

[2]The effort to modernize NATO's infrastructure is known in the NATO Communication and Information Agency as the Digital Endeavor. See this page of the Agency: https://www.ncia.nato.int/ NewsRoom/Pages/2019.03.019.aspx.

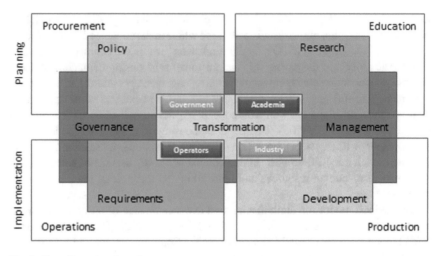

Fig. 1 Transformational quadrants

Fig. 2 Mind-map for change of IRM

The Initial cost related to all activities for implementation of the new capability (planning, R&D, procurement, training, etc.) will increase at the beginning, especially with procurement of new equipment. After the initial growth these costs will decrease during the setup, testing and validation.

The old capability has to be maintained during the transition phase until the full implementation of the new one. This may cause some additional costs. The O&M for new capability will be less low than suggested costs for the old capability, because of

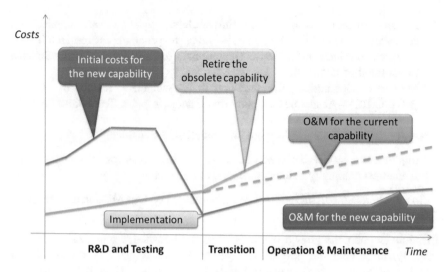

Fig. 3 Cost—Time implications of the digital transition

the sharing of the assets and activities within different projects, improved processes and use of good governance practices [4].

In order Information Reseource Management Organization (IRMO) to be successful the Academy needs establishment of Chief Information Officer/Chief Information Security Officer (CIO/CISO) institution for the Academic Community [5] and with the recent developments around GDPR[3] there is a need for (personal) Data Protection Officer (DPO).

The establishment of CIO/CISO can be organized through the following possible steps:

Digitalization is to be considered as a change management effort:

2020—preparation

1. Stock taking project for IT assets/services in BAS/academic community and CIO certification program preparation.

2021—consolidation

1. Service Catalogue and costing;
2. CIO for different organizations in the community and CIO Council with CIO certification program;
3. Architecture of the IT infrastructure/applications.

2022–23—rationalization

---

[3]GDPR—General Data Protection Regulation (https://gdpr.eu/).

1. IT Modernization (improve the infrastructure and software applications according to the architecture agreed)—along the programs for research infrastructure and centers under the EU Operational Program "Education and Science for an Intelligent Growth";
2. Network Operations Center/Computer Incident Response Center (NOC/CIRC)—Academic Computer Emergency Response Team (CERT)/NOC.

2024—optimization (incl. Mandate for unified ICT provider)

1. Implement a service based and customer funded model for IT support of the academic community as a model for the public sector.

In this context the IRMO in BAS could be defined with its Mission and Vision as follows.

Mission: Support the growth of the BAS as the knowledge organization of 21st centurty, preserving the best of 150 years of success and leading the digital transformation of Bulgaria as part of Digital Europe. in the EU Digital Decade

Vision: Start as virtual, project based organization with CIOs' Council of BAS bodies under the Council of Directors. Federate the existing digital infrastructure to create a data rich environment for research and knowledge management to maximize the efficiency of using available computing power (including the HPC assets), specialized digital input/output periphery and specialized research/scientific appliances. Immprove the access to the European digital infrastructure for research and knowlededge management and enhance Bulgariana contribution to the European digital space for research. Enable the work of scientific staff and guarantee cyber security of the digital research environment.

In parallel implement the best practices for the BAS management system to reduce time and efforts of the sciencitfic staff dealing with the administrative tasks and radically improve management processes in support of high quality reasearch environment in line with the government effort to establish e-Government environment for the Public Administration.

Implementation of the Vision includes measures on 3 levels:

1. Governance and management (processes and organization)—establishing the function of CIO/CISO (DPO/GDPR officer) for most effective, efficient and cyber resilient development of the e-Infrastructure of the BAS and its bodies.
2. Technology—optimize the network and establishing effective academic NOC/ academic CERT for professional management of e-Infrastructure.
3. Human capital—improve the professionalism of IT personnel in BAS and effective use of outsourcing on competitive basis to professional IT companies to impement decisions talen on governance level and optimize the use of available technology

This transformation is a typical change management task to be followed with the continuous improvement process [6] until the next turning point in technology, processes, organization or human capital appears.

Key changes to be considered in support of digital transformation are as follows:

- Identification of BAS bodies CIOs and establishment of CIO Council for BAS with supporting change management project;
- Architecture and Federated Mission Networking process (mission is BAS to operate as a Knowledge-based organization);
- Security—Academic CERT, Academic NOC/CIRC;
- Catalogue as an Unified list of Services (with costs):

  - Data Storage;
  - Computers;
  - Printing;
  - Scanning;
  - Internet access;
  - E-mail;
  - Portal management;
  - Document Handling System (DHS);
  - Tasker Tracker Enterprise (TTE);
  - BAS App Store.

- Transition from asset management to service management;
- Transition to costed services/paid services and related ICT investment management;
- Professionalize CIO/administrators and unify the service provision organization.

Road map for change will cover following activities:

1. Map existing situation;
2. Identify resources, services, customers, cost;
3. Consolidate through federation;
4. Rationalize around services;
5. Optimize against cost and value of the services.

Digital Transformation for effectiveness, efficiency and savings/resilience will address several key aspects as depicted on Fig. 4.

Therefore, a Global Change/Improvement Plan could be discussed in following 3 steps:

- Establish in the BAS a virtual Academic Communication and Information (C&I) Organization, based on the best practices from NCIA (incl. Academic CIRC/CERT and National C&I Academy with focus on CIO certification);
- Support the improvement process for C&I organization for the security sector in line with the best practices in NATO/EU, based on projects between MoD/MoI and BAS including and the Defense Institute;
- Support the improvement of C&I organization for the Government, based on projects SA e-Government—BAS.

In order to achieve an agreement on the digital transformation a discussion is needed on the following topics:

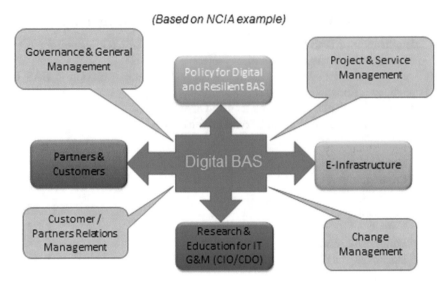

Fig. 4 Aspects of digital transformation of BAS

1. Do we need change in IRM area for BAS?
2. Could we agree on CIO/Council of CIOs concept?
3. Do we want to start a project (internally or funded by external good governance program) for improvement of IRM in BAS?
4. What could be the mandate for ICT institutes and laboratories in order to prepare the project (including commitment of human resources from different BAS bodies)?
5. What is our level of ambition in improving IRM?
6. Could we agree on NATO C&I Organization as a model for IRM improvement in BAS?
7. Do we want to present IRM improvement project in BAS as a model for IRM improvement in public administration?

In the framework of this general model the following 4 key issues for improvement need to be considered: project management, service management, funding, innovation and personnel management. The change could be managed as a portfolio of transformational projects [7].

3 Projectized Model for the Development of ICT Systems

Development of ICT systems as part of research projects is quite ineffective, because every project is limited as a funding and as aims to create a unique system to support the research goals of the specific project. The Academy can establish ICT program or at least a portfolio of ICT projects [8]. Thus, on one hand will provide required

Table 1 Matrix of Research Projects (RPs) and ICT Projects (IPs)

	RP$_1$	RP$_2$	RP$_3$	...	RP$_m$
IP$_1$		X		...	X
IP$_2$	X	X		...	
...	...	...	...	...	...
IP$_n$			X	...	X

services for the specific research project, where funding is provided. On the other hand, this will assure compliance with the general architecture for the ICT system which will support the academic institution as a whole in long term.

In such a way the Academy will have a matrix of Research Projects (RPs) and ICT Projects (IPs), where RPs will provide Requirements and Funding (RF) to IPs, but IPs will contribute to the overall architecture of the academic ICT system. It means one RP could be served by several IPs and one IP project could serve several RPs (see Table 1). With this matrix structure the Academy could improve its effectiveness, efficiency and realize overall cost savings. In addition, the architecture could focus on secure by design approach, providing cyber resilience of the resulting infrastructure.

Requirements and Funding (RF) will be integrated on annual basis within BAS as a sum of RF$_i$ planned within different RPs. Thus, the Academy could plan and use budget to fund several IPs to meet the RPs requirements and activities throughout the year (and beyond). This could be established potentially as a service with respective Service Level Agreement (SLA).

This approach gives opportunity to consolidate the ICT team and focus their professional efforts on effective and efficient implementation of the set of IPs—{IP} for optimal support to the set of RPs—{RP} over the year.

Implementation of this projectized approach to ICT support of course will require quite mature approach to the RPs as well, especially well defined ICT requirements in their project plans with clear costing from the RPs budget. Taking in consideration these prerequisites, the Academy have to stress that well established Portfolio, Program, and Project Management (P3M) system [9] will be used both for RPs and IPs.

Development of P3M system could start with the IPs, where more unified approach is possible, especially if a well-established framework of best practices is used—as an example—PRINCE II. Such an approach could facilitate development of a professional Project Management Office (PMO) that could be used in the future for support of the RPs' P3M system. PMO have to be not just an operational body to support the effective project management, but also have to be a training team to develop professional Project Managers (PM) for the organization (and outside, if it is decided). Introducing Portfolio Management and Program Management is a logical next step in the development of an effective projectized organization.

Project management will include all the procurement activities and especially in an academic environment with dominantly public funding, there is a requirement to establish proper acquisition system [10] to serve this large and diverse projectized organization.

4   Transition to Service Provision—Internal and External

With increase of the maturity level of the ICT system and its governance/management the Academy could consider the transition to service oriented model. In this case RP "buy" services (IS) for a fee with a SLA between RP (research team) and internal (or external) ICT organization for the academic institutions. The ICT organization manages projects for improvement of the ICT services and adding new services, funded by these fees from the SLAs.

It is certain that there always will be a need for massive investment in new equipment for new services. These investments might come from a big customer such as the Ministry of Education and Science (MoES), for example. This will define mixed model of funding for the ICT organization.

The transition to service provision is an evolutionary process to include [11]:

1. Definition of services;
2. Costing of services;
3. Development of Catalogue of services;
4. Implementation of SLAs—costed as a whole or based on costed as set of services/fee for service information;
5. Organizing of Service Desk to support users according to the SLA arrangements;
6. Establishing of user/customer satisfaction program for continuous improvement of the services;
7. Introducing investment program for the new services.

Such a transition for sure will require certain level of maturity of the projectized model, described above with better planning from the point of view of the {RPs} of the required services and their respective levels. Thus, the RPs will be able to negotiate SLAs for the next year well before the end of the current year giving time for the ICT support team to organize the SLAs implementation.

Because the journey from distributed asset management to consolidated service management is quite challenging the establishment of roadmap with well-defined maturity levels is required [8].

5   Funding Models for ICT Support

The most important aspect for each funding model is to keep well-defined records of expenses and their relations to the expected results. There are two crucial conditions to have an appropriate solution for expenses tracking: (a) a good project management for investment in new capabilities and an effective service management are established, and (b) the acquired equipment is used to provide services with defined level of performance to the users [12].

In principle, there are different funding sources for ICT [13]. The usual sources can be identified as follows:

1. Budget (state or local) for all IT activities;
2. ICT items in RPs budgets of one or several organizations, provided a pool of resources or specifically dedicated to IP subproject;
3. Dedicated IPs with state or local funding;
4. Trust funds for ICT projects/services;
5. Donations, dedicated for ICT projects/services, incl. ICT vouchers;
6. Others.

Other aspect of funding is the transition from input to output costing. The output costing is also called Unit or Single Costing. One established practice is to define the internal structure of expenses of the ICT (internal or external) organization of the academic institution or directly to define standardized costs rates, based on previously accepted assumptions about total costs, unit's costs, quantity, quality, personnel, etc. Having these standardized rates, it is easy to provide for each project or service transparency and actual expenses monitoring for external services and materials [14].

In this model it is important to differentiate between two types of customers— internal, eligible to receive funds from state investments, made for example by MoES for the infrastructure and external customers. Second type of customers is not covered by state investment during the design, development and transition of the service, so they will have lower priority and in order to use the services the fee imposed to these customers need to include justifiable share of the investment made for the development of the service.

Of course there is a third type of customers, who could be partially subsidized, but they still have to be considered as an external customer.

It is important to introduce the priority levels of the customers, or their requirements in order to define the mechanism of using limited capacity in competitive environment and to define this through the funding mechanism.

Funding is related to optimization of the process for effectiveness, efficiency and savings—a study that is presented in a separate paper.

6 Selection and Development of People, Innovation Management and Technology

The most critical resource in any ICT organizations are people. The academic ICT organization is not an exception. Therefore, the above approach that provides opportunity to develop ICT service center have to create an attracting environment for highly qualified personnel that could serve all RPs through IPs and IS. In this case one additional benefit is the improvement of cyber resilience at lower cost, because of consolidation of the ICT infrastructure and personnel.

Special focus is needed on hiring, continuous professional development, retaining of people, research based education and training, including PhD education. This study is also presented in a separate paper [15].

Critical for addressing this challenge is the level of knowledge on IRM Change management in the following areas:

1. Successful models of change management;
2. *Mission, Vision and Strategy* of the effective and efficient IRM and a mandate for change for 5 years;
3. Governance, funding of change and business case;
4. Leadership: Selection of the professional leader of change, management team and oversight of mandate implementation;
5. Planning and management of successful change program and projects/portfolio;
6. Personnel management, selection, assessment, development/rotation;
7. Risk management and the role of consultants in different scenarios;
8. External and internal audit missions and corrective plans;
9. The role of technology and innovation—the challenge of CTO (Chief Technology Officer);
10. Strategic communications and cultural change;
11. The most important challenge—the challenge of the Chief Customer Officer (CCO)—customer centric change management, development of partnerships with customers, industry and sister organizations.

BAS IRM improvement needs a change management team development course focused on the following topics:

1. BAS as a knowledge based public organization in support to the country/society;
2. CIO roles and responsibilities in IRM: IT Strategy;
3. IT Architecture and Federated Mission Networking;
4. BAS Apps;
5. Information Assurance;
6. IT Shared Services models;
7. CTO roles and responsibilities: IT taxonomy and Innovation management;
8. COBIT 5 for IT Governance;
9. Management Successful Programs (MSP);
10. Prince II for IT project management;
11. ITIL V3 for IT service management;
12. Customer funding models;
13. Service Based Organization;
14. Customer (Internal, External) Relations and Partnership Development;
15. BAS IT investment management for new capabilities/services;
16. IT Acquisition;
17. BAS IT Service Catalogue management;
18. BAS IT SLA development and management;

19. Service Operations Management;
20. Enterprise Business Application/Enterprise Resource Planning (EBA/ERP) for IRM/financial management of break-even type (to be able to reach break-even point in 3 year period for example—typical for non-for-profit organizations under governance level supervision of funding organizations).

Besides people, the technology and innovation area is the other key factor for success, related to effective, efficient and resilient use of ICT. It requires special focus on innovation—innovation hub for digital transformation and cyber resilience with required ranges and BEST (Basic Environment for Simulation and Training). This area is the topic of an additional research, which results will be presented in a separate paper.

7 Governance and Management as an Integration and Change Management Tool

When the set of projects, services, people and funding sources are well defined, it is mandatory to establish an appropriate system for governance and management in order to achieve effectiveness, efficiency and resilience of the ICT system [1].

In the specific case of the Bulgarian Academy of Sciences there are more than 40 research institutes and laboratories with large autonomy. Each of these organizations are running between 20 and 100+ projects, providing internal and external services and operating its own ICT system. It is obvious that the task of digitalization goes through processes of consolidation, rationalization and optimization of the current ICT arrangements. Successful execution of these processes will provide the position for the Academy to address the common challenges of digitalization and cyber resilience. Successful and effective ICT organization will assure higher level of effectiveness and success for the Academy, as a whole. These challenges require governance and management arrangements for the complex network environment within the Academy. The level of complexity becoming even higher if external relations are added to the model—relationships with MoES, Government as a whole, industry and society as well as foreign/international partners.

At the same time the complexity, scope and academic nature of the environment is inceptive for experimentation and innovation to develop and validate solutions in the area of digitalization and cyber resilience that could be transferred to public administration, industry and society as a whole.

The presented results of the initial research on applicability of known best practices in the Academy are mostly on management level. An additional research will be required for governance level. This was one of the reasons to establish a Consultative Council of effective, efficient and cyber resilient management of the ICT resources to the Chairman of the Academy.

The key areas of decision making for the governance framework are as follows:

1. Level of ambition for digitalization of the Academy and contribution to the digitalization of the country;
2. Mandate for the digitalization of the Academy;
3. Vision of the ICT organization to support the digitalization;
4. Business model, organizational development and in particular personnel development;
5. Innovation management;
6. Funding;
7. Strategy for digitalization and the implementation plan.

The initial research during the last 2 years provides evidence to propose the model for digitalization of the Bulgarian Academy of Sciences as discussed above.

The successful implementation of this approach has a potential to be replicated to more challenging areas, such as digitalization of the public administration, industry (case by case) and other aspects of country's social and economic life.

The idea that the administration itself, one or another university or private academic organization could fill this gap, if the Academy is not able to digitalize itself and to provide academic support for the digitalization of the country, has a low probability of success, because of:

- Scope of the challenge;
- Complexity, comprehensiveness;
- Human resources and knowledge level required.

If the Academy of Sciences is not leading in the process of digitalization and cyber resilience, there is a high probability the academic support for this effort in public administration and industry to come from abroad and it will marginalize the Academy in contrary to the mandate given by its founders 150 years ago.

8 Conclusions

Summarizing this initial research, it central focus is the proposal to embrace the model for digitalization and cyber resilience of the Academy of Sciences. The validation and following adaptation of the model for the much more complex process of digitalization of the public administration and the industry are considered as a high priority joint task between the process participants.

The main challenge is to agree and to fund the Digitalization and Cyber Resilience Strategy, supported by the 3 year Change Management Plan in order to design **optimal processes** and to develop **organizational structure** around these processes with required **technology** and **people** for effective and efficient operation.

The optimization of the processes is the most stable fundament for the change, based on agreement and commitment to be implemented for the benefit of the Academy, not for the protection of private interests.

The very first step is the nomination of Chief Digital Officer (CDO or just CIO/CISO) with a mandate for digital transformation and cyber resilience of the Bulgarian Academy of Sciences or to select such a person based on the agreed mandate and required competencies. The selection should not be limited to Bulgarian citizens only, if the goal is to be successful and competitive in the digital era.

The implementation of the initiative for digital transformation and cyber resilience of the Bulgarian Academy of Sciences could lay the fundaments for the national programs on digital transformation and skills, cyber security and trust—3 of the key areas of the Digital Europe program, together with High Performance Computing and Artificial Intelligence.

It will require from the very begging the initiative for digitalization and cyber resilience of the Academy of Sciences to be aligned with the Digital Europe and Horizon Europe developments in order to benefit from and to contribute to these EU intergovernmental level programs. Currently, the legal framework for these initiatives in Europe is also in process of development.

Most advanced is the work in the area of European wide governance and management of the European Cybersecurity Industrial, Technology and Research Competence Centre (the 'Competence Centre'), supported by Network of National Coordination Centres coordinating the Cybersecurity Competence Community (see Fig. 5). In this context first step in implementation could be driven by establishment of Academic CERT to participate in the Cybersecurity Competence Community under the National Coordination Centre of Bulgaria to be nominated in 2021. Initial model

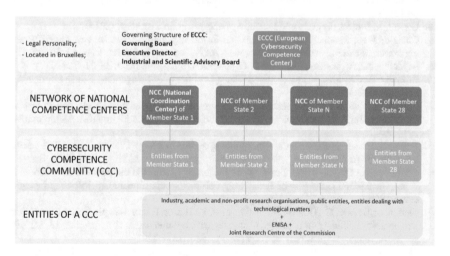

Fig. 5 Organization of the EU network for cyber security research and technology development (envisioned in the draft proposal for European Parliament regulation (This Regulation establishes the European Cybersecurity Industrial, Technology and Research Competence Centre (the 'Competence Centre'), as well as the Network of National Coordination Centres, and lays down rules for the nomination of National Coordination Centres as well as for the establishment of the Cybersecurity Competence Community))

for such a consolidation [16] is developed as part of the study, described here, used currently on the work performed for the ECHO project.

Acknowledgements This work was supported by the ECHO project, which has received funding from the European Union's Horizon 2020 research and innovation programme under the grant agreement no. 830943.

References

1. Shalamanov, V.: Institution building for it governance and management. Inf. Secur. Int. J. **38**, 13–34 (2017)
2. Doz, Y., Kosonen, M.: Fast Strategy: How Strategic Agility Will Help You Stay Ahead of the Game. Wharton School Publishing (2008)
3. Ribes, D., Finholt, T.: The long now of technology infrastructure: articulating tensions in development. J. Assoc. Inf. Syst. **10**(5), 375–398 (2009). https://doi.org/10.17705/1jais.00199
4. Kearns, G.S., Lederer, A.L.: A resource-based view of strategic IT alignment: how knowledge sharing creates competitive advantage. Decis. Sci. **34**(1), 1–29 (2003). https://doi.org/10.1111/1540-5915.02289
5. Weill, P., Ross, J.W.: IT Governance: How Top Performers Manage IT Decision Rights for Superior Results. Harvard Business School Press (2000)
6. Persse, J.R.: Process Improvement Essentials: CMMI, ISO 9001, Six Sigma; O'Reilly (2006)
7. Durbin, P., Doerscher, T.: Taming. Change Within Portfolio Management: Unify Your Organization, Sharpen Your Strategy and Create Measurable Value. Greenleaf Book Group Press (2010)
8. Carroll, J.: Project Program and Portfolio Management, In easy steps (2014)
9. Couture, J.-R.: Reconciling operational and financial planning views in a customer-funded organization: making customer-funding work for NC3A. Inf. Secur. Int. J. **38**, 63–69 (2017)
10. Szydelko, A.: Acquisition in a large IT organization. Inf. Secur. Int. J. **38**, 71–76 (2017)
11. Brewster, E., Griffiths, R., Lawes, A., Sansbury, J.: IT Service Management. British Information Society (2012)
12. Chih, Y.-Y., Zwikael, O.: Project benefit management: a conceptual framework of target benefit formulation. Int. J. Project Manage. **33**(2), 352–362 (2015). https://doi.org/10.1016/j.ijproman.2014.06.002
13. Ballinger, P.: Lessons from a customer-funding regime in a large IT organization. Inf. Secur. Int. J. **38**, 49–62 (2017)
14. Robinson, M.: Output-Driven Funding and Budgeting Systems in the Public Sector. Working Paper, Queensland University of Technology, Feb 2002. eprints.qut.edu.au. https://eprints.qut.edu.au/535/
15. Matern, S., Savova, G., Goleva, D., Shalamanov, V.: Human Factor in Digitalization and Cyber Resilience of Public Administration, Computer and Communications Engineering, 13, 2/2019, pp. 3–14. Technical University of Sofia (2019). ISSN:1314-2291
16. Shalamanov, V., Penchev, G.: Academic Support to Cyber Resilience: National and Regional Approach, Computer and Communications Engineering, 13, 2/2019, pp. 73–80. Technical University of Sofia (2019). ISSN:1314-2291

Information Sharing and Cyber Situational Awareness

Design Science Research Towards Ethical and Privacy-Friendly Maritime Surveillance ICT Systems

Jyri Rajamäki

Abstract Maritime surveillance is essential for creating maritime awareness and saving lives. With new technologies, privacy and ethics in surveillance will be of special concern. On the other hand, processing of personal data in surveillance is regulated by the General Data Protection Regulation (GDPR) and/or by the 'Police-Justice Directive' 2016/680. GDPR encourages applying a Data Protection Impact Assessment (DPIA) to identify and minimize data protection risks as the initial step of any new project. This design science research (DSR) shows how DPIA is adapted in the MARISA project and tries to be a step towards new meta-artifacts and useful methods for the design and validation of privacy requirements engineering approaches into maritime surveillance ICT systems. This DSR also highlights the importance of DSR cycles in designing new artifacts and processes for building these artefacts.

Keywords Design science · Maritime surveillance · Impact assessment · Open source intelligence · OSINT · Privacy · GDPR directive

1 Introduction

Maritime surveillance as well as search and rescue (SAR) operations in EU are based on various international legislation, EU-level legislation and local legislation, including United Nations Convention on the Law of the Sea (UNCLOS), International Maritime Organization (IMO) and International Convention for the Safety of Life at Sea (SOLAS). As regards surveillance activities, a coastal State has the exclusive right to undertake monitoring and surveillance within its territory including territorial sea. A coastal State also has the exclusive right to undertake monitoring and surveillance in connection with the economic exploitation and exploration of its Exclusive Economic Zone. Furthermore, all states have the implied right to undertake monitoring and surveillance in the high seas, but not to the extent of interfering

J. Rajamäki (✉)
Laurea University of Applied Sciences, Espoo, Finland
e-mail: jyri.rajamaki@laurea.fi

© The Author(s), under exclusive license to Springer Nature Switzerland AG 2021
T. Tagarev et al. (eds.), *Digital Transformation, Cyber Security and Resilience of Modern Societies*, Studies in Big Data 84,
https://doi.org/10.1007/978-3-030-65722-2_7

with the exercise of the freedom of the high seas by ships flying a foreign flag. On the other hand, all public safety organizations, emergency services, etc. which process personal data and even sensitive data, are subject to the General Data Protection Regulation (GDPR). In law enforcement operations, also the Directive 2016/680 'on the protection of natural persons with regard to the processing of personal data by competent authorities for the purposes of the prevention, investigation, detection or prosecution of criminal offences or the execution of criminal penalties, and on the free movement of such data' (hereafter, the "Police-Justice Directive") regulates processing of personal data.

The ongoing MARitime Integrated Surveillance Awareness (MARISA) project [1] funded by the Horizon 2020 programme, focuses on four major objectives: (1) create improved situational awareness with a focus on delivering a complete and useful comprehension of the situation at sea; (2) support the practitioners along the complete lifecycle of situations at sea, from the observation of elements in the environment up to detection of anomalies and aids to planning; (3) ease a fruitful collaboration among adjacent and cross-border agencies operating in the maritime surveillance sphere (Navies, Coast Guards, Customs, Border Polices) in order to pull resources towards the same goal, leading to cost efficient usage of existing resources; and (4) foster a dynamic eco-system of users and providers, allowing new data fusion services, based on a "distilled" knowledge, to be delivered to different actors at sea by the integration of a wide range of data and sensors.

GDPR and Police-Justice Directive apply in the MARISA context. GDPR applies during the development of the MARISA Toolkit and its operational trials. Whereas in the operational end-use of the MARISA Toolkit in the future either GDPR or Police-Justice Directive applies based on the authority and the purpose for which the MARISA Toolkit is used. Therefore, both regulations must be taken into consideration when designing the MARISA Toolkit. This design science research (DSR) tries to be a step towards new meta-artifacts and useful methods for the design and validation of privacy requirements engineering approaches into maritime surveillance ICT systems.

In contrast to behavioral science, DSR aims to provide four general outputs: (1) constructs, (2) models, (3) methods, and (4) instantiations [3]. Figure 1 shows how the DSR framework is applied in this paper. After this introduction, Sect. 2 presents the maritime surveillance environment, the MARISA Toolkit and its privacy requirement. Section 3 deals with the present knowledge base with regard to privacy and surveillance as well as data life cycles (personal data, big data). Section 4 presents designing of ethical and privacy-friendly MARISA services and Sect. 5 how these are evaluated in the MARISA project. Finally, Sect. 6 concludes the paper.

2 Maritime Surveillance and the MARISA Toolkit

The Maritime Common Information Sharing Environment (CISE) seeks to further enhance and promote relevant information sharing between authorities involved

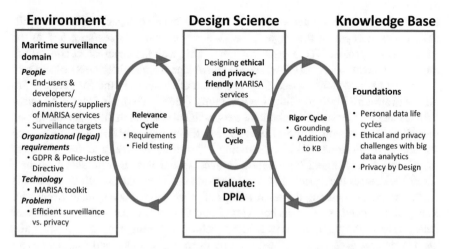

Fig. 1 Design science research framework of the study (modified from [2])

in maritime surveillance from coastguards and navies to port authorities, fisheries controls, customs authorities and environment monitoring and control bodies. The EUCISE2020 project [4] has taken the level of collaboration forward, in putting operational authorities together at an unprecedented scale to define the largest European test bed for data and information exchange. DG MARE Test Project CoopP on cooperation in execution of various maritime functionalities at sub-regional or sea-basin level in the field of integrated maritime surveillance has investigated information exchange needs, barriers, benefits and technologies by analyzing use cases, agreed at the level of large user community.

The MARISA project fosters faster detection of new events, better informed decision making and achievement of a joint understanding of a situation across borders and allowing seamless cooperation between operating authorities and on-site/at sea/in air intervention forces. Its solution is a toolkit that provides a suite of services to correlate and fuse various heterogeneous and homogeneous data and information from different sources, including Internet and social networks. The project also aims to build on the huge opportunity that comes from using the open access to "big data" for maritime surveillance: the availability of large to very large amounts of data, acquired from various sources ranging from sensors, satellites, open source, internal sources and of extracting from these amounts through advanced correlation improves knowledge.

2.1 Technology

The MARISA Toolkit provides new means for the exploitation of the bulky information silos data, leveraging on the fusion of heterogeneous sector data and taking benefit of a seamless interoperability with the existing legacy solutions available across Europe. In this regard, the data model and services specified in the EUCISE2020 project [4] are exploited, combining with the expertise of consortium members in creating security intelligence knowledge from a wide variety of sources.

Level 1 addresses the aspect of "Observation of elements in the environment" to build and enrich a Maritime Situation Awareness (MSA). The main focus is on establishing enhanced information about the geographical position of the observed objects providing Data Fusion services, such as "Multi Sensor/Target/Common Operating Picture (COP) Fusion", "Object Clustering", "Maritime route extraction", "Density maps" and "Multilingual Information Extraction and Fusion from social media". Level 2 addresses the aspect of "Comprehension of the current situation" to provide useful information among the relationships of objects in the maritime environment. The goal is to detect suspicious behavior of maritime entity (particular and irregular patterns) and infer the real vessel identity (fishing, polluting, smuggling) providing Data Fusion services, such as "Business Intelligence", "On-Demand Activity Detection", "Behavior Analysis", "Anomaly Detection & Classification" and "Alarm Generation". Level 3 addresses the aspect of "Projection of future states" to predict the evolution of a maritime situation, in support of rapid decision making and action. The focus is on predicting future behavior (time, place and probability of type of activity) and mission planning support based on predicted behavior of vessels in the region of interest providing Data Fusion services, such as "Predictive Analysis" and "Mission Planning" [5].

The "MARISA User Application" level includes all the computing facilities to let MARISA End Users to visualize results of the MARISA services in a set of different graphical and statistical presentations, based on a Web Based approach. For each end-user community of interest (generic, data fusion expert and MSA operators) different representation of MARISA Data Fusion Products will be made available, based on access privileges assigned to them. MSA Presentation Web Console enables generic user to access, analyze and visualize maritime entities in textual (dashboard) or graphical views, using a Web browser as a client. The situational awareness of the maritime domain will be provided through a fused maritime picture based on a WebGIS and reference detailed cartographic map of a selected Area of Interest (AoI). This service also includes the capability to monitor the Maritime Situation to detect abnormal behavior and highlight alarms. Data Fusion Expert Console addresses the aspect of "Man in the loop" as defined in MARISA Level 4 services. Data Fusion experts will be able to have a range of interactive content types available, in order to refine data sources and data fusion products coming from MARISA processing. A web based application approach will be pursued. System Administration Console will be primary devoted to address the general management activities of the MARISA system. The console will also be used to profile and assign privileges to generic end users and operational systems when accessing data fusion and HCI services [5].

The "MARISA Networking and Integration Services" level includes computing components. Access Control Services manage the access to MARISA Data Fusion products and deal with the ability of MARISA to identify, record and manage users' identities and their related access to all the services made available by the toolkit.

They include (a) Identity and Access management services to identify all the users connecting to the toolkit's services and to assure that access privileges are granted according to defined security policies, and all individuals and systems are properly authenticated, authorized and audited, (b) User Profiling service record and assign privileges to all users (human/device/process) connecting to MARISA. Data Source Interfaces (I/F) Services gather data, information and services from external sources, such as End User Legacy Systems & Assets, Free & Open Internet Sources, Simulation Sources as well as some assets directly provided by MARISA such as Satellite Data, Signal Analysis Devices, Automatic Identification System (AIS) Sources. The sources feeding the MARISA Toolkit are expected to be: Maritime data (e.g. AIS Network, System Tracks, Mission Plans, etc.); Satellite data (e.g.: COSMO-SkyMed SAR data, Sentinel-1, Sentinel-2, commercial optical missions, etc.); Intelligence data (e.g.: OSINT, Signal Analysis) [5].

2.2 Legislative Background

The European Data Protection Reform Package consists of two texts: GDPR and the Police-Justice Directive. In addition to this Package, there are guidelines [6] from supervisory authorities and Article 29 Working Party ("WP29"), recently replaced by the European Data Protection Board ("EDPB"), as well as Member States' national protection laws which adapt and complete the GDPR, and transpose the Police-Justice Directive. Data Protection refers to legislation that is intended to (1) protect the right to privacy of individuals (all of us), and (2) ensure that Personal Data is used appropriately by organizations that may have it (Data Controllers). Personal data is any information that can be used to identify a natural person (Data Subject), such as name, date of birth, address, phone number, email address, membership number, IP address, photographs, etc. Some categories of information are defined as 'special categories of personal data' (e.g., religion, ethnicity, sexual orientation, trade union membership, medical information) and they require more stringent measures of protection. Also, criminal data and children's data need additional protection.

2.2.1 Privacy by Design

Privacy by Design (PbD) is one of the key requirements in that reform. PbD is an approach to systems engineering approach intended to ensure privacy protection from the earliest stages of a project and to be taken into account throughout the whole engineering process. The PbD concept is closely related to the concept of privacy enhancing technologies (PET) published in 1995 [7]. PbD framework was published in 2009 [8]. The concept is an example of value sensitive design that takes human values into account in a well-defined manner throughout the whole process. GDPR also requires Privacy by Default, meaning that the strictest privacy settings should be the default.

2.2.2 Data Protection Impact Assessment

To satisfy its requirement, GDPR encourages organizations to undertake a Data Protection Impact Assessment (DPIA) to identify and minimize data protection risks as the initial step of any new project. The 'controller' is responsible for ensuring that the DPIA is carried out if necessary (Article 35(2)). Carrying out the DPIA may be done by someone else, inside or outside the organization, but the controller remains ultimately accountable for that task. When a processing activity is likely to result in a high risk to the rights and freedoms of natural persons, the controller shall, prior to the processing, carry out an assessment of the impact of the envisaged processing operations on the protection of personal data. The high risk to the rights and freedoms of natural persons, of varying likelihood and severity, may result from personal data processing which could lead to physical, material or non-material damage, in particular [6]:

(a) *Evaluation or scoring*: where personal aspects are evaluated, in particular analyzing or predicting aspects concerning performance at work, economic situation, health, personal preferences or interests, reliability or behavior, location or movements, in order to create or use personal profiles.

(b) *Automated-decision making with legal or similar significant effect*: where processing aims at taking decisions on data subjects producing "legal effects concerning the natural person" or which "similarly significantly affects the natural person".

(c) *Systematic monitoring*: processing used to observe, monitor or control data subjects, including data collected through "a systematic monitoring of a publicly accessible area" where data subjects may not be aware of who is collecting their data and how it will be used. The WP29 interprets "systematic" as meaning one or more of the following characteristics (see the WP29 Guidelines on Data Protection Officer 16/EN WP 243): occurring according to a system; pre-arranged, organized or methodical; taking place as part of a general plan for data collection; carried out as part of a strategy.

(d) *Sensitive data*: this includes special categories of data as defined in Article 9 as well as personal data relating to criminal convictions or offences as defined in Article 10. The fact that personal data is publicly available may be considered as a factor in the assessment if the data was expected to be further used for certain purposes.

(e) *Data processed on a large scale*: where processing involves a large amount of personal data and affects a large number of data subjects. The GDPR does not define what constitutes large-scale, recommends that the following factors be considered when determining whether the processing is carried out on a large scale: the number of data subjects concerned, either as a specific number or as a proportion of the relevant population; the volume of data and/or the range of different data items being processed; the duration, or permanence, of the data processing activity; the geographical extent of the processing activity.

(f) *Datasets that have been matched or combined*: where two or more data processing operations performed for different purposes and/or by different data controllers in a way that exceed the reasonable expectations of the data subject.

(g) *Data concerning vulnerable data subjects*: where personal data of vulnerable natural persons, especially of children are processed. There is an increased power imbalance between the data subject and the data controller, meaning the individual may be unable to consent to, or oppose, the processing of his or her data.

(h) *Innovative use of technology*: where solutions technology can involve novel forms of data collection and usage, possibly with a high risk to individuals' rights and freedoms. Indeed, the personal and social consequences of the deployment of a new technology may be unknown.

(i) *Data transfer across borders outside the European Union*: where export/import of personal data is performed with a non EU country.

(j) *Prevention of data subjects from exercising their rights*: where data subjects might be deprived of their rights and freedoms or prevented from exercising control over their personal data.

Keeping in line with the risk-based approached embodied by the GDPR, carrying out a DPIA is not mandatory for every processing operation. A DPIA is only required when the processing is "likely to result in a high risk to the rights and freedoms of natural persons" (Article 35(1)). The WP29 considers that the more criteria are met by the processing, the more likely it is to present a high risk to the rights and freedoms of data subjects, and therefore to require a DPIA. As a rule of thumb, a processing operation meeting less than two criteria may not require a DPIA due to a lower level of risk, and processing operations which meet at least two of these criteria will require a DPIA.

2.2.3 Mandatory Legal Requirements for the MARISA Toolkit

With the European Data Protection Reform requirements, Privacy by Design and Privacy by Default are mandatory requirements for the MARISA Toolkit and for all MARISA services. The MARISA Toolkit is considered to have a processing activity with a 'high risk', because:

- it contains systematic monitoring of Twitter messages;
- it processes data on a large scale via its Big Data architecture;
- it deals with datasets that have been matched or combined (data fusion);
- it contains innovative use of technology.

This means that the Data Protection Impact Assessment is to be carried out.

3 Knowledge Base

As Hevner and Chatterjee [2] state design science draws from a vast knowledge base of scientific theories and engineering methods that provides the foundations for rigorous DSR. This section defines the state of the art in the application domain of the research, which is privacy in surveillance and management of data throughout its lifecycles.

3.1 Privacy and Surveillance

New surveillance technologies became omnipresent in our everyday live. While early research focused on functionality of these technologies, e.g., face recognition or violence detection, latterly also privacy and transparency related work has been made [9]. While this research helps us to design systems that combine functionality and privacy, only little understanding is present how the people under surveillance will react to the new systems; average citizens do not understand technological details and they are unable to distinguish between systems with varying privacy protection [9]. Privacy in surveillance is a special concern when, for example, using drones and surveillance cameras, with automated border control, and when collecting and analyzing big data. In addition, the impact of new surveillance technologies on the fundamental rights of asylum seekers and refugees as well as the increased responsibility this more effective situational awareness brings (under international refugee law and the Search and Rescue regime: duty to render assistance) have all been debated by numerous scholars [10]. Surveillance has a bad reputation in most countries. Many surveys for understanding the acceptance of surveillance were made in special places (airports, public transport and shopping malls), but their outcome depends on recently happened events, e.g., a terrorists attack or a reported misuse of a video sequence and the underlying factors are not considered and no generic model for the acceptance exists [9].

Koops et al. [11] considers the challenge of embedding PbD in open source intelligence (OSINT) carried out by law enforcement. Ideally, the technical development process of OSINT tools is combined with legal and ethical safeguards in such a way that the resulting products have a legally compliant design, are acceptable within society (social embedding), and at the same time meet in a sufficiently flexible way the varying requirements of different end-user groups. Koops et al. [11] use the analytic PbD framework and they discusses two promising approaches, revocable privacy and policy enforcement language. The approaches are tested against three requirements that seem suitable for a 'compliance by design' approach in OSINT: purpose specification; collection and use limitation and data minimization; and data quality (up-to-datedness) [11]. For each requirement, they analyze whether and to what extent the approach could work to build in the requirement in the system. They demonstrates that even though not all legal requirements can be embedded fully in

OSINT systems, it is possible to embed functionalities that facilitate compliance in allowing end-users to determine to what extent they adopt a 'privacy by design' approach when procuring an OSINT platform, extending it with plug-ins, and fine-tuning it to their needs. Therefore, developers of OSINT platforms and networks have a responsibility to make sure that end-users are enabled to use PbD, by allowing functionalities such as revocable privacy and a policy enforcement language [11]. Even though actual end-users have a responsibility of their own for ethical and legal compliance, it is important to recognize that it is questionable whether all responsibility for a proper functioning and use of OSINT platforms can be ascribed to the end-users; and some responsibility for a proper functioning of OSINT framework in practice also lies with the developers of the platform and individual components [12].

3.2   Personal Data Life-Cycles

Legal frameworks and standards, such as GDPR (see Table 1) and the Global Privacy Standard (GPS) [13], govern the processing of personal data, but there is a disconnect between policy-makers' intentions and software engineering reality [14]. The Abstract Personal Data Lifecycle (APDL) model was developed to serve as a model for personal data life-cycles, distinguishes between the main operations that can be performed on personal data during its lifecycle by outlining the various distinct activities for each operation outlining these activities in relation to the GPS principles with the aim of governing the behavior of these operations [15]. The separation (1) helps support the manageability and traceability of the flow of personal data during its lifecycle, (2) ensures and demonstrates compliance with legal frameworks and standards, (3) reflects the extent to which the flow of personal data is appropriate in terms of involved actors and their assigned roles and responsibilities, and (4) facilitates the identification of data processing activities that may lead to privacy violations or harms [15].

Big data offers great promise and opportunity in terms of the development of more efficient and effective government programs, but it presents serious privacy challenges, such as unwarranted surveillance activities, inaccurate results or predictions, the use of biased data, and unauthorized secondary uses of personal information [16]. Data management throughout the lifecycle of the data is difficult especially with the proliferation of cloud computing and the increasing needs in analytics for big data. Table 2 present an audit framework for data lifecycles in a big data context [17].

Table 1 GDPR framework throughout the lifecycle of the data

Data life-cycle	GDPR factor	GDPR principles
Capture	What you are allowed to capture? How you may do so? What you must tell the person in advance? What you must get from them (their permission)?	Data Minimization; only ask for what is needed Privacy Notices; clearly inform what, why, who and where Data Subject Rights; state the persons rights under the legislation Obtain Consent; consent must be freely given and explicit for the purpose or purposes
Store	How you must store it? Where it can be stored? Obligations of third parties? What happens if you lose it?	Safe and Secure; information must be stored appropriately e.g. locked cabinets/password protected files) Restricted Access; only authorised persons should have access to it Data Inventory; information captured should be recorded Subject Access Requests; must be in a position to provide ALL information held Contracts with Data Processors; any third parties must have GDPR contracts in place Data Breaches; processes to detect, report and investigate Data Breaches must be in place
Use	What you can use it for? What you can't use it for?	Appropriate use; must be for the purpose(s) originally stated Consent; must have person's consent or a lawful basis for processing it Manage Consent; individuals have the right to revoke consent for part or all of the processing, this must be managed Restricted; profiling or automated decision making are restricted International Transfers; any processing that occurs outside EU must have been communicated to person at time of data capture and must have additional safeguards in place

(continued)

Table 1 (continued)

Data life-cycle	GDPR factor	GDPR principles
Destroy	How long you can keep it for? When you must destroy information?	Retention Period; retention periods must be documented and justified and data must be destroyed after its useful retention period has expired Right to Erasure; must be erased upon request from person Portability; must be provided in standard format Third Party Copies; all copies of information must be deleted including those held by third parties. Systems like Whatsapp can be an issue here due to the lack of control over the personal data held within it
Assess	How to ensure compliance?	Data Protection by Design and by Default; all relevant projects or initiatives must consider impacts on privacy from the outset Data Protection Impact Assessment (DPIA); must be conducted for new technology, profiling, large scale processing, or engagement of a new third party data processor Documentation; decisions and rationale for decisions around Data Protection should be documented

Table 2 Data lifecycle phases with their category in a Big Data context

Category	Phases	An example of audit questions (totally 2–8/phase)
operational phases	Collection	Are controls performed to ensure the quality of collected data?
	Integration	Are there rules and policies for integrating distributed data?
	Filtering	Are there filtering rules?
	Enrichment	Is there an enrichment plan?
	Analysis	Is this phase adapted for Big Data?
	Visualization	Is the visualization flexible?
	Storage	Are the means by which the data is stored mixed (Cloud, private external DBs)?
	Destruction	Do data arrive at a given moment to deletion?
	Archiving	Is there an archiving policy based on specific parameters or business rules?
Management phases	Management	Is management transversal throughout operational phases?
	Planning	Does Planning concern all operational phases?
Support phases	Security	Is there a respect for privacy and anonymity?
	Quality	Is there quality control in all phases?

4 Designing Ethical and Privacy-Friendly MARISA Services

The implementation of privacy-by-design in the MARISA Toolkit is an overall requirement or constraint for the development of the whole MARISA project [5]. The MARISA Toolkit has two relevant data sources: (1) data coming from the sensors, and (2) data coming from OSINT/Social Media. *Data from Sensors*: These sensors are embodied in the operational environment of the Legacy Systems. In these environments, owned by Participating Member State governmental entities, we can suppose that the data are used on the basis of need-to-know and need-to-share. Thus, the observance of the privacy of the data can be taken for granted. *Data from Open Sources*: This case is more problematic, since the origin of the data is not controlled for any public entity. Nevertheless, here there are two possibilities: (1) System performing in a classified environment (as could be the case in managing EU-Restricted data). Here the data coming from open sources enters, by means of a cross-domain exchange devices, in a highly regulated environment, where again the privacy of the data managed can be taken for granted, on the basis of need-to-know and need-to-share. (2) System performing in an unclassified environment (this will be the most common case) [5].

The MARISA Toolkit is built on the top of a big data infrastructure that provides the means to collect external data sources and operational systems products and to organize and exploit all the incoming data as well as all the data produced by the various services. Table 3 summarize some relevant big data related ethical and privacy challenges.

Figure 2 shows the two major streams of data protection in the MARISA project: compliance with new regulatory framework and exploitation of opportunities given by the development of peculiar expertise on the topic.

The key objectives of Data Protection can be summarized as follows:

- Lawfulness, Fairness, Transparency;

Table 3 Ethical and privacy challenges in a Big Data context

Big data phase	Challenge
Collection	Active: data owner will give the data to a third party Passive: data are produced by data owner's online actions (e.g., browsing) and the data owner may not know about that the data are being gathered by a third party
Storage	Clouds: cloud customer doing anything more than storing encrypted data must trust the cloud provider
Analysis	Machine learning techniques including neural networks run in two phases: training phase and prediction phase quality of predictions is absolutely dependent on examples used for the training phase ML systems are only as good as the data sets that the systems trained and worked with Does not directly touch the individual and may have no external visibility
Use	Ethical issue comes especially with automated policing
Destroy	How be sure that has destroyed all data in all redundant data storages in multiple physical locations?

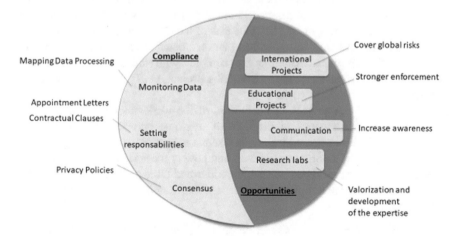

Fig. 2 The two sides of Data Protection in the MARISA project

- Purpose Limitation (Use only for one or more specified purposes);
- Data Minimisation (Collect only the amount of data required for the specified purpose(s));
- Accuracy (Ensure data is kept up to date, accurate and complete);
- Storage Limitation (Kept for no longer than necessary for the specified purpose(s));
- Integrity and Confidentiality (Processed ensuring appropriate security of data);
- Accountability (essential not only to be compliant, but to be able to demonstrate compliance).

5 Evaluation

Evaluation is a key element of DSR. When the artifact has been built, the next phase is evaluation in terms of functionality, completeness, consistency, accuracy, performance, reliability, usability, fit with the organization, and other relevant quality attributes [2]. Section 5.1 presents how the new artifacts (the MARISA services) are introduced into the application domain and how the field tests are carried out. Section 5.2 deals with evaluations during the internal design cycles.

5.1 Relevance Cycle: From Use Cases to Operational Trials

The MARISA project follows the Systems Engineering (SE) approach instructed by the International Council on Systems Engineering (INCOSE) [18]. The role of the systems engineer encompasses the entire life cycle for the system-of-interest [19]. The MARISA project life-cycle is limited to the Concept and Development Stages, as the Exploratory Research has been already performed by Consortium members and is part of their background [19]. One task of the Development Stage is to verify and validate the system, i.e. to confirm that the specified design requirements are fulfilled by the system and that the system complies with stakeholders' requirements in its intended environment. In the MARISA project, verifying and validation are carried out via operational trials.

For the MARISA project, the definition of the use cases have been made in previous European projects in the field of maritime surveillance (e.g., CoopP, EUCISE 2020). The selected use cases provide scenarios demonstrating how the information sharing environment is used and how to meet the user's requirements. The use cases cover all seven user communities and three processes describing the overall performance of how the information sharing system works. The three process levels are:

(1) *Baseline Operations*: this level describes "Everyday monitoring of events in the maritime domain", or "Behavior monitoring". The purpose of this process is to ensure the lawful, safe and secure performance of maritime activities.

Furthermore, to detect anomalies (detection of possible non-compliance) and other triggers/intelligence to improve decision making for the use of response capabilities (e.g. targeting of inspections). This level also contains "simple" response to single incidents or actions within the maritime domain—everyday operations [19].

(2) *Targeted Operations*: the "Targeted operations" level describes operations planned in advance towards a specific activity. The purpose of this process is to react to or to confront specific threats to sectorial responsibilities as discovered in risk analysis/intelligence gathering processes. They will give support to operational decision-making when employing operational assets [19].

(3) *Response Operations*: response to major incidents, events or accidents. The purpose would be to respond to events affecting many actors across sectors and borders and with a potentially major impact on, e.g., the environment and economy [19].

The following use cases, among the whole set defined within the EUCISE2020 project, will be exercised in MARISA Project: (1) *Use Case 13b*: Inquiry on a specific suspicious vessel (cargo related); (2) *Use Case 37*: Monitoring of all events at sea in order to create conditions for decision making on interventions; (3) *Use Case 44*: Request any information confirming the identification, position and activity of a vessel of interest; 4) *Use Case 70*: Suspect Fishing vessel (small boat) is cooperating with other type of vessels; (5) *Use Case 93*: Detection and behavior monitoring of illegal, unreported and unregulated (IUU) vessels listed by Regional Fishery Management Organisations (RFMOs) [19].

From use cases a continuous link with the user needs were established in order to verify the matching between the preliminary trials objectives and the user requirements collected with the user communities. By using the systems engineering approach, the operational scenarios were defined through a certain number of parameters and the whole set of possible values which can be verified during the execution of the trials. The parameters are: (1) incident type; (2) geographic characteristics of the trial area; (3) meteo-marine conditions; and (4) the traffic conditions and target types. The selected incident types (i.e. the operational situation in which the MARISA services could provide additional information) are human trafficking and smuggling; Maritime Situation Exchange and Assessment Service (MSEAS) for safety and security; irregular immigration; and safety. The geographical characteristics: i.e. the characteristics to which the trial area refers. An application domain can refer to geographical areas small or large, wide or thin, or related to the possibility to track vessels along routes. For instance, in Maritime Border Surveillance context, the interest is more focused on vessels heading orthogonally with respect to borders than on vessels sailing in parallel. The meteo-marine conditions: i.e. currents, winds, waves, temperature, etc. Meteo-marine conditions are relevant for two reasons. First of all, they can have impacts on tools performances (e.g. in the discrimination of targets with respect to the clutter). Secondarily, they define the scenarios in which MARISA services will be asked to work, and so they have to be defined in order to have a maximum added value. The traffic conditions and target types: i.e. ships

densities (high/low), ship dimensions (majority of small/big ships), targets' characteristics (e.g. in terms of target motion type and speed), vessels' equipment properties
(e.g. the presence of on-board ship reporting systems), vessels preferred routes (for
anomalies detection), seasonal traffic variation, etc. [19].

Finally, the trails were described considering the specific goal, the area to test,
the end-users and MARISA partners which will be involved, and the use cases to
be tested. During the trials, a particular focus will be made on the validation of the
users' assets availability, the constraints to data availability in relation to MARISA
nodes installation, and the needed input data types.

Altogether five operational trials for the MARISA Toolkit validation, each
covering a different area and involving different partners, have been defined [19]: (1)
Northern Sea Trial (maritime situation exchange and assessment service for safety
and security, end user is Dutch Coast Guard); (2) Iberian Sea Trial (irregular immigration, end users are Guardian Civil and Portuguese Navy); (3) Strail of Bonifacio
Sea Trial (safety and illegal immigration, end users are PMM and Italian Navy); (4)
Ionian Sea Trial (human trafficking between Corfu and Italy, end users are Greek
Ministry of Defence and Italian Navy); and (5) Aegian Sea Trial (human trafficking
and smuggling, end user is Greek Ministry of Defence).

5.2   Design Cycle: DPIA

As stated in Sect. 2.2, a DPIA is a process designed to describe the processing, assess
its necessity and proportionality and help manage the risks to the rights and freedoms
of natural persons resulting from the processing of personal data by assessing them
and determining the measures to address them [6]. DPIAs are important tools for
accountability, as they help controllers not only to comply with requirements of
the GDPR, but also to demonstrate that appropriate measures have been taken to
ensure compliance with the Regulation. In other words, a DPIA is a process for
building and demonstrating compliance [6]. DPIAs should be relatively unexpensive
to implement with sufficient resources and tools [20]. However, while there is advice
on the legal requirements for DPIA and the elements of what practitioners should do
to undertake a DPIA, there has been little prescription about how security and privacy
requirements engineering processes map to the necessary activities of a DPIA, and
how these activities can be tool-supported [20].

Coles et al. [20] have studied existing Privacy Requirements Engineering
approaches and tools that support carrying out DPIAs. The existing approaches
capture the elements that would be needed by a DPIA, but two barriers need to
be overcome before such approaches are ready for security and practitioners to use
in DPIAs [20]: (1) more prescription is needed to indicate what tools and techniques map to different stages of a DPIA, and (2) such steps need to be adequately
tool-supported, such that data input in one step can be used to support reasoning
and analysis in others. Their main contributions [20]: (1) existing Requirements
Engineering techniques associated with Integrating Requirements and Information

Security process framework can be effective when supporting the different steps needed when carrying out a DPIA, but there is no one-to-one mapping between requirements and techniques, and several techniques might be needed to support a single step; (2) demonstration how an exemplar for Security Requirements Engineering tools supports and helps reason about potential GDPR compliance issues as a design evolves; and (3) they present a real example where their approach assessed the conceptual design of a medical application without an initial specification, and only the most preliminary of known functionality. They show that the use of this approach and the Requirements Engineering techniques in general, are effective in discovering additional functionality, and envisaging different forms of intended and unintended device use [20].

DPIA can be conducted by meeting relevant stakeholders, identifying potential privacy issues and agreeing ways to mitigate the risk of issues occurring. It must be documented and retained. The processor of MARISA personal data is the MARISA consortium jointly, based on the MARISA Grant Agreement and Data Sharing Agreement. All MARISA DPIAs were carried out by the help of CNIL's PIA software [21]. Figure 3 presents DPIA phases.

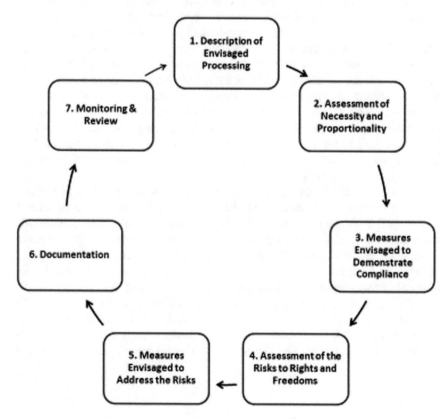

Fig. 3 Data Protection Impact Assessment process phases

As Sect. 2.1 presents, the MARISA Toolkit collects data from sensor and open sources. The first MARISA DPIA was created prior the first operational trial. It covers the services in which persons can be identified directly or indirectly included AIS data on vessels (>indirect identification), data base of historic incidents (>indirect identification) and personal data on MARISA end-users (>direct identification). AIS is a maritime technical standard developed by the International MAritime Organization (IMO). It is a radio technology combining GPS, VHF and data processing technologies to enable the exchange of relevant information in a strictly defined format between different entities.

The second MARISA DPIA covers three services (Twitter service, GDELT service and OSINT service) that collect open source information. Their main target is to extract and integrate maritime related safety and security events. OSINT service mainly collects its information via Twitter service and DGELT service. From data collection point of view, MARISA GDELT service may not have privacy concerns because professional journalists should have taken that issue into account when making news. In Twitter, several technical features and tweet-based social behaviors occur that might compromise privacy. Tweets are complex objects that, in addition to the message content, have many pieces of associated metadata, such as the username of the sender, the date and time the tweet was sent, the geographic coordinates the tweet was sent from if available, and much more [22]. "Most metadata are readily interpretable by automated systems, whereas tweet message content may require text processing methods for any automated interpretation of meaning" [22]. "Direct Messages" are the private side of Twitter and "retweeting" is directly quoting and rebroadcasting another user's tweet. Someone might unintentionally or intentionally retweet private tweet to a public forum. Other behaviors include mentioning another user in one's tweet that is, talking about that user. According to Rumbold and Wilson [23], when one puts any information in the public domain—whether intentionally or not—one does not waive one's right to privacy, but one can only waive one's right to privacy by actually waiving it.

The MARISA DPIAs conducted with the CNIL's PIA software show that privacy and data protection is in order in the MARISA project at the moment, but more DPIA related R&D is needed (e.g. maturity level improvements). Foreseeable privacy risks in the future are: (1) how to handle personal data in tweets (in a content of tweet, before commercialization); (2) the mosaic effect in data fusion; and (3) big data analytics by AI. A DPIA is a constant process which continues after the termination of the MARISA project. The MARISA project should offer a tool, to carry out this for the future administrators and end-users of the toolkit. Thus far, the CNIL's PIA software is a good tool in Big Data dimensions: (1) data generation and collection and (4) technology and infrastructure behind data. On the other hand, it covers weakly the dimensions (2) data analysis (i.e. mosaic and transparency), and (3) use of data.

6 Conclusions

Following the design science research paradigm [2], the major contribution of this research is to highlight the importance of three inherent design science research cycles (see Fig. 1). These cycles are applied in designing of new artifacts and processes for building these artefacts. The new artifacts of this study are ethical and privacy-friendly maritime surveillance services. During the MARISA project's internal *design cycle*, ethical and privacy issues of new developed maritime surveillance services are evaluated by Data Protection Impact Assessment and redesigned when needed. The MARISA project's *relevance cycle* ties new innovative artifacts to the maritime domain; it sets design goals and (privacy and ethical) requirements, and via operational trials determines whether additional iterations of the relevance cycle are needed. The *rigor cycle* provides past Knowledge Base (KB) to the MARISA project and brings additions to the KB as a result of the research.

Privacy by Design (PbD) and a Data Protection Impact Assessment (DPIA) as concepts are well-known and recently many research papers has been published on this area. However, it turns out that there is not much standardization in how to actually apply PbD throughout the whole engineering process. On the other hand, new software tools are released to make the data protection impact assessment more practical and to foster collaboration between stakeholders, and in this study we have applied a free software developed by CNIL.

Ethical issues concerning OSINT are diverse and evolving. Their impact on the MARISA Toolkit concern both technology, user processes and the business/governance model. Even though international regulatory guidelines are available, specific allowances, prohibitions and exceptions mainly stem from national legislation. European Data Protection Reform partly harmonizes data protection regulation in EU member states, but still leaves the possibility for variation on the national level. Big challenge in OSINT is coping with the mosaic effect. Data protection sets strong requirements on MARISA technology utilizing various data sources and performing data fusions on various levels. Another challenge concerns the reliance of automated analysis: How can data fusion algorithms that are reliable and transparent for the end-user be developed? As dos Passos [24] argues, associated with OSINT, big data is about being able to map behavior and tendencies. However, data science is needed in OSINT because of the lack/low quality of big data, to find the correct answers, capture the correct data and to have the correct perception of how to proceed throughout the process. Current academic and public debates entertain the notion of shifting the emphasis from data collection to data analytics and data use. There are scholars who underline the need for "algorithmic accountability" [25]. It can therefore be expected that the legal requirements concerning OSINT and big data may develop in this direction. Therefore, to separate the ethics of data collection from the ethics of the processing and use of data is essential.

Acknowledgements Acknowledgement is paid to the MARISA Maritime Integrated Surveillance Awareness project. This project is funded by the European Commission through the Horizon 2020 Framework under the grant agreement number 740698. The sole responsibility for the content of this paper lies with the authors. It does not necessarily reflect the opinion of the European Commission or of the full project. The European Commission is not responsible for any use that may be made of the information contained therein.

References

1. MARISA: MARISA Maritime Integrated Surveillance Awareness (2018). [Online]. Available: https://www.marisaproject.eu/
2. Hevner, A., Chatterjee, S.: Design Research in Information Systems: Theory and Practice. Springer Science and Business Media, New York (2010)
3. March, T., Smith, G.: Design and natural science research on information technology. Decis. Supp. **15**(4), 251–266 (1995)
4. EUCISE2020:EUropean test bed for the maritime Common Information Sharing Environment in the 2020 perspective [Online]. Available: https://www.eucise2020.eu/
5. MARISA:D3.2 MARISA services description document (2018)
6. ARTICLE 29 DATA PROTECTION WORKING PARTY,Guidelines on Data Protection Impact Assessment (DPIA) and determining whether processing is "likely to result in a high risk" for the purposes of Regulation 2016/679 (2017)
7. Hustinx, P.: Privacy by design: delivering the promises. Identity Inf. Soc. **3**(2), 253–255 (2010)
8. Cavoukian, A.: Privacy by Design: The 7 Foundational Principles. Information and Privacy Commissioner of Ontario, Ontario (2011)
9. Krempel, E., Beyerer, J.: TAM-VS: a technology acceptance model for video surveillance. In: Privacy Technologies and Policy, pp. 86–100. Springer, Cham (2014)
10. Rajamäki, J., Sarlio-Siintola, S., Simola, J.:Ethics of open source intelligence applied by Maritime Law Enforcement authorities. In: Proceedings of the 17th European Conference on Cyber Warfare and Security ECCWS (2018)
11. Koops, B., Hoepman, J., Leenes, R.: Open-source intelligence and privacy by design. Comput. Law Secur. Rev. **29**, 676–688 (2013)
12. Guest Editorial: Legal aspects of open source intelligence—results of the VIRTUOSO project. Comput. Law Secur. Rev. **29**, 642–653 (2013)
13. Global Privacy Standard: International Data Protection Commissioners Conference (2006). [Online]. Available: https://thepublicvoice.org/2006/12/global-privacy-standard.html
14. Alshammari, M., Simpson, A.:A UML Profile for Privacy-Aware Data Lifecycle Models. Lecture Notes in Computer Science, vol. 10683 (2017)
15. Alshammari, M., Simpson, A.: Personal Data Management for Privacy Engineering: An Abstract Personal Data Lifecycle Model (2017). [Online]. Available: https://www.cs.ox.ac.uk/publications/publication10942-abstract.html
16. Information and Privacy Commissioner of Ontario,Your privacy and big data (2017). [Online]. Available: https://www.ipc.on.ca/privacy-individuals/your-privacy-and-big-data/
17. Arass, M., Tikito, I., Souissi, N.: An Audit framework for data lifecycles in a big data context. In:2018 International Conference on Selected Topics in Mobile and Wireless Networking (MoWNeT), pp. 1–5 (2018)
18. INCOSE:SE Handbook
19. MARISA:D2.7 MARISA operational scenarios and trials (2017)
20. Coles, J., Faily, S., Ki-Aries, D.: Tool-supporting data protection impact assessments with CAIRIS. In:2018 IEEE 5th International Workshop on Evolving Security & Privacy Requirements Engineering (ESPRE), pp. 21–27 (2018)
21. Commission Nationale de l'Informatiqueet des Libertés,May 2018 updates for the PIA tool.[Online]. Available: https://www.cnil.fr/en/may-2018-updates-pia-tool

22. Glasgow, K.: Big data and law enforcement: Advances, implications, and lessons from an active shooter case study. In: Application of Big Data for National Security, pp. 39–54. Butterworth-Heinemann, Waltham (2015)
23. Rumbold, B., Wilson, J.:Privacy rights and public information. J. Polit. Philos. (2018)
24. dos Passos, D.: Big Data, Data Science and their contributions to the development of the use of open source intelligence. Syst. Manage. **11**, 392–396 (2016)
25. Broeders, D., Schrijvers, E., van der Sloot, B., van Brakel, R., Hoog, J., Ballin, E.: Big data and security policies: towards a framework for regulating the phases of analytics and use of Big Data. Comput. Law Secur. Rev. **33**, 309–323 (2017)

Maritime Surveillance and Information Sharing Systems for Better Situational Awareness on the European Maritime Domain: A Literature Review

Ilkka Tikanmäki, Jari Räsänen, Harri Ruoslahti, and Jyri Rajamäki

Abstract This chapter concentrates on the principles and practices of the different ways of conducting maritime surveillance, and compares the advantages and disadvantages of different solutions. European maritime authorities aim at increasing situational awareness, and it is unlikely that any single system would fulfil all the tasks needed in maritime surveillance. To ensure safety and security on the maritime domain requires good situational awareness by the Member States of the European Union. The purpose of this study is to investigate maritime surveillance and information sharing systems and how they can improve maritime safety and security. The research questions are *how maritime surveillance solutions enhance security and safety* and *how information sharing enhances awareness on the cyber situation in the European maritime domain?* Current maritime surveillance solutions include coastal radars, satellite surveillance, boat patrolling, and both manned and unmanned aerial surveillance. The common information sharing environment for the EU maritime domain (CISE) is used for common situational awareness between European maritime authorities. Though an extensive and objective comparison of naval surveillance systems, based on public sources, is challenging because information from different sensors and providers is difficult to compare, the authors discuss advantages and barriers for various maritime surveillance systems.

I. Tikanmäki (✉) · J. Räsänen · H. Ruoslahti · J. Rajamäki
Security and Risk Management, Laurea University of Applied Sciences, Espoo, Finland
e-mail: ilkka.tikanmaki@laurea.fi

J. Räsänen
e-mail: jari.rasanen@laurea.fi

H. Ruoslahti
e-mail: harri.ruoslahti@laurea.fi

J. Rajamäki
e-mail: jyri.rajamaki@laurea.fi

I. Tikanmäki
Department of Warfare, National Defence University, Helsinki, Finland

© The Author(s), under exclusive license to Springer Nature Switzerland AG 2021
T. Tagarev et al. (eds.), *Digital Transformation, Cyber Security and Resilience of Modern Societies*, Studies in Big Data 84,
https://doi.org/10.1007/978-3-030-65722-2_8

117

Keywords Cybersecurity · Early warning · Information sharing · Maritime
surveillance · CISE · Project RANGER

1 Introduction

Maritime surveillance is essential for creating maritime awareness, in other words
'knowing what is happening at sea'. Integrated maritime surveillance is about
providing authorities interested or active in maritime surveillance with ways to
exchange information and data. Support is provided by responding to the needs
of a wide range of maritime policies—irregular migration/border control, maritime
security, fisheries control, anti-piracy, oil pollution, smuggling etc. Also the global
dimension of these policies is addressed, e.g. to help detect unlawful activities in
international waters. Sharing data will make surveillance less expensive and more
effective. Currently, EU and national authorities responsible for different aspects of
surveillance e.g. border control, safety and security, fisheries control, customs, environment or defence, collect data separately and often do not share them. As a result,
the same data may be collected more than once. A common information-sharing
environment (CISE) is currently being developed jointly by the European Commission and EU/EEA member states with the support of relevant agencies such as the
European Fisheries Control Agency (EFCA). It will integrate existing surveillance
systems and networks, and give all those authorities concerned access to the information they need for their missions at sea. The CISE will make different systems
interoperable so that data and other information can be exchanged easily through the
use of modern technologies.

European maritime authorities aim at increasing situational awareness, and it is
unlikely that any single system would fulfill all the tasks needed in maritime surveillance [1]. By combining different surveillance systems (e.g. radar, satellite, Remotely
Piloted Aircraft (RPAS), boat patrols) it is possible to create a more comprehensive
Maritime Situation Picture (MSP).

As part of the RANGER project [2], this study assists in identifying which systems
and platforms could be possible alternatives for obtaining maritime surveillance.
Combining multiple solutions are needed to augment and even replace existing
systems and solutions. The purpose of this study is to investigate maritime surveillance systems, and how they can improve maritime safety and security. The research
questions for this study are:

1. How maritime surveillance solutions enhance security and safety on the European
   maritime domain?
2. How information sharing enhance cyber situation awareness on the European
   maritime domain?

The next Section presents Maritime Surveillance solutions currently in use; coastal
radars, satellite surveillance, boat patrolling, aerial surveillance etc. Also, their best

features and weaknesses are presented. Section 3 deals with cybersecurity information sharing governance structures. Section 4 introduces the results of the study, and finally, Sect. 5 concludes the Chapter.

2 Existing Maritime Surveillance Methods

The information and details of maritime surveillance systems presented in this chapter are based on public sources, mostly websites and commercial brochures and reports. The information found is analyzed by the best knowledge of surveillance systems and technical solutions of the authors by placing the data into a table to identify benefits and barriers of the different surveillance systems.

2.1 Maritime Surveillance Systems

To ensure safety and security at sea having good situational awareness becomes mandatory for each coastal nation. Maritime Surveillance systems are designed to assist authorities in controlling sea borders and traffic on critical sea areas. There are multiple manufacturers on the market, and the solutions that they offer range from systems that can are built piece by piece according to customer needs to complete surveillance systems.

Furuno provides surveillance systems that are designed for coastal surveillance, port surveillance and ground surveillance. For example, their maritime solution FF-Coast is designed to be used for the surveillance of coastal, port and critical areas. The FF-Coast system produces a situational traffic picture, reliably and in real time, to increase safety and security in the above-mentioned areas. One advantage mentioned by Furuno is that several kinds of sensors can be connected to the FF-Coast system for surveillance purposes. The number of these sensor stations are not limited per workstation. Sensors, which can best be applied in each area, can be selected to establish a cost-efficient solution [3].

Depending on the radar model and antenna type the radar system can track and control fast-moving small targets (6 m or larger) at sea level at distances from 3 to 20 km. In normal cases, the number of tracked targets are not a limited. A camera system can, depending on the chosen camera models, reach similar levels of efficiency. The alarm zone, type and generation management of abnormal behavior, collision risk or other predetermined factor can be customized based on customer needs [3].

Kongsberg has developed the Norcontrol Coastal Surveillance System, which creates real-time Common Operational Picture/Recognized Maritime Picture (COP/RMP) by providing detection, classification and identification of cooperative and non-cooperative vessels. The aim is to provide information in time to task the assets required for the reaction [4].

According to Kongsberg the Norcontrol Surveillance System can assist coastal authorities to (a) Improve security of the maritime domain and coastline; (b) Detect vessels at long ranges; (c) Identify and classify vessels; (d) Highlight vessels of interest; (e) Visualize land-based reaction forces with the maritime picture; (f) Prevent ships from entering dangerous, sensitive, prohibited or restricted areas; (g) Prevent illegal immigration, drug trafficking and smuggling; (h) Detect oil spills with radar and/or satellite images; (i) Coordinate Search and Rescue and oil spill clearance/containment operations; and (j) Integrate Blue Force Tracking [4].

Raytheon SMARTBLUE is the latest command and control (C2) system, engineered by Raytheon Anschütz. It has been specifically designed for maritime situational awareness, collision avoidance, asset protection and security. SMARTBLUE is based on open "software architecture". According to Raytheon, its customers benefit from the flexibility of selecting from a broad range of surveillance sensors depending on their specifications and budget. Additional safety, environmental and security systems can be integrated into SMARTBLUE to improve efficiency and reduce security related and environmental risks [5].

The European Union (EU) aims to enhance pre-operational Information Sharing between the maritime authorities of its Member States. Projects EUCISE2020 and MARISA are examples of milestones in the roadmap to implement the European Common Information Sharing Environment—CISE—that "represents the European test-bed for the maritime Common Information Sharing Environment in the 2020 perspective" to promote innovation and increase efficiency, quality, responsiveness and coordination of European maritime surveillance operations. The ideology behind this development is that authorities share much of the information that they have for the benefit of other actors on the maritime domain [6, 7].

2.2 Radars

Radar surveillance is one of the main methods for maritime surveillance. Radar surveillance distance was limited to the horizon before radars that use the surface wave were developed. The HF band together with surface wave technology enable the detection of maritime targets behind the horizon. Due the frequency band short range radars are needed for coastal surveillance and able for example the Vessel Traffic Services. In this chapter sort overview of existing radars is presented. On the Over the Horizon (OTH) market, there are four systems for maritime surveillance which use High Frequency Surface Wave (HFSW) technology. Diginext Stradivarius, BAE HFSWR, ELM 2270 and Raytheon HFSWR [2, 8–10].

All these four OTH radars use the same High Frequency Surface Wave principle to detect maritime targets behind the horizon. Each of these four radars have their own advantages, which the customer must evaluate against their specific needs or requirements, but other technical differences between the systems could not be found due to the limited information in public sources. All these above-mentioned radars are in operational use. New technologies will be used as they are developed and new versions/generations enter the markets.

There are several manufacturers and different types of radars on the short range radar markets. Terma SCANTER 2000, 4000 and 5000 systems, GEM Elettronica Sentinel series radars, Sentinel 50, 100, 200 and 400 systems, Hensoldt SBS-700, SBS-800 and SBS-900 radar systems, Furuno Electric Co, Ltd all have surveillance systems, which are designed for coastal, port and ground surveillance, The ELM-2226 ACSR (Advanced Coastal Surveillance Radar) [8–10]. The above-mentioned radar systems are just some examples of the solutions that are available on the markets. Typically these systems claim to be essential tools to detect (a) smugglers in very fast boats (b) illegal immigrants traveling in small, slow boats (c) boats or jet skies with hostile intentions, e.g. piracy (d) illegal fishing and (e) Search and Rescue operations.

2.3 Patrol Boats

This section focuses on vessels that have electrical surveillance capabilities (radar, optical etc.).

The effectiveness of surface vessel patrols for maritime surveillance is related to the size and capabilities of the vessel. Often, the main surveillance system is the navigational radar. The key producers of vessel-mounted versions are the same that develop surveillance radars and systems. Navigational radars have better range accuracy and resolution, but smaller surveillance areas.

For surveillance purposes, patrol boats usually have navigational radars, AIS systems, and optoelectronic systems. Larger vessels may also have air surveillance radars and capabilities for underwater surveillance. Navigational radar control units are often capable of sensor fusion (sensors on-board the vessel), and at a minimum level, AIS targets are displayed on the radar operator screen [11].

Surface vessel mounted systems are capable of surveillance areas that vary from a few nautical miles to 96 nautical miles (NM), by a surveillance system that has an X-band radar for surface and air targets [12]. The maximum surveillance distances are for air targets only. Additionally, the total coverage of surveillance is related to vessel route and speed. With a cruising speed of, for example, 17 knots (KN), a patrol vessel can cover approximately 400 NM in 24 h and thus, depending on the tracking area of its surveillance system, antenna height, and target size it can cover a mathematical surveillance area from 30,000 to 80,000 km^2.

The main advantage of patrol boat surveillance is that it is capable of visual identification of interesting targets that its on-board or other sensors have detected. This increases the importance of patrol boat surveillance. The operational costs of patrol boat surveillance are high compared to radar surveillance only. Solely the personnel costs are high as a vessel patrolling for 24/7 needs at least two watch crew.

2.4   Satellites

Space assets are an important tool for strengthening EU-wide capacity to protect its maritime security interests, including maritime surveillance. Several EU agencies have integrated satellite technology in their maritime surveillance activities.

The European Commission entrusted on 2015 European Maritime Safety Agency (EMSA) the operation of the European maritime satellite surveillance component to the Copernicus security service. According to the agreement, EMSA will use space data from Copernicus Sentinel 1 and other satellites combining it with other marine data sources, for the efficient monitoring of marine areas of interest [13].

According to the European Commission [14, p. 2] "The goal of the Copernicus Maritime Surveillance Service, managed by the European Maritime Safety Agency (EMSA), is to support its users by providing a better understanding and improved monitoring of activities at sea that have an impact on maritime safety and security, fisheries control, marine pollution, customs and general law enforcement as well as the overall economic interests of the EU".

EMSA offers the satellite-based CleanSeaNet solution for oil spill and vessel detection services in near real-time. CleanSeaNet bases on Synthetic Aperture Radar (SAR) satellite images that provide worldwide coverage of maritime areas night and day, and independent from fog and cloud cover. Data is processed to images and analyzed for oil spill, vessel detection and meteorological variables. Optical satellite imagery can also be obtained on request, depending on the situation and user needs. Each coastal state has access to CleanSeaNet through interfaces, which allow them to view the ordered images [15].

Electro optical (EO) sensors cover a 400 km by 400 km area in medium resolution and an analysis is ready in under 30 min. The Earth observation data center (EODC) has the capacity to acquire satellite images that are from areas 500 km wide and up to 1,600 km long. CleanSeaNet is mainly meant for oil-spill detection and Search and Rescue operations providing detailed views of accident areas [15].

2.5   Remotely Piloted Aircraft Systems (RPAS)

According to the International Civil Aviation Organization (ICAO) a Remotely Piloted Aircraft Systems (RPAS) are "a remotely piloted aircraft, its associated remote pilot station(s), the required command and control link and any other components as specified in the type design" [16]. RPAS offers a wide range of applications e.g. for surveillance, disaster or environmental monitoring and border control. [17].

Globally there are over 2,000 different types of RPAS and over 660 manufacturers in nearly 60 countries [18]. There are several different ways to classify RPAS, and the following Table 1 presents one example of those.

The abbreviations in the above Table 1 identify three types of RPAS. MALE stands for Medium Altitude Long Endurance, HALE stands for High Altitude Long Endurance and UCAV for Unmanned Combat Aerial Vehicle.

Table 1 RPAS classification example

	Short range	Medium range	MALE	HALE	UCAV
Weight (kg)	50–250	150–500	500–1500	2500–5000	1500–10,000
Ceiling Alt (m)	3000	5000	8000	20 000	10,000+
Operation time (h)	3–6	6–10	24–48	24–50	5–18
Distance (km)	30–70	70–200	>500	>2000	>2000

Table 2 Comparison of fixed and rotary wing RPAS [21]

Fixed wing	Rotary wing
Large intelligence area—high altitude	Smaller intelligence area—low altitude
Low power consumption—long operating time	High power consumption—short operating time
High payload take-off mass	Small payload take-off mass
Fast moving from one waypoint to another	Slow moving from one waypoint to another
Good resistance to wind load	Worse resistance to wind load
High space requirement on the ground—runway or parachute	Small space requirement on the ground—VTOL
Laborious integrating a new payload	Easier integrating a new payload
De-icing systems are possible	No de-icing systems in use
Low fault sensitivity	Higher fault sensitivity—more moving parts

EMSA provides maritime surveillance services with RPAS for authorities belonging to EU Member states, Iceland, Norway and the European Commission. These services are by request and free of charge to EU Member States, candidate countries and European Free Trade Association (EFTA) countries [19].

RPAS services can, for example, include (a) monitoring of marine pollution and emissions, (b) detection of illegal fishing, drug trafficking and illegal migration, and (c) search and rescue (SAR). RPASs equipped, for example, with optical and infrared cameras, radar, gas sensors and AIS sensors. Aircraft currently available have a durability of 6–12 h. There are three types of RPAS for different operational purposes: (1) Medium size with long endurance; (2) Larger size with long endurance and a comprehensive set of sensor capabilities and; (3) A Vertical-Take-Off-and-Landing (VTOL) [19].

In 2018, Frontex launched RPAS testing for border control in Greece, Italy and Portugal to monitor the European Union's external borders. Frontex examines the monitoring capabilities for Medium Altitude Long Endurance (MALE) and evaluates their cost effectiveness and robustness [20].

RPAS systems can simply be divided into two categories based on the principle of their retention in the air. Table 2 presents comparison of fixed wing and rotary wing RPAS features.

Table 3 Comparison of Lockheed Martin 74 K and 420 K airships [24]

	Lockheed Martin 74 K	Lockheed Martin 420 K
Length (m)	35	64
Payload (kg)	500	1000
Max. ceiling altitude (m)	1500	4600
Operation time	30 days/150 km (Line-of-sight, LOS)	30 days/240 km (Line-of-sight, LOS)
Max wind (m/s)	33	33
Radar horizon	N/A	275 km

2.6   Lighter Than Air (LTA) Systems

Lighter than air systems are usually divided into three types: airship (or often zeppelin), aerostat, and hybrid airship. Airships are powered and steerable aircraft that are inflated with a gas lighter than air. Airships are divided into three main types (by keeping them in shape): (1) rigid bodies (aluminum body) like zeppelin; (2) semi-rigid hull; and (3) unbraced [22]. The term zeppelin is a generic trademark that originally referred to airships manufactured by the German Zeppelin Company, and airships may commonly be referred as Zeppelin. Aerostat remains in the air using buoyancy or static lift. Hybrid airships enable the delivery of heavy cargo and personnel to areas away from infrastructure [23, 24].

The remote-control location of lighter than air systems can be positioned, for example, on-board a surface vessel. Their operating times are not limited by fuel or battery capacity. Operation does not require flying, and payload control is mainly stationary. These floating aircraft can be lifted to heights of 100–150 m with payloads up to 18 kg [22, 24]. Typical airtimes are between 20 and 30 days with 5–7 types of sensors: Electro Optical (EO), infrared (IR), Synthetic Aperture Radar (SAR), Laser RF, and Laser illumination. Acoustic sensors are often included for weapon and projectile detection [25].

Table 3 gives an example of two Aerostats/Airships and their features.

2.7   Maritime Patrol and Surveillance Aircraft (MPA / MSA)

Maritime patrol aircraft (MPA) are a fixed wing aircraft that operate for long durations over sea and coastal areas. Typically, MPA are fitted with radar for surface ship movement detection and tracking, infrared cameras (Forward Looking Infrared, FLIR) [26–28].

Maritime Surveillance Aircraft (MSA) provide maritime surveillance solutions designed for Search and Rescue (SAR), anti-piracy patrols and coastal and border security. MSA use proven technologies to provide multi-mission surveillance capabilities [29]. The main difference between MPA and MSA is that MPA is normally armed and MSA is unarmed.

There are multiple MPA manufacturers on the market producing different type of MPA. Two types of MPA has been selected as a reference in this article: Saab GlobalEye and P-8A Orion. The Saab GlobalEye presents the newest technology and innovations in Airborne Early Warning and Control (AEW&C) solutions. The GlobalEye provides a multi mission solution for air, land and maritime surveillance. The sensors of GlobalEye include Erieye ER (Extended Range) radar, Seaspary 7500E maritime surveillance radar, electro-optical sensor, ESM/ELINT, AIS and IFF/ADS-B. All the sensors are connected to command and control system which can also be operated remotely from a land based command centre [27, 30].

The P-8A Orion belongs to the MPA category and it is mainly designed for anti-submarine warfare (ASW). For the ASW operations, the P8 is equipped with an active multi-static and passive acoustic sensor system, inverse synthetic aperture radar, new electronic support measures system, new electro-optical/infrared sensor and a digital magnetic anomaly detector. The AN/APY-10 radar system is a multi-mission maritime and overland surveillance radar. The P-8A radar is capable of performing long-range surface search and target tracking, periscope detection, ship imaging and classification using synthetic aperture radar (SAR) and inverse synthetic aperture radar (ISAR) [31].

The challenge of using the MPA are their costs. Flight costs are high even if the original cost of MPA and their surveillance systems are not taking into account. Several MPA and crews per unit are needed for continuous 24/7/365 surveillance, due to crew rest regulations during peacetime operations.

However, the MPA is an effective tool for supporting maritime surveillance when discussing of accuracy, areal coverage and selection of surveillance equipment.

3 Cybersecurity Information Sharing Governance Structures

Almost all the business areas are using networked systems or services and the services provided by globally interconnected, decentralized IT systems and networks, the cyberspace, play a prominent role in our world. Cyberspace reaches all corners of human access and encompasses all interconnected devices into one large virtual entity. To understand the complexity and issues associated with cybersecurity, one must be knowledgeable about the evolution and growth of cyberspace, and the fact that cyberspace is mostly unregulated and uncontrolled [32]. Cyber threats, cyber-attacks, or more commonly intrusions, might affect to the continuity of business in all sectors. The dilemma of digitalisation poses the requirement for comprehensive situational awareness in cyber security as a backbone for decision-making. Dependence on these services requires the high-level security of cyberspace that can be ensured by a broad cooperation of different organisations. Information sharing is a vital component of cyber risk management, and has benefits in both preventing incidents, and managing them when they do occur. Actors sharing or exchanging information related to cyber intrusions would use it as an early warning information for immediate intrusion mitigation and threat response activities. Of course information sharing can also be useful after an incident. "Zero Day Attacks" are attacks that exploit previously unknown vulnerabilities. Reporting these incidents can help spread the word to others and enable them to prepare. Reporting incidents to trade associations, regulators, and others may also provide access to mitigation measures [33]. Franke's and Brynielsson's [34] systematic review of the literature with regard to cyber SA found that one way of gaining increased cyber SA is to exchange information with others, and Table 4 summarises their findings in that area. Successful and efficient cooperation cannot be achieved without a similar level of information exchange between the actors, and their IT systems that requires interoperability of these systems [35]. Information exchange receives much attention in the national

Table 4 Articles with regard to cyber SA information exchange

Article	Content
Klump and Kwiatkowski [37]	An architecture for information exchange about incidents in the power system
Hennin [38]	Sharing of information about suspicious IP addresses
Brunner et al. [39]	Principled problems as they ponder the trade-off between the increased awareness gained by sharing data and the loss of privacy entailed. Combining peer-to-peer networking and traceable anonymous certificates, they propose a collaborative and decentralized concept for an exchange platform
National Coordinator for Security and Counterterrorism [40]	The Netherlands find "information-exchange between the various players" to be "of the utmost importance" for fighting cybercrime
Australian Government, Attorney-General's Department [41]	The Australian government strives to foster "more intensive trusted information exchanges with high risk sectors to share information on sophisticated threats", aiming primarily at telecommunications, banking and finance, and owners of industrial control systems
Cyber Security Strategy Committee, Ministry of Defence [42]	Estonia highlights the importance of exchanging expert information within the frameworks of the international network of national CERTs, the network of government CERTs, Interpol, Europol and organizations dealing with critical information infrastructure protection

strategies. Information related to cyber threat is often sensitive and might be classified, so when that information is shared with other organisations, there is a risk of being compromised [36].

Information sharing among industry peers, and with government agencies, can allow a company to identify possible vulnerabilities in their systems, anticipate attacks, and provide access to software patches and other mitigation tools. Some reports indicate that as much as 85% of successful cyber breaches are in part preventable in that they exploit known vulnerabilities for which software patches have been available for at least a year [33]. There are different types of cybersecurity-related information that could be shared to improve cybersecurity defences and incident response. Munk [35] divides this information into four major groups: information related to events, to vulnerabilities, to threats, and other information. Sedenberg's and Dempsey's [43] division includes incidents (including attack methods), best practices, tactical indicators, vulnerabilities, and defensive measures. According to them, organizations are engaged in sharing tactical indicators ("indictors of compromise", IOCs). IOCs are artifacts that relate to a particular security incident or attack, such as filenames, hashes, IP addresses, hostnames, or a wide range of other information. Cybersecurity defenders may use IOCs forensically to identify the compromise or defensively to prevent it [43].

Sedenberg and Dempsey [43] identify seven different cyber information sharing models in the U.S. that are summarised in Table 5. Their taxonomy of cybersecurity information sharing structures may help illustrate how different design and policy choices result in different information sharing outcomes. Based on the governance models described, they identified a set of factors or determinants of effectiveness that appear in different cybersecurity information sharing regimes [43].

Always, when dealing with information exchange and sharing, the main question is "trust" [44]. The lack of trust in information propagation is the key to a lack of robust security [35]. Lack of trust is the primary reason cyber vulnerability and threat data is not shared within and between the public and private sectors [45]. Sedenberg and Dempsey [43] identify that trust within cybersecurity information sharing must be bidirectional, meaning that (1) the sharing entity needs to trust that the information will not be used against it for regulatory or liability purposes, obtained by adversaries and exploited against it as a vulnerability, or disclosed publicly to hurt the reputation of the sharer; and (2) the recipient of information needs to trust the integrity of the information shared. Also, reciprocity is important; parties need to trust that other participants will contribute roughly equivalent information [43].

Reporting to law enforcement and government agencies is required in some industries, and can help public servants "connect the dots" and identify patterns that suggests further attacks (including physical attacks) are likely, or can help authorities identify the perpetrators [33]. In the U.S., the Cybersecurity Information Sharing Act (CISA) attempts to alleviate trust burdens that accompany sharing private sector information with the government, by limiting public disclosure through Freedom of Information Act (FOIA) and by offering protections against liability and regulation. Sedenberg and Dempsey [43] found no evidence to Indicate that CISA has

Table 5 Taxonomy of information sharing models [43]

Classification	Organizational units	Example organizations	Governance types
Government-centric	Government operated; private sector members can be corporations, private sector associations (e.g., ISACs), non-profits (e.g., universities), or individuals	DHS AIS; US-CERT ECTF; FBI's e-guardian ECS	Federal laws and policies Voluntary participation rules range from open sharing subject to traffic light protocol or FOUO (for official use only) to classified information restrictions (ECS)
Government-prompted, industry-centric	Sector or problem specific	ISACs; ISAOs	Sector or problem specific voluntary participation generally organized as non-profits, use terms of service or other contractual methods to enforce limits on re-disclosure of information
Corporate-initiated, peer-based (organizational level)	Specific private companies	Facebook ThreatExchange; Cyber Threat Alliance	Reciprocal sharing; closed membership; information controlled by contract (e.g., ThreatExchange Terms and Conditions)
Small, highly vetted, individual-based groups	Individuals join, take membership with them through different jobs	OpSec Trust; secretive, adhoc groups	Trust based upon personal relationships and vetting of members; membership and conduct rules
Open-source sharing platforms		Spamhaus Project	Information published and open to all; no membership but may be formed around community of active contributors and information users; one organization may manage platform infrastructure
Proprietary products	Organization or individuals participate by purchasing the product	AV and firewall vendors	Information via paid interface; responsibility and security management still in house

(continued)

Table 5 (continued)

Classification	Organizational units	Example organizations	Governance types
Commercialized services	Organizations purchase service	Managed Security Service Providers	Outsourcing of security

succeeded in encouraging increased cybersecurity information sharing, and their research highlights some of the limitations of the statute's approach.

4 Results

This section is guided by the research questions of this paper. The first sub-section discusses the use of platforms and sensors (radar and alternative surveillance solutions) to enhance European maritime safety and security. The second sub-section deals with cyber information sharing in Maritime domain.

4.1 Platforms and Sensors to Enhance Maritime Safety and Security

One main advantage for patrol boat surveillance is its capability to visually identify any interesting target that is detected by the sensors of the boat. This increases the meaning of patrol boat surveillance.

One main obstacle to the use of satellite imagery is the considerable time difference between the satellite crossing and the transmission of satellite imagery data to Frontex and the National Coordination Centers (NCC). There still remains considerable potential for improvement, such as adding value to satellite imagery, as their results are needed within minutes or at least hours. Another problem is that, in civilian authority applications (e.g. maritime surveillance) access to high-resolution satellite imagery is restricted, which does not allow full use of existing capabilities [46].

Satellite systems are an essential instrument for consolidating the capacity of the EU to secure its maritime security interests and maritime surveillance. The benefit of using satellites are considered to be improved efficiency combined with lower operating costs. Traditional control methods (e.g. on-board inspections) will not be abandoned but van be simplified and centralized through the introduction of new technologies [47].

RPAS may be used, for example, for (1) Pollution monitoring and response; (2) Real time ship emission checks; (3) Search and Rescue (SAR); (4) Fisheries control; (5) Customs Control; (6) Border Control; and (7) Law enforcement. Equipped with suitable sensors, RPAS can detect marine pollution when carrying out targeted or routine marine surveillance operations, or RPAS can confirm the pollution that was initially detected by other resources (e.g. satellite) and collect water and pollution samples [48].

Real-time ship emissions control is an apparent task for aircraft that have the capability of flying through the emission plumes of interesting ships. Because there are no people on board, RPAS is the perfect resource to perform this such a task, and meet the operational needs. For the Maritime community, RPAS bring new abilities to the implementation of legislation on emissions from ships and carry out such duties [48].

For Search and Rescue organizations, RPAS is a tool that can improve search capabilities in large SAR areas for improved durability on manned aircraft and faster response times compared to vessels. In adverse weather conditions, RPAS can perform better than manned aircraft or surface vessels, as the associated risk to onboard personnel can be reduced. SAR operations may be conducted in off-shore locations with little or no communication coverage (EMSA 2016). "RPAS can be used as communication relay platforms to support the coordination of resources involved in the search and rescue operations" [48, pp. 4–5].

RPAS flight endurance, range and hidden features are important when targeting and tracking fishing vessels that may be involved in illegal activities such as fishing activities in restricted areas. The main advantage for Fisheries control authorities in using the RPAS would be the detection of illegal activities, which nowadays, with existing means, is still difficult to in a cost-effective manner.

4.2 Cyber Information Sharing in Maritime Domain

Cyberspace in the maritime domain comprises ports and harbours, shipping, offshore facilities, and autonomous ships, and the satellites that keep these systems connected to the deepest depths of the ocean where autonomous underwater vehicles navigate [32]. The global maritime system—including all civilian, commercial, and military ship traffic—is a system of systems, in which each system can be described as a set of components and the communication pathways between those components [49]. The maritime transportation system is increasingly a target of cyberattacks [49]. The ECHO project's maritime sector use case focuses on the commercial ship that is itself a complex cyber-physical system (CPS) with a large variety of communication systems for crew, passengers, external sources, and internal operations. According to Kessler et al. [49], the ship CPS includes:

- Bridge Navigation Systems (e.g., GPS, Electronic Chart Display and Information System [ECDIS], AIS, LRIT)
- External Communication Systems (e.g., satellite communications, FleetBroad-band, Internet)
- Mechanical Systems (e.g., main engine, auxiliary engine, steering control, ballast management)

- Ship Monitoring and Security Systems (e.g., closed-circuit television, Ship Security Alert System [SSAS], access control systems, sensors)
- Cargo Handling Systems (e.g., valve remote control systems, level/pressure monitoring systems)
- Other specialized networks (e.g., Combat Command & Control Systems on warships, Entertainment Systems and Point-Of-Sale terminals on passenger vessels; Vessel Management Systems on commercial fishing vessels).

The maritime industry has a long history of success in risk management. While physical and personnel risks are relatively easy to identify, cyber risks pose a unique challenge [33]. In modern ships, IT technology and operational technology (OT) on board are networked and highly integrated, so in order to maintain the naval surviv-ability main aspects (Susceptibility, Vulnerability, Recoverability), the underlying IT Infrastructure must be designed to assure the cyber security triad (availability, confi-dentiality and integrity) of any information and IT service, application, industrial control. The starting point is *a cyber-risk assessment* of the IT infrastructure, of the organization and of the available operators' skill, in order to evaluate the risk posed by the cyber threats or change on the services in all the possible operational condi-tions and finds, in each case, the most appropriate strategy of prevention, control and reaction. The scope of the risk management must encompasses all digital systems on vessels. These systems can be divided in two main categories: (1) the IT networks, the hardware and software dedicated to manage and to exchange information; and (2) the Operational Technology (OT) networks, the hardware and software dedicated to detecting or causing changes in physical processes through Industrial Control Systems which direct monitor and control the physical devices such as engines, rudder, valves, conveyors, pumps, etc.

When cyber risks are recognized, the organization can select mitigation strate-gies to reduce that risk. Policy enforcement controls required for risk mitigation that include Technical Cyber Security Controls and Procedural controls. The Cyber Security policy adopted should be defined and distributed over five different levels: Secure by Design, Access Control Management, Proactive Protection, Continuous Threat Monitoring and Disaster Recovery Procedure.

5 Conclusions

An extensive and objective comparison of naval surveillance systems, based on public sources, is not possible because the information producers provide vary too much. The advantage of using RPAS in general maritime surveillance (e.g. customs control, border control and law enforcement) to gain intelligence and surveillance data of suspicious vessels, is its ability to provide a communications relay for offshore tasks, and offer longer operating times, and range, as compared to manned resources. The main benefit of using RPAS from an operational point of view is that RPAS offer greater operational flexibility over conventional resources. Enhanced operational performance united with cost benefits is a tempting alternative for end-users. RPAS can complement and/or replace existing resources in a cost-effective way. Lighter than air systems offer the possibility of remote control from onboard marine vessels, with long operating times, which are not limited fuel, battery capacity, or pilots. Maritime Patrol Aircraft (MPA) are effective for surveillance, as they can provide accurate areal coverage and select surveillance equipment. The capability of the P-8 Poseidon, for example, enables to survey and compose the maritime situation of the critical parts of the Mediterranean Sea during a target period by flying in the international airspace. New capabilities as GlobalEye's remotely used surveillance systems increase the cost-efficiency of MPAs.

Traditionally, Maritime Situation Picture (MSP) has been created by using and combining data collected from a variety of technologies and platforms, such as manned aircraft (helicopters/airplanes), Earth observation (satellite systems), land-based infrastructure and patrol vessels. Modern technologies and systems provide innovative key features that can potentially provide additional sources of information and performance. They can bridge the gap between satellite-based information and locally acquired information. RPAS is one example as a tool that can reinforce existing resources and/or replace them in a more cost-effective way. However, no single solution can replace all existing systems and solutions.

Information sharing limitations on the maritime domain could be divided in at least in technical and organisational limitations. The actualized CISE network would not support classified information sharing. An organisational limitation is based on the observation that maritime authorities have outsourced network controlling and therefore co-operation might be limited between the actors. On the other hand CISE network itself and the traffic inside the network has to be controlled by the Members States and whenever a cyber-threat is found in one MS it should be informed to the other MS's. In other words, it is mandatory for CISE operational phase on 2020 to start building up the cyber information-sharing network among the maritime authorities. A wide scale of open or undiscussed issues of cyber information exchange exists among maritime CISE consortium. The common understanding or agreement which data model should be used for sharing has not been determined so far as well as the information type, which will be shared.

Acknowledgements This study has received co-funding from the European Union's H2020 research and innovation programme under grant agreement no 700478 Radars for long distance maritime surveillance and SAR operations (RANGER).

References

1. Tikanmäki, I., Ruoslahti, H.: Increasing cooperation between the European Maritime Domain authorities. Int. J. Environ. Sci. **2**, 392–399 (2017). ISSN: 2367–8941
2. RANGER: Radars for long distance maritime surveillance and SAR operations. The European Union's H2020 research and innovation programme under grant agreement no 700478 (2016)
3. Furuno Finland: Maritime surveillance system—FF-Coast. https://www.furuno.fi/eng/survei llance/surveillance_systems/ff_coast/. Accessed 15 Sept 2019
4. Kongsberg: Coastal Surveillance. https://www.kongsberg.com/kda/about-us/knc-systems/knc-coastal-surveillance/. Accessed 1 Oct 2019
5. Anschütz, R.: Maritime Surveillance and Offshore Safety/Security. https://www.raytheon-ans chuetz.com/products-systems/c2surveillance-solutions/maritime-surveillance-and-offshore-safety-security/. Accessed 5 Sept 2019
6. EUCISE2020.: EUropean test bed for the maritime Common Information Sharing Environment in the 2020 perspective. https://www.eucise2020.eu/. Accessed 22 Sept 2019
7. MARISA: Improving maritime surveillance knowledge and capabilities through the MARISA toolkit. https://www.marisaproject.eu/. Accessed 27 Oct 2019
8. BAE Systems: HF Over-The-Horizon Surface Wave Radar (HFSWR). https://products. rfsworld.com/rfs-and-bae-systems-collaborate-on-advanced-high-frequency-surface-wave-radar-coastal-surveillance-s,62,1,pressreleases,570.html. Accessed 12 Oct 2019
9. Israel Aircraft Industry: ELM-2270 OTH Over-The-Horizon HF Coastal Surveillance Radar. https://www.iai.co.il/p/elm-2270-oth. Accessed 23 Sept 2019
10. Raytheon: Next Generation High Frequency Surface Wave Radar (HFSWR). https://www.ray theon.com/capabilities/products/hfswr. Accessed 3 Sept 2019
11. Raymarine: See and be seen. https://www.raymarine.com/ais.html. Assessed 8 Nov 2019
12. Radartutorial "Scanter 2001", "Scanter 6000." https://www.radartutorial.eu/19.kartei/07. naval/. Assessed 8 Nov 2019
13. Copernicus: Copernicus security service. https://www.copernicus.eu/en/services/security. Accessed 20 Oct 2019
14. European Commission: Security service. https://www.copernicus.eu/sites/default/files/docume nts/Copernicus_Security_October2017.pdf. Accessed 20 Oct 2019
15. European Maritime Safety Agency: Celebrating the CleanSeaNet Service: A ten year anniversary publication (2017)
16. International Civil Aviation Organization: Remotely Piloted Aircraft System (RPAS) concept of operations for international IFR operations. https://www.icao.int/safety/UA/Documents/ ICAO%20RPAS%20Concept%20of%20Operations.pdf. Accessed 4 Nov 2019
17. European Defence Agency: Remotely Piloted Aircraft Systems—RPAS. https://www.eda. europa.eu/what-we-do/activities/activities-search/remotely-piloted-aircraft-systems---rpas. Accessed 2 Nov 2019
18. Blyenbugh & Co.: RPAS: The global perspective. https://uvs-international.org/yearbook-des cription/. Accessed 18 Oct 2019
19. European Maritime Safety Agency: Remotely Piloted Aircraft Systems (RPAS). https://www. emsa.europa.eu/operations/rpas.html. Accessed 17 Sept 2019
20. Frontex: Frontex begins testing unmanned aircraft for border surveillance. https://frontex.eur opa.eu/media-centre/news-release/frontex-begins-testing-unmanned-aircraft-for-border-sur veillance-zSQ26A. Accessed 3 Oct 2019
21. Insta ILS: Insta ILS presentation 19 Sep 2018.
22. Cooper, R.: The return of the airship. The Week, February 19, 2019. https://theweek.com/art icles/824018/return-airship. Accessed 1 Nov 2019
23. Air Power Development Centre: Lighter than air and hybrid airships. Pathfinder, Issue 224, July 2014
24. Lockheed Martin: Lighter than air: Aerostats & Stratospheric Airships. https://www.lockhe edmartin.com/en-us/products/unmanned-aerostats-airships-and-lighter-than-air-technology. html. Accessed 2 Nov 2019

25. Office of the Under Secretary of Defense for Research and Engineering. Lighter than air vehicles. Report: OMB No. 0704-0188 (2012)
26. Boeing: Maritime Surveillance Aircraft Quick Facts. https://www.boeing.com/defense/mar itime-surveillance/maritime-surveillance-aircraft/index.page. accessed 23 Oct 2019
27. Saab 2019: Performance beyond limits. https://saab.com/air/airborne-solutions/airborne-sur veillance/globaleye/?gclid=EAIaIQobChMIuqnB94Db5QIVio2yCh3MRApeEAAYASAAE gKYYvD_BwE. Accessed 8 Nov 2019
28. Defence iQ: Maritime Patrol Aircraft. https://www.defenceiq.com/glossary/maritime-patrol-aircraft. Accessed 24 Oct 2019
29. S. Writers: Boeing maritime surveillance aircraft demonstrator completes first flight. UPI Space Daily, 2014. https://search-proquest-com.nelli.laurea.fi/docview/1505377451?accoun tid=12003. Accessed 6 Sept 2019
30. Lentoposti: Saab esitteli uuden GlobalEye AEW&C-monitoimivalvontakoneen (The roll-out of Saab new AEW&C surveillance system GlobalEye) (2018). https://www.lentoposti.fi/artikk elit/saab_esitteli_uuden_globaleye_aewc_monitoimivalvontakoneen. Accessed 28 Oct 2019
31. Military.com: P-8A Poseidon. https://www.military.com/equipment/p-8a-poseidon. Accessed 23 Oct 2019
32. Koola, P.M.: Cybersecurity—a systems perspective. In:Dynamic Positioning Conference, Marine Technology Society, pp. 1–12 (2018)
33. Tucci, A.E.: Cyber risks in the marine transportation system.In: Clark, R., Hakim, S. (eds.) Cyber-Physical Security. Protecting Critical Infrastructure at the State and Local Level, pp. 113–131. Springer (2017)
34. Franke, U., Brynielsson, J.: Cyber situational awarenesse: a systematic review of the literature. Comput. Secur. **46**, 18–31 (2014)
35. Munk, S.: Interoperability services supporting information exchange between cybersecurity organisations. AARMS **17**(3), 131–148 (2018)
36. Kokkonen, T.: Anomaly-based online intrusion detection system as a sensor for cyber security situational awareness system. University of Jyväskylä, Jyväskylä (2016)
37. Klump, R., Kwiatkowski, M.: Distributed IP watchlist generation for intrusion detection in the electrical smart grid.In: Moore, T., Shenoi, S. (eds.) Critical Infrastructure Protection IV. ICCIP 2010. IFIP Advances in Information and Communication Technology, vol. 342, pp. 113–126. Springer, Berlin (2010)
38. Hennin, S.: Control system cyber incident reporting protocol. In:IEEE International Conference on Technologies for Homeland Security, pp. 463–468 (2008)
39. Brunner, M., Hofinger, H., Roblee, C., Schoo, P., Todt, S.: Anonymity and privacy in distributed early warning systems. In:CRITIS 2010: Critical Information Infrastructures Security, pp. 81–92 (2010)
40. National Coordinator for Security and Counterterrorism, Netherlands, National Cyber Security Strategy 2 (2013)
41. Australian Government, Attorney-General's Department: Cyber Security Strategy (2009)
42. Cyber Security Strategy Committee, Ministry of Defence:Cyber Security Strategy (2008)
43. Sedenberg, E.M., Dempsey, J.X.: Cybersecurity Information Sharing Governance Structures: An Ecosystem of Diversity, Trust, and Tradeoffs, 31 May 2018. [Online]. Available: https:// arxiv.org/abs/1805.12266. Accessed 30 March 2019
44. Rajamäki, J., Knuuttila, J.: Cyber security and trust: tools for multi-agency cooperation between public authorities. In:Proceedings of the 7th International Conference on Knowledge Management and Information Sharing—KMIS, pp. 397–404 (2015)
45. Harwood, M.: Lack of Trust Thwarts Cybersecurity Information Sharing. Security Management (2011)
46. Seiffarth, O.: EUROSUR and Copernicus—a positive example of how to create synergies at EU level. In: BRIDGES—Window on Copernicus—Discover the Security dimension of Copernicus (2013)
47. Bosilca, R.-L.: The use of satellite technologies for maritime surveillance: an overview of EU initiatives. Incas Bull. **8**(1/2016), 153–161 (2016). ISSN 2066-8201

48. European Maritime Safety Agency: User-benefit analysis of RPAS operations in the maritime domain. Executive summary (2016)
49. Kessler, G., Craiger, J., Haass, J.: A taxonomy framework for maritime cybersecurity: a demonstration using the automatic identification system. Int. J. Marine Navig. Safety Sea Transp. **12**(3), 429–437 (2018)

Comparing Cybersecurity Information Exchange Models and Standards for the Common Secure Information Management Framework

Jussi Simola

Abstract Cyber threats have increased in spite of formal economic integration in the world. Decision-makers and authorities need to respond to the growing challenge of cyberthreats by increasing cooperation. Information is one of the main facilities when the objective is to prevent hybrid threats at EU level and between the western countries. The main purpose of the study is to find out separating and combining factors concerning existing cyber information sharing models and information management frameworks in western countries. The aim is also to find out crucial factors, which affect the utilization of a common Early Warning System for the ECHO stakeholders. The main findings are that unclear allocation of responsibilities in national government department departments prevents authorities from fighting together against cyber and physical threats. Responsibilities for developing cybersecurity have been shared among too many developers. Operational work concerning cyber threat prevention between European public safety authorities should be more standardized, with more centralized information management system. When the purpose is to protect the critical infrastructure of society, public safety organizations in European Union member states need proactive features and continuous risk management in their information systems. The sharing of responsibilities for standardization concerning information management systems and cyber emergency procedures between authorities and international organizations is unclear.

Keywords Information sharing · Early warning · Standards · ECHO project

1 Introduction

The purpose of this paper is to support European ECHO Early Warning Solution developers, European politicians and end users but also provide features of existing information sharing models to identify and to take into consideration territorial, organizational, managerial, legal and societal dimensions of the existing information

J. Simola (✉)
University of Jyväskylä, Mattilanniemi 2, Jyväskylä, Finland
e-mail: jussi.hm.simola@jyu.fi

sharing solutions, models and frameworks. The research will comprise new database for the Echo Early Warning System concept. E-EWS aims at delivering a security operations support tool enabling the members of the ECHO network to coordinate and share information in near real-time. Echo Early Warning System will provide a mechanism for EU partners to share incident and other cybersecurity relevant data to partners within the ECHO network.

The sub-research's question focused on how it is possible to integrate US-related cyber information sharing models to Europe. Within E-ECHO consortium, there is a need to protect information sharing, information management and practices. The purpose is to propose initial risk management framework for the common early warning system. There are territorial and cultural differences between The United States of America and European Union, but technological solutions create new kind of opportunities within EU member countries to reach the same situation as USA have concerning proactive intrusion detection systems. The research needs equivalences of the concepts and other variable factors in other territory—in the area of European Union.

USA is the main actor in the field of information exchange in the western world. Therefore it is important to notice information sharing frameworks and models that are already in use in global level. There are many similarities concerning legislation and technical solutions between the unions and organizations, but also differences. It is important to separate predictive and preventive purposes, because legislation differ between the countries. Despite of the formal legislative dimension, agencies of The United States of America has enough resources to act proactively and use predictive functions in cyber space. According to they have capability already and legislative implementation for the new cybersecurity features is under the progress. This research belongs to European network of Cybersecurity centres and competence Hub for innovation and Operations project, which is part of the Horizon2020 program. The rest of this paper is divided as follows. Section 2 proposes central concepts. Section 3 handles background of the cyber information sharing. Sections 4 handles legislation and regulation. Section 5 handles relevant standards. Section 6 presents Method and Process. Section 7 handles information sharing models and frameworks. Section 8 presents findings. Section 9 presents conclusion about the research.

2 Central Concepts

CERT (Computer Emergency Response Team)

An organization that provides incident response services to victims of attacks, including preventive services (i.e. alerting or advisory services on security management). The term includes governmental organizations, academic institutions or other private body with incident response capabilities. (European Union Agency for Cybersecurity (ENISA) [12]. The EU Computer Emergency Response Team (CERT-EU) was set up in 2012 with the aim to provide effective and efficient response to information security incidents and cyber threats for the EU institutions, agencies and bodies.

Critical Infrastructure protection (CIP) Critical Information Infrastructure Protection (CIIP)

Critical infrastructure refers to the structures and functions which are necessary for the vital functions of society. They comprise fundamental physical facilities and structures as well as electronic functions and services. Critical infrastructure (CI) includes Energy production, transmission and distribution networks, ICT systems, networks and services (including mass communication), financial services, transport and logistics, water supply, construction and maintenance of infrastructure, waste management in special circumstances. Transforming the nation's aging electric power system into an interoperable smart grid enabling two-way flows of energy and communications. That smart network will integrate information and communication technologies with the power-delivery infrastructure [4, 28] According to Secretariat of the Security Committee [39].

Critical Information Infrastructure means any physical or virtual information system that controls, process, transmits, receives or stores electronic information in any form including data, voice or video that is vital to the functioning of critical infrastructure. Those interconnected information systems and networks, the disruption or destruction of which would have a serious impact on the health, safety, security, or economic well-being of citizens, or on the effective functioning of government or the economy [32].

Cyber-Physical Systems (CPS)

Cyber-physical systems integrate computing and communication capabilities with monitoring and control of entities into the physical world. In CPS, embedded computers and networks monitor and control the physical processes. CPS are enabling next generation "smart systems" like advanced robotics, computer-controlled processes and real-time integrated systems [25].

Cyber Threats in Critical Infrastructure

These threats can be initiated and maintained by a mixture of malware, social engineering, or highly sophisticated advanced persistent threats (APTs) that are targeted and continues for a long period of time. Channel jamming is one of the most efficient ways to launch physical-layer DoS attacks, especially for wireless communications. According to National Institute of Standards and Technology [32], National Institute of Standards and Technology [34].

ENISA

The European Union Agency for Network and Information Security (ENISA) is a centre of network and information security expertise for the EU, its member states, the private sector and Europe's citizens. ENISA works with these groups to develop advice and recommendations on good practice in information security [6].

Information Security Management System (ISMS)

An Information Security Management System (ISMS) describes and demonstrates an organization's approach to Information Security (and privacy management). It includes how people, policies, controls and systems identify, then address the opportunities and threats revolving around valuable information and related assets.

The European Cyber Security Organisation (ECSO)

It represents the contractual counterpart to the European Commission for the implementation of the Cyber Security contractual Public–Private Partnership (cPPP). ECSO members include a wide variety of stakeholders such as large companies, SMEs, research centres, universities, end-users, operators, clusters and association as well as European Member State's local, regional and national administrations, countries part of the European Economic Area (EEA) and the European Free Trade Association (EFTA) and H2020 associated countries.

Information Exchange

According to ISO/IEC 27002 Information exchange should base on policies, procedures and agreements (e.g. non-disclosure agreements) concerning information transfer to/from third parties, including electronic information sharing (e.g., messaging).

Information Sharing and Analysis Centers (ISACs)

ISAC is collaboration community created for sector-specific national or international information sharing. Information Sharing and Analysis Centers are trusted entities to foster information sharing and good practices about physical and cyber threats and mitigation. The ISAC could support the implementation of new European legislation (e.g. NIS Directive) or support economic interests [7].

Information Sharing and Analysis Organization (ISAO)

An ISAO is any entity or collaboration created or employed by public- or private sector organizations, for purposes of gathering and analysing critical cyber related information in order to better understand security problems and interdependencies related to cyber systems to ensure their availability, integrity, and reliability [43].

North Atlantic Treaty Organization (NATO)

NATO is a 70 years old security alliance of 28 full member countries from North America and Europe. NATO's primary goal is to protect the Allies' security by political and military means. NATO is the principal security instrument of the transatlantic community. The security of North America and Europe are permanently tied together with allies. NATO enlargement has furthered the U.S. goal of a Europe whole, free, and at peace [42].

Risk Assessment Framework (RAF)

According to National Institute of Standards and Technology [35], the purpose of risk assessments is to inform decision makers and support risk responses by

(a)   Identifying relevant threats to organizations or threats directed through organizations against other organizations;
(b)   Identifying internal and external vulnerabilities;
(c)   Impact to organizations that may occur given the potential for threats exploiting vulnerabilities and
(d)   Likelihood that harm will occur. The result is a determination of risk.

Risk Management Framework (RMF)

Comprehensive risk management process by NIST, which Integrate the risk Management Framework into the system development lifecycle.

Standards ISO 27000 family

This family of 27000 standards provide fundamental bases for the definition and implementation of an Information Security Management System (ISMS) [31] (JRC TAXONOMY). The Security Measurement Index is based on ISO 27000 international standards and input from an advisory board of security professionals. It consists benchmarking tools for assessing organizations' security practices, a global assessment of IT and a basis for developing security measurement best practices to help make cybersecurity more effective and efficient [22].

Among ISO 27000 family, target audience comprise e.g. personnel of risk management. Personnel as skilled lead auditors are needed to grant certification [13].

Standard ISO/IEC 27010:2015 (ISO/IEC 2700 family)

Is a key component of trusted information sharing is a "supporting entity", defined as "A trusted independent entity appointed by the information sharing community to organise and support their activities, for example, by providing a source anonymization service" [18].

Tactics, Techniques, and Procedures (TTPs)

The behaviour of an actor. A tactic is the highest-level description of this behaviour, while techniques give a more detailed description of behaviour in the context of a tactic, and procedures an even lower level, highly detailed description in the context of a technique (National Institute of Standards and Technology [33].

Threat Information

Any information related to a threat that might help an organization protect itself against a threat or detect the activities of an actor. Major types of threat information include indicators, TTPs, security alerts, threat intelligence reports, and tool configurations [33].

3 Cooperation Within the USA, NATO and EU

The Department of Homeland Security (DHS) is the U.S. Federal Government focal point of the U.S. cyber information-sharing ecosystem. It is responsible for the government's operational responses to major cybersecurity incidents, analyzing threats and exchanging critical cybersecurity information with the owners and operators of critical infrastructures and trusted worldwide partners. DHS as part of U.S Government and NATO (North Atlantic Treaty Union) have developed advanced situational awareness systems within cyber ecosystem. NATO is developing a Cyber Rapid Reaction Team (RRT) that protect its critical infrastructure. U.S. Cyber Command's Cyber Protection Teams (CPT's) creates security for all states in USA. NATO does not have an inherent cyber offensive capability, as the U.S Cyber CPT.

NATO CCD COE's mission is to enhance cooperation and information sharing between NATO member states and NATO's partner countries in the area of cyber defence by virtue of research, education and consultation. The Centre has taken a NATO-oriented interdisciplinary approach to its key activities, including academic research on selected topics relevant to the cyber domain from the legal, policy, strategic, doctrinal and/or technical perspectives, providing education and training, organizing conferences, workshops and cyber defence exercises, and offering consultations upon request [37]. NATO does not have own cyber weapons against cyber-attacks [41]. The U.S.-led alliance established an operations centre on Aug. 31.2018 at its military hub in Belgium and the U.S.A, Britain, Estonia and other allies have since offered their cyber capabilities [3]. NATO's CYOC (CYOC Cyber Operations Center) is under development, and it will provide coordination and integration fuctions for allies.

The MITRE Corporation is a private, not-for-profit organization that manages and operates federally funded research and development centers (FFRDCs) that support United States (U.S.) government sponsors. FFRDCs serve as long-term strategic partners to the government, providing objective guidance in an environment free of conflicts of interest. MITRE has substantial experience as a trusted, independent third party providing secure stewardship, sharing, and transformational analyses of sensitive information in USA [2].

3.1 Background of Information Sharing in EU

In 2009 ENISA (European Network and Information Security Agency) defined information exchange as follows: An information exchange is a form of strategic partnership among key public and private stakeholders. The common goal of the information exchange is mostly to address malicious cyber-attacks, natural disasters and physical attacks. The drivers for this information exchange are the benefits of member countries working together on common problems and gaining access to information, which is not available from any other sources [12].

The European Commission presented the cybersecurity strategy of the European Union in 2013. It sets out the EU approach on how to best prevent and respond to cyber disruptions and attacks as well as emphasizes that fundamental rights, democracy and the rule of law need to be protected in the cyber-atmosphere. Cyber resilience as one of the strategic priorities. That means effective cooperation between public authorities and the private sector is crucial factor [7].

The European Public-Private Partnership for Resilience (EP3R) was established in 2009 and was the very first attempt at Pan-European level to use a Public-Private Partnership (PPP) to address cross-border Security and Resilience concerns in the Telecom Sector. After the EP3R the main principles for setting up a PPP ecosystem in Europe are to provide legal basis of cooperation. It is also important to ensure open communication between public and private sector. Involvement of Small and Medium Enterprises (SMEs) in the process of PPP building is also crucial, since they are the backbone of the European economy [11, 14].

3.2 Information Exchange in Law Enforcement

How to prevent criminal activities has been one of the main question when public safety authorities have tried to solve a common problem within EU countries. Hague Programme and Stockholm Programme introduced the principle of availability as the guiding concept for information exchange of law enforcement. Information that is available to law enforcement authorities in one Member State should be made accessible to law enforcement authorities or public safety authorities in other Member States [27].

Regulations and Policy Documents; European Regulation and policy documents were considered as sources for legal definitions and to cover the gaps left by the vocabularies extracted from standards when dealing with non-technical definitions [27].

Law enforcement authorities can use Schengen Information Systems (SIS) to consult alerts on wanted persons etc. both inside the EU and at the EU external border. The SIS improve information exchange on terrorist suspects and efforts Member States of EU invalidate e.g. the travel documents [27].

The European Commission has adopted a Communication on the European Information Exchange Model (EIXM). The instruments covered by EIXM allows other to exchange automatically fingerprints, DNA and vehicle registration data (Prum decision). Swedish decision sets out how information should be exchange between EU Member States [27].

Europol supports Member States of the European Union as the information hub for EU law enforcement. Its Secure Information Exchange Network Application (SIENA) enables authorities to exchange information with each other, with Europol, and with a number of third parties. Europol's databases help law enforcement from different countries to work together by identifying common investigations, as well as providing the basis for strategic and thematic analysis [27].

4   Legislation and Regulation Concerning Information Exchange in USA and Europe

4.1   Regulation in the USA

The White House designated the National Coordinating Center for Communications (NCC) as Information Sharing and Analysis Center (ISAC) for telecommunications in accordance with presidential Decision Directive 63 in 2000 (President's National Security Telecommunications Advisory Committee (NSTAC) [38].

The communications Information Sharing and Analysis Center (Comm-ISAC) incorporating dozens of organisations. It has facilitated the exchange of information among industry and government participants regarding vulnerabilities, threats, intrusions and anomalies affecting the telecommunications infrastructure.

The exchange of information between the EU and the US has been regulated among other things, as follows; The European Commission and the U.S. Government reached a political agreement on a new framework for transatlantic exchanges of personal data for commercial purposes named the EU-U.S. Privacy Shield. The European Commission adopted the EU-U.S. Privacy Shield on July of 2016 [8].

The framework protects the fundamental rights of anyone in the EU whose personal data is transferred to the United States as well as bringing legal clarity for businesses relying on transatlantic data transfers.

The EU-U.S. Privacy Shield based on the principles: Obligations on companies that handle data. (a) The U.S. Department of Commerce will conduct regular updates and reviews of participating companies to ensure that companies follow the rules they submitted themselves to. (b) Clear safeguards and transparency obligations on U.S. government access: The US has given the EU assurance that the access of public authorities for law enforcement and national security is subject to clear oversight mechanisms. (c) Effective protection of individual rights: citizen who thinks that collected data has been misused under the Privacy Shield scheme will benefit from several accessible dispute resolution mechanisms. It is possible for a company to resolve the complaint by itself or give it to The Alternative Dispute resolution (ADR) to be resolved for free. Citizens can also go to their national Data Protection Authorities, who will work with the Federal Trade Commission to ensure that complaints by EU citizens are investigated and resolved [8]. The Court of Justice of the European Union issued a judgement declaring as invalid the European Commission's Decision (EU) 2016/1250 on the adequacy of the protection provided by the EU-U.S. Privacy Shield Framework is no longer a valid mechanism to comply with with EU data protection requirements when sharing personal data from the European Union to the United States [45]. Participated organizations of the Privacy Shield program are required to re-certify to the Department of Commerce annually. The Department will remove an organization from the Privacy Shield List if it voluntarily withdraws from the Privacy Shield or if it fails to achieve its annual re-certification to the Department. An organizations's removal from the list means it may no longer claim that it benefits from the Privacy Shield.

4.1.1 Freedom of Information Act (FOIA)

The Freedom of Information Act (FOIA) has provided the public the right to request access to records from any federal agency. The FOIA requires agencies to proactively post online certain categories of information, including frequently requested records. It is often described as the law that keeps citizens in the know about their government. Federal agencies are required to disclose any information requested under the FOIA unless it comprises under one of nine exemptions which protect interests such as personal privacy, national security, and law enforcement. Any person can make a FOIA request (Office of Information Policy (OIP) [36].

4.1.2 Cybersecurity Information Sharing Act (CISA)

CISA authorizes companies to monitor and implement defensive measures on their own information systems to counter cyber threats. CISA provides certain protections to encourage companies voluntarily to share information about "cyber threat indicators" and "defensive measures" with the federal government, state and local governments, and other enterprises and private entities. These protections comprise protections from liability, non-waiver of privilege, and protections from FOIA disclosure, although, importantly, some of these protections apply only when sharing with certain entities. Qualifying these protections requires that, the information sharing must comply with CISA's requirements, including regarding the removal of personal information [16].

4.2 Regulation in the European Union

The list of the most relevant regulation taken into consideration in EU level.

4.2.1 NIS Directive

ENISA, Europol/EC3 and the EDA are three agencies active from the perspective of NIS, law enforcement and defines respectively. These agencies have Management Boards where the Member States are represented and offer platforms for coordination at EU level [10].

DIRECTIVE (EU) 2016/1148 OF THE EUROPEAN PARLIAMENT AND OF THE COUNCIL of 6 July 2016 concerning measures for a high common level of security of network and information systems across the Union (NIS directive). The NIS Directive (see EU 2016/1148) is the first piece of EU-wide cybersecurity legislation. The goal is to enhance cybersecurity across the EU. The NIS directive was adopted in 2016 and subsequently, because it is an EU directive, every EU member state has started to adopt national legislation, which follows or "transposes' the directive. EU directives give EU countries some level of flexibility to take into account national circumstances, for example to re-use existing organizational structures or to align with existing national legislation [5]. The European Parliament resolution on the European Union's cyber Security Strategy states e.g. that the detection and reporting of cyber-security incidents are central to the promotion of information networks Sustainability in the Union [26].

The NIS Directive consist three parts:

1. National capabilities: EU Member States must have certain national cybersecurity capabilities of the individual EU countries, e.g. they must have a national CSIRT, perform cyber exercises, etc.
2. Cross-border collaboration: Cross-border collaboration between EU countries, e.g. the operational EU CSIRT network, the strategic NIS cooperation group, etc.
3. National supervision of critical sectors: EU Member states have to supervise the cybersecurity of critical market operators in their country: Ex-ante supervision in critical sectors (energy, transport, water, health, and finance sector), ex-post supervision for critical digital service providers (internet exchange points, domain name systems, etc.).

4.2.2   General Data Protection Regulation

The EU General Data Protection Regulation (GDPR) harmonize data privacy laws across Europe, to protect and empower all EU citizens' data privacy and to reshape the way organizations across the region approach data privacy. GDPR applies to all businesses offering goods and/or services to the EU. That means that the organizations do not have to reside in the EU area or even in Europe, if you are holding private information about an EU citizen whom you provide services, GDPR applies [9]. The Regulation introduces stronger citizens' rights as new transparency requirements. It strengthens the rights of information, access and the right to be forgotten. The law is technology neutral and applies to both automated and manual processing if the data is organized in accordance with pre-defined criteria [9]. It also does not matter if the data is stored in an IT system through video surveillance, or on paper. In all these cases personal data is subject to the protection requirements set out in the GDPR.

5   Relevant Standards Concerning Cyber Secure Information Sharing

What is Data protection and relationship between 27000 and 29000 family standards?

Data protection is the basic legal right of all individuals to protect their own personal information. Personal information is any information relating to an identified or identifiable person. The purpose of data protection is to indicate when and under what conditions personal data may be processed. Organizations processing personal da ta are required to take reasonable steps to protect it [15].

How should personal data be processed?

The processing of personal data or privacy issues is subject to requirements in several different laws. The processing of personal data must be confidential and secure. The processing of personal data according to the principles is only for a specific and legitimate purpose. Privacy Policy—Consent and Freedom of Choice. Legality and definition of purpose. Limitation of data collection. Restriction of data processing. Restriction on Use, Storage and Disposal SFS-ISO / IEC 2910 [15].

Important standards of data protection

The 29000 series contains standards that fundamentally govern privacy, although the 29000 series contains a very wide variety, most of which have nothing to do with privacy issues. The 27000 series describes the standards related to the security management method, some of which also directly concern data protection. The 27000 Series management template can be used to implement a data-driven environment, which is a prerequisite for data protection [15]. As Fig. 1 illustrates, information security consist of CIA (Confidentially, Integrity and Availability) features. Confidentiality means that information is only accessible to those entitled to it.

Integrity or correctness of information means that the information must be true and correct. Availability means that information is available when you want to use the data of the data subject. The right to privacy or the rights of the data subject required by data protection cannot be fulfilled without the implementation of the data security attributes as mentioned above. For example, the data subject has the right to know who has accessed the data stored in the register. This requires confidentiality and integrity.

Figure 2 presents relationships between the elements of data protection.

Figure 2 presents relationships between the elements of data protection and standards (modified from SFS 2018 publication).

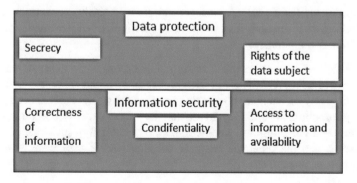

Fig. 1 Privacy elements (CIA)

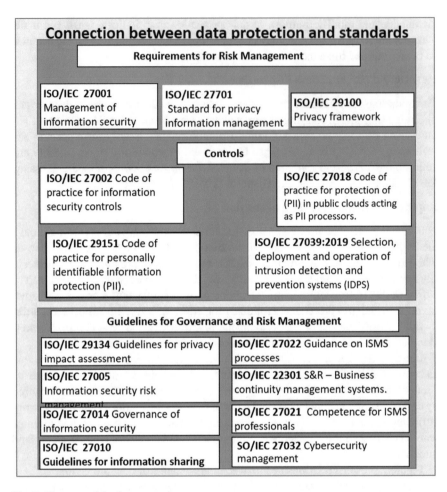

Fig. 2 Elements of the data protection

According to ISECT [23] risk management, ISO/IEC 27005 is a remarkable standard which propose ongoing process consisting of a structured sequence of activities, some of which are iterative:

- Establish the risk management context (e.g. the scope, approaches or methods to be used and relevant policies and criteria such as the organization's risk tolerance)
- Quantitatively or qualitatively assess means identify, analyze and evaluate relevant information risks, taking into account the information assets, threats, existing controls and vulnerabilities to determine the likelihood of incidents or incident scenarios, and the predicted business consequences if they were to occur, to determine a "level of risk".

- Manage and modify by using information security controls, retain or "accept", avoid and/or share with third parties the risks appropriately, using those "levels of risk" to prioritize them;
- Keep partners informed throughout the process; and Monitor and review risks, risk treatments, obligations and criteria on an ongoing basis, identifying and responding appropriately to significant changes [23].

ISO/IEC 29134:2017 [19] gives guidelines for a process on privacy impact assessments and a structure and content of a PIA report. It is applicable to all types and sizes of organizations, including public companies, private companies, government entities and not-for-profit organizations. ISO/IEC 29134:2017 is relevant to those involved in designing or implementing projects, including the parties operating data processing systems and services that process PII [19].

According to requirements for system management ISO/IEC 29100:2011 provides a privacy framework which specifies a common privacy terminology; defines the actors and their roles in processing personally identifiable information (PII); describes privacy safeguarding considerations; and provides references to known privacy principles for information technology. It is applicable to natural persons and organizations involved in specifying, procuring, architecting, designing, developing, testing, maintaining, administering, and operating information and communication technology systems or services where privacy controls are required for the processing of PII [17].

ISO/IEC 27001 formally specifies an Information Security Management System (ISMS). It is a suite of activities concerning the management of information risks (called "information security risks" in the standard). The ISMS is an overarching management framework through which the organization identifies, analyzes and addresses its information risks. The ISMS ensures that the security arrangements are fine-tuned to keep pace with changes to the security threats, vulnerabilities and business impacts—an important aspect in such a dynamic field, and a key advantage of ISO27 family's flexible risk-driven approach. "Statement of Applicability" (SoA) is not explicitly defined, it is a mandatory requirement. SoA refers to the output from the information risk assessments and in particular the decisions around treating those risks. The SoA may, i.e. take the form of a matrix identifying various types of information risks on one axis and risk treatment options on the other and show how the risks are to be treated in the body, and perhaps who is accountable for them. It usually references the relevant controls from ISO/IEC 27002 but the organization may use a completely different framework such as NIST SP800-53, the ISF standard, BMIS and other [24].

Management methods and controls

Management consists ISO/IEC 29151:2017 and ISO/IEC 27002:2013. ISO/IEC 29151:2017 establishes control objectives, controls and guidelines for implementing controls, to meet the requirements identified by a risk and impact assessment related to the protection of personally identifiable information (PII). ISO/IEC 29151:2017 is applicable to all types and sizes of organizations acting as PII controllers (as defined in ISO/IEC 29100), including public and private companies, government entities and not-for-profit organizations that process PII [21].

ISO/IEC 27002:2013 gives guidelines for organizational information security standards and information security management practices including the selection, implementation and management of controls taking into consideration the organization's information security risk environment(s). It is designed to be used by organizations that intend to: select controls within the process of implementing an Information Security Management System based on ISO/IEC 27001; implement commonly accepted information security controls; develop their own information security management guidelines [20].

Continuity management and relationship to the Cyber-Physical System

ISO/IEC 22301:2019 set frames to the Security and resilience. It consists requirements for business continuity management systems. It represents how to manage business continuity in an organization [1]. This standard based on leading business continuity specialists opinions and supplies the framework for managing business continuity in an organization [1]. Other relevant standards are listed on the Fig. 3.

6   Method and Process of the Research

Case study illustrates the attempt to produce a profound and detailed information about the object under research. The materials collected for this case study based on scientific publications, official documents, collected articles and literary material. The research is focused on how it's possible integrate USA- related information sharing models in European level. Yin [44] identifies five components of research design for case studies: (1) the questions of the study, (2) its propositions, if any; (3) its unit(s) of analysis; (4) the logic linking the data to the propositions; and (5) the criteria for interpreting the findings. This case study is carried out with the guidance of Yin [44].

There are country-specific differences, institutional differences, legislative differences in legislation, etc. The purpose is to categorize things into their own groups. Some information sharing models and information management frameworks are simple diagrams, some are ready-made templates, and some information sharing models have concrete instruments and tools. The purpose of the analysis is to find out about the functionalities, useful standards and features of information sharing systems in the EU, USA and NATO. Outcome of the research is combined proposal of information sharing model and initial risk management framework.

7 Definition of Information Sharing Goals

According to National Institute of Standards and Technology [33] the organization should establish goals and objectives that describe the desired outcomes of threat information. These objectives will help guide the organization through the process of scoping its information sharing efforts, joining sharing communities and providing ongoing support for information sharing activities.

According to Skopik et al. [40] primary dimensions of security information sharing can be divided as follows: (a) Cooperation and coordination economic need for coordinated cyber defense. There exists variety of classification of information that are viable for a wide range of stakeholders: indicators of compromise, technical vulnerabilities, zero-day exploits, social engineering attacks or critical service outages. (b) Legal and Regulatory Atmosphere: information sharing requires a legal basis. Therefore, the European Union and its Member States and the US, have already done a set of directives and regulations. (c) Standardization Efforts means enabling information sharing, standards and specifications need to standardize that are compliant with legal requirements (e.g. NIST, ENISA, ETSI and ISO). (d) Regional and International Implementations means taking these standards and specifications, organizational measures and sharing structures need to be realized, integrated and implemented. CERTs and national cyber security centers work on this issue. (e) Technology Integration into Organizations means sharing protocols and management tools on the technical layer need to be selected and set into operation.

7.1 Identify Internal Sources of Cyber Threat Information

CORA (Cyber Operations Rapid Assessment) methodology was developed to study issues and best practices in cyber information sharing. In addition, it consists as an engagement tool for assessing and improving threat-based security defenses. CORA identifies five major areas of cyber security where the proper introduction of threat information can have tremendous impact on the efficacy of defenses: External Engagement—Tools and Data Collection—Tracking and Analysis—Internal Processes—Threat Awareness and Training.

The TICSO gather cyber threat intelligence and information from a variety of sources including open source reporting by researchers and consultants, government and law enforcement sources [USCERT, INFRG], fee-for-service threat Intel feeds from vendors and industry sector and regional threat sharing communities such as ISACs and ISAOs. The TICSO focuses collection efforts on the most relevant information by defining prioritized intelligence requirements (PIR), and continuously evaluating the quality of intelligence from different sources in terms of relevance, timeliness, and accuracy (MITRE Corporation).

A first step in any information sharing effort is to identify sources of threat information within an organization. According to National Institute of Standards and

Technology [33]. The process of identifying threat information sources includes the following sections:

(a) Identify sensors, tools, data feeds, and repositories that produce threat information and confirm that the information is produced at a frequency, precision, and accuracy to support cybersecurity decision-making.
(b) Identify threat information that is collected and analyzed as part of an organization's continuous monitoring strategy.
(c) Locate threat information that is collected and stored, but not necessarily analyzed or reviewed on an ongoing basis.
(d) Identify threat information that is suitable for sharing with outside parties and that could help them more effectively respond to threats. Examples of selected Internal Information Sources [33].

7.2   Comparing Features of the Information Sharing Models

There are several different information sharing models in the world. The most important thing was to choose such cyber information sharing models that are widely used in the European Union countries, USA and NATO. It is not necessary to compare all models or frameworks because availability of information varies a lot. Usually the information-sharing model is incomplete frame that is believed to solve all the problems concerning cyber security. As Table 1 illustrates five different type of models has chosen to more detailed review.

8   Findings

Mechanism type of the ISAC concerns the overall structure that is used to exchange information. This type of mechanism often has a central hub that receives data from the participants. The hub can redistribute the incoming data directly to other members, or it can provide value-added services and send the updated information or data to the members. The hub may act as a "separator" that can facilitate information sharing while protecting the identities of the members. One of the main tasks of ISACs is sharing information on intrusions and vulnerabilities. These types of information are usually troublesome; therefore, companies often decide to keep silent. ISAC hub system relies on the functionality of the hub, which makes the system vulnerable to delays and systemic failures [29]. Important information is often unnecessary to achieve, delays in information sharing can reduce the benefits of the information-sharing hub mechanism. In post to all model information is shared among stakeholders. MITREs model is one kind of hybrid information sharing model. It is a partner for helping private or public organizations stand-up and run information sharing exchanges. Mechanism of MITRE use automated processing of information. This work has enabled security automation in vulnerability management, asset

Table 1 Examples of information sharing models

Organization //Name //System/model or framework type	Main tasks/ features	Special tasks or info	Major areas of cyber impacts	Instruments
MITRE// CORA // Assessment of cyber operations (not-for-profit organization)	Developed for to study issues and best practices in cyber information sharing. It serves as an engagement tool for assessing and improving threat-based security defences	Based on NIST Special Publication 800–150: Guide to Cyber Threat Information Sharing	External Engagement Tools and Data Collection Tracking and Analysis Internal Processes Threat Awareness	indicators scan networks and systems—Reporting new indicators about attacks on its own networks
MITRE// TISCO// Threat-Informed Model	It collects cyber threat intelligence and information from a variety of sources including open source reporting by researchers and consultants		External Engagement Tools and Data Collection Tracking and Analysis Internal Processes Threat Awareness	Sensors monitoring attack activity such as phishing email addresses and URLs of malicious sites, host-based indicators
ENISA// ISAC// Member driven organization model Country-focused ISAC - International ISAC -	Sharing knowledge about incidents with the member organizations and prevent/ respond to the incidents which occur (ISAC is a fast way to get all the knowledge and way of networking and meeting people from different organizations	ISAC gives the public sector access to knowledge about the cybersecurity level in critical sectors. It provides information about threats and incidents. (close cooperation with the industry, public entities get better understanding of the private sector)	(a) Some information can be shared widely with all members. (b) The shared information is more detailed in internal circle. c) use of the (TLP) to share information	web portal/platform (following a specific template) and encrypted emails

(continued)

Table 1 (continued)

Organization //Name //System/model or framework type	Main tasks/ features	Special tasks or info	Major areas of cyber impacts	Instruments
ENISA// PPP// Cooperative model	Access to public funds. Opportunity to influence national legislation and obligatory standards. Access to public sector knowledge and confidential information (EU legislation, fighting against cybercrime)	Helps to achieve resilience in the cyber ecosystem. PPP Increase the trust between public–public–private. it allows to have better information and proactive attitude in case of crisis	Incident handling and crisis management, Information exchange, Early warnings, Technical evaluation, Defining standards etc.	Help desk helps PPP's members. PPP does not consist real-time instruments against cyberattacks
NIST// Framework//	NIST FW targeting on risk management, procedures and privacy preservation aspects	The guidelines included in the ISO/IEC27010 standard, it is oriented toward the protection of the data exchanged in the information sharing process	Techniques standards and protocols for systems monitoring, threat detection, vulnerability inventory and incident exchange	Framework adds consist different kind of tools, but only framework does not offer protection for shared information or information for incident handling process

management, and configuration management though the Security Content Automation Protocol program. Members of MITRE do not share information. Each participant sends its sensitive data to MITRE, and MITRE works diligently to ensure that member data is kept confidential [29].

There is a need to develop Public–Private information-sharing models in EU level because public safety organizations of the Department of the Homeland Security in USA are capable to handle external threats more effectively. There are international organizations which have formulated co-operational working environment such a way that western world could operate for the common purpose. International organizations like UN (United Nations) and NATO are the connecting factors concerning harmonization of information sharing procedures in the EU and USA and between them, not forgetting NATO. In this author's view, the so-called "triangle" should be called a "square."

The requirements of the system integrity means that it's impossible to separate information system -related standards from the information sharing methods when the purpose is to design common cyber ecosystem for the western world. Interoperability should be coordinated through standards as Fig. 3 illustrated.

Cyber-physical system allows to protect critical infrastructure because of the automated functionalities. E.g., in a finance sector it is not possible to protect it without interfering with the activities of the attacker. Automated physical actions mean Physical functionalities e.g., in finance sector and/or cyber-defence functionalities against the attacks but everything must be reverted to existing standards. Privacy impact (PI) is crucial element in all situations when the purpose is to develop system which handle

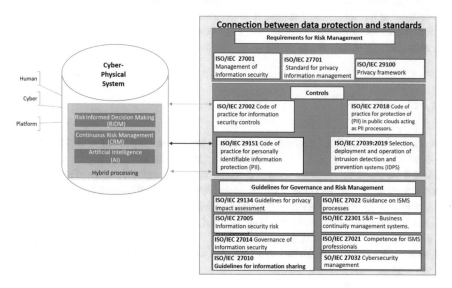

Fig. 3 Relationship between CPS and continuous risk management system

privacy identifiable information. PI could result from the processing of Privacy Identifiable information (PII). According to ISO/IEC 29134:2017 (International Organization for Standardization (ISO) [19] a Privacy Impact Assessment (PIA) is a tool for addressing the potential impacts on privacy of a process information system, programme, or device. It will inform to all participants which have to take actions in order to treat privacy risk. PIA is ongoing process and report may include documentation about measures taken for risk treatment, measures may arise from the use of the ISMS.

At a general level, collaboration between cyber-physical system and continuous risk management requires collaboration between these elements. In the traditional sense, three levels can be found; human; platform layer and cyber layer as figure illustrates, but that's not enough. Proposed framework require to take into account standards and information management when purpose is to develop common early warning solution for the western allies.

At the technical level, the challenge of semantic interoperability is that information systems should automatically understand the concepts arising from the actions of people and organisations. Therefore, it is important to create a common risk management framework for both. It is possible to connect different kind of decision-making strategies to the cyber physical framework as proposal illustrates above. Legislation and regulation must be the fundamental basis for all functions and operations.

This means that fundamental frame of the cyber-physical system based on legislation, rules and standards. E.g., higher-level EWS should be structured from the view of "regulation". The operations of the system must be based on rules and standards. Semantic interoperability means that an information system is able to combine the information it receives from different sources and process it in a way that preserves the meaning of the information. E.g., there are business-related differences concerning sector-specific stakeholders of the ECHO consortium.

9   Conclusions

Separate functionalities between the EU member states are not only problem. When the common goal is to improve Cyber Situational Awareness, it is important to deepen the cooperation between western stakeholders. Major problem of information sharing models is related lack of real-time cyber information management between participants. There is essential problem with features of information sharing models. When the purpose is to protect vital functions of society, public safety organizations in European Union member states needs proactive features in their information systems. A shared common cyber situational awareness means that real time communication links between the states must exist.

Legislation is not only factor, which affects to completely secure cyber-ecosystem. Developed systems need coherent standardization, common management system and governance model. The USA and its public safety cyber defense organizations has ability to combat cyberattacks, which have made against vital functions, but also make counter-attacks [41]. It is one of the most important features in protecting the western world. Cooperation and collaboration in triangle EU-NATO-USA is therefore particularly important. In addition The United Nations acts as the fourth element. Utilizing the best features of the information sharing models will ensure procedures of continuity management. It is therefore important to place EU countries in the right context. Legislation has been harmonized, but occasional is to trust organization's functionalities. Common continuous risk management system helps to handle the databases concerning privacy issues. Lack of standardization may cause obstacles when the aim is to catch cyber criminals or find out state level actor that has caused a cyber or hybrid attacks.

It is a fundamental problem that, as the geographical area of the European Union expands, it does not have the capability to prevent hybrid-threats. Controlled governance model for the EWS and common standardization concerning information management systems and cyber emergency procedures between authorities, and international organizations helps to achieve common situational awareness inside the western world. It is not enough that every country tries to tackle cyberthreats separately. There is a need for a jointly controlled information exhange framework for the EU countries and credible counter opeartion tools for counter-attack operations that must be connectable to another defense mechanism. Nato is setting up a joint coordination center against cyberattacks by 2023, but NATO will also neeed centralized mechanism to defend allies against cyber-threats.

References

1. Advisera Expert Solutions: What is ISO 22301? [Homepage of Advisera Expert Solutions] (2019). [Online]. Available: https://advisera.com/27001academy/what-is-iso-22301/. 28 Aug 10
2. Bakis, B., Wang, E.D.: Building a National Cyber Information-Sharing Ecosystem. MITRE Corporation (2017)
3. Bigelow, B.: The Topography of cyberspace and its consequences for operations. In: 10th International Conference on Cyber Conflict 2018, NATO CCD COE Publications (2018)
4. Department of Homeland Security (DHS): Blueprint for a Secure Cyber Future—The Cybersecurity Strategy for the Homeland Security Enterprise. DHS (2011)
5. ENISA: NIS Directive [Homepage of European Union Agency for Network and Information Security] (2019-last update), [Online]. Available: https://www.enisa.europa.eu/topics/nis-dir ective [6/2019]
6. ENISA: Position Paper of the EP3R Task Forces on Trusted Information Sharing (TF-TIS). European Union Agency for Network and Information Security, Greece (2013)
7. ENISA & ITE: Information Sharing and Analysis Centres (ISACs) Cooperative models. European Union Agency for Network and Information Security, Greece (2017)
8. European Commission: EU-U.S. Privacy Shield: Stronger Protection for Transatlantic Data Flows. Brussels (2016)
9. European Commission: General Data Protection Regulation (EU) 2016/679. Regulation edn. Brussels (2016)
10. European Commission: Joint Communication To The European Parliament, The Council, The European Economic And Social Committee And The Committee Of The Regions. European Commission, Brussels (2013)

11. European Union Agency for Cybersecurity (ENISA): Public Private Partnerships (PPP) Cooperative models. European Union Agency for Network and Information Security, Greece (2017)
12. European Union Agency for Cybersecurity (ENISA): Good Practice Guide—Network Security Information exchanges. ENISA, Greece (2009)
13. European Union Agency for Network and Information Security (ENISA): Smart grid security certification in EUROPE. ENISA, Greece (2014)
14. European Union Agency for Network and Information Security (ENISA): EP3R 2013—Position Paper of the EP3R Task Forces on Trusted Information Sharing (TF-TIS). European Union Agency for Network and Information Security, Greece (2013)
15. Finnish Association for Standardization SFS RY: Information technology. Safety. Information security management systems. Privacy Standards. SFS (2018)
16. Harvard Law School Forum on Corporate Governance and Financial Regulation: Federal Guidance on the Cybersecurity Information Sharing Act of 2015 [Homepage of The President and Fellows of Harvard College] (2016). [Online]. Available: https://corpgov.law.harvard.edu/2016/03/03/federal-guidance-on-the-cybersecurity-information-sharing-act-of-2015/. 11 Oct 2019
17. International Organization for Standardization (ISO): ISO/IEC 29151:2017 Information technology—Security techniques—Code of practice for personally identifiable information protection [Homepage of ISO] (2018), [Online]. Available: https://www.iso.org/obp/ui/#iso:std:iso-iec:29151:ed-1:v1:en
18. International Organization for Standardization (ISO): International Standard ISO/IEC 27010:2015. Standard edn. Switzerland (2015)
19. International Organization for Standardization (ISO): ISO/IEC 29134:2017 Guidelines for privacy impact assessment (2017). Available: https://www.iso.org/standard/62289.html
20. International Organization for Standardization (ISO): ISO/IEC 27002:2013 Security techniques—Code of practice for information security controls [Homepage of ISO] (2013), [Online]. Available: https://www.iso.org/standard/54533.html
21. International Organization for Standardization ISO: ISO/IEC 29100:2011 information technology—Security techniques—Privacy framework [Homepage of ISO] (2018), [Online]. Available: https://www.iso.org/standard/45123.html2019
22. International Telecommunication Union: Global Cybersecurity Index (GCI) 2018. ITU, Switzerland (2018)
23. ISECT: ISO/IEC 27005:2018 Information technology—Security techniques—Information security risk management (third edition [Homepage of IsecT Limited] (2018), [Online]. Available: https://www.iso27001security.com/html/27005.html
24. ISECT: ISO/IEC 27001 Information security management systems—Requirements [Homepage of IsecT Limited] (2017), [Online]. Available: https://www.iso27001security.com/html/about_us.html
25. Lee, E.A., Seshia, S.A.: Introduction to Embedded Systems, A Cyber-Physical Systems Approach, 2 edn. (2015)
26. Lehto, M., Limnéll, J., Kokkomäki, T., Pöyhönen, J., Salminen, M.: Kyberturvallisuuden strateginen johtaminen Suomessa. 28. Valtioneuvoston kanslia, Helsinki (2018)
27. Migration and Home Affairs: Information exchange [Homepage of European Commission] (2019), [Online]. Available: https://ec.europa.eu/home-affairs/what-we-do/policies/police-cooperation/information-exchange_en. [06/2019, 17/06/2019].
28. Ministry of the Interior: National Risk Assessment. Ministry of the Interior, Helsinki (2018)
29. MITRE: Cyber Information-Sharing Models: An Overview. MITRE Corporation (2012)
30. MITRE Corporation: Cyber Operations Rapid Assessment (CORA): A Guide to Best Practices for Threat-Informed Cyber Security Operations | The MITRE Corporation. Available: https://www.mitre.org/sites/default/files/publications/pr_15-2971-cyber-operations-rapid-assessment-best-practices_0.pdf [3/20/2016, 2016]
31. Nai-Fovino, I., Neisse, R., Lazari, A., Ruzzante, G., Polemi, N., Figwer, M.: European Cybersecurity Centres of Expertise Map—Definitions and Taxonomy. Publications Office of the European Union, Luxemburg (2018)

32. National Institute of Standards and Technology: Framework for Improving Critical Infrastructure Cybersecurity. 1.1. NIST (2018)
33. National Institute of Standards and Technology: Guide to Cyber Threat Information Sharing. NIST Special Publication 800–150. National Institute of Standards and Technology, Gaithersburg (2016)
34. National Institute of Standards and Technology: Guidelines for Smart Grid Cybersecurity—Volume 2 privacy and the Smart Grid. U. S. Department of Commerce (2014)
35. National Institute of Standards and Technology: Guide for Conducting Risk Assessments. 800–30. U.S. Department of Commerce, Gaithersburg (2013)
36. Office of Information Policy (OIP): What is FOIA? [Homepage of U.S. Department of Justice] (2019), [Online]. Available: https://corpgov.law.harvard.edu/2016/03/03/federal-guidance-on-the-cybersecurity-information-sharing-act-of-2015/ [10/11, 2019].
37. Pernik, P., Wojtkowiak, J., Verschoor-Kirss, A.: National Cyber Security Organisation: United States. CCDCOE, Tallinn (2016)
38. President's National Security Telecommunications Advisory Committee (NSTAC): Report to the President on the National Coordinating Center. Department of the Homeland Security (2006)
39. Secretariat of the Security Committee: Finland's cyber security strategy—government resolution. Ministry of Defense (2013)
40. Skopik, F., Settanni, G., Fiedler, R.: A problem shared is a problem halved: a survey on the dimensions of collective cyber defense through security information sharing. Comput. Secur., 154–176 (2016)
41. Smeets, M.: NATO Allies Need to Come to Terms with Offensive Cyber Operations [Homepage of Lawfare] (2019), [Online]. Available: https://www.lawfareblog.com/nato-allies-need-come-terms-offensive-cyber-operations [11/19, 2019].
42. U.S. Mission to NATO: About NATO (2019). Available: https://nato.usmission.gov/our-relationship/about-nato/
43. White, G., Lipsey, R.: ISAO SO Product Outline. ISAO Standards Organization (2016)
44. Yin, R.K.: Case Study Research, Design and Methods, 5th edn. Sage, Thousand Oaks, CA (2014)
45. Court of Justice of the European Union: The Court of Justice invalidates Decision 2016/1250 on the adequacy of the protection provided by the EU-US Data Protection Shield (2020)

Cyber Situational Awareness in Critical Infrastructure Organizations

Jouni Pöyhönen, Jyri Rajamäki, Viivi Nuojua, and Martti Lehto

Abstract The capability related to cybersecurity plays an ever-growing role on overall national security and securing the functions vital to society. The national cyber capability is mainly composed by resilience of companies running critical infrastructures and their cyber situational awareness (CSA). According to a common view, components of critical infrastructures become more complex and interdependent on each other and, as a consequence, ramifications of incidents multiply. In practice, the actions relate to developing better CSA and understanding of a critical infrastructure organization. The aim is to prepare for incidents and their management in a whole-of-society approach. The arrangement is based on drawing correct situation-specific conclusions and, when needed, on sharing critical knowledge in the cyber networks of society. The target state is achieved with an efficient process that includes a three-leveled (strategic, operational and technical/tactical) operating model related to the organization's decision-making. The cyber environment is dynamic and hence especially the strategic agility is required when preparing for incidents. The pervasive incidents targeting society are a challenging cyber environment when it comes to the critical reaction speed required by the situation management.

Keywords Critical infrastructure · Cybersecurity · Information sharing · Situational awareness · Vital societal functions

J. Pöyhönen (✉) · V. Nuojua · M. Lehto
University of Jyväskylä, Jyväskylä, Finland
e-mail: jouni.a.poyhonen@jyu.fi

V. Nuojua
e-mail: viannuoj@jyu.fi

M. Lehto
e-mail: martti.j.lehto@jyu.fi

J. Rajamäki
Laurea University of Applied Sciences, Espoo, Finland
e-mail: jyri.rajamaki@laurea.fi

1 Introduction

The national cybersecurity capability is vital for overall security of society and securing its crucial functions. Mostly, the national cybersecurity capability consists of resilience of private critical infrastructure (CI) companies, and of the cyber situational awareness (CSA) they constantly maintain. CI can be described as a three-leveled system of systems (Fig. 1), efficient and appropriate operations can be targeted at its three levels, from bottom to the top: power grid, data transmission network and services [14]. It is generally supposed that CI becomes more complex and its parts are increasingly dependent on each other, and that way, the ramifications of the incidents can be multiple compared with the original impact. The operation of CI and the threats having an impact on it are not limited only on organizations or administrative borders [17].

An efficient incident management requires tight collaboration between the management, situational awareness (CA) and communication. A good management requires: (1) unquestionable managerial responsibility, casting of different operators and decision-making ability of the ministerial authority, (2) building of situation awareness (situational understanding, evaluation of situational development), (3) crisis communication, (4) information sharing, and supporting technical solutions, (5) business continuity management, and (6) co-operation.

This study is a continuum of the research 'Cyber strategic management in Finland' [11], in which one task was to formulate management proposals for the management of nationally pervasive incidents concerning cyber environment. A good CSA has an essential impact on the incident management. The collection of research data was created on the basis of open theme interviews with a material-based content analysis, a document analysis and international comparison information. All three levels of the CI system of systems (see Fig. 1) were represented. There were altogether 40 interviewees from 25 private or public Finnish organizations, who were leaders or persons responsible for the information/cyber security of their organizations. The observations, presentations and models presented in this article, are based on qualitative analysis of this data.

The research questions of this study are:

1. How the cyber situational awareness of a CI organization can be developed?
2. What kind of cyber situational awareness challenges exist in CI?

In Finland, the significance of the private businesses is emphasized in the operation of critical infrastructure, since approximately 80% of the operations can be estimated

Fig. 1 Plain structure of critical infrastructure

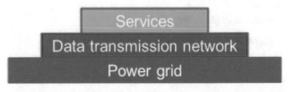

to belong to their responsibility. During the research, six private businesses were interviewed, as well as public authorities, such as the National Cyber Security Centre Finland (NCSC-FI) and the National Emergency Supply Agency (NESA). Section 2 deals with the need for SA, related decision-making levels, and the SA models applied in this study. In Sect. 3, the formation of situational awareness in Finnish critical infrastructure organizations are explored. Section 4 discusses the challenges of SA in critical infrastructure. Section 5 concludes this study.

2 Fundamentals of Situational Awareness

In order to function, every organization needs information about its environment and events, and also about their impact on its own operation. An appropriate and fast SA is based on correct information and evaluations, and it is emphasized in case of incidents when very pervasive decisions must be made quickly. In order to make correct solutions, decision-makers have to know the base for their decisions, consequences how the others react to them and what risks the decisions include. For that reason, decision-makers must have sufficient SA and understanding on all the operational levels, which enables a timely decision-making and operation. SA and understanding require collaboration and expertise, which enables the comprehensive monitoring of the operational environment, data analysis and aggregation, information sharing, recognition of the research needs and network management. The information systems must enable the systematic use of information sources and collaboration, and the flexible sharing of situation information related to it [12].

The organizations' and decision-makers' formation of SA is supported by the situation awareness arrangements. In general, SA means the description of the dominant circumstances and the operational preparedness of different operators aggregated by the specialists, the happenings caused by an incident, its background information and the evaluations concerning the development of a situation. In addition, data analysis based operational recommendations may be related to SA. The general view is constituted by utilizing a networked operational model based on different sources. The process consists of data acquisition, information aggregation, classification and analysis, and of a timely and efficient sharing of the analyzed information with those in need. The surrounding data space is organized such that the information is understood correctly, and that the operators have a chance to get the information important to their operation [12].

The pervasive incidents targeting society are a challenging cyber environment when it comes to the critical reaction speed required by the situation management. Advanced persistent threats are unfamiliar attacks to the traditional protection ways and can proceed quickly, when a fast information sharing and good SA play an important role in the incident management. In a worst-case scenario, the delegation of responsibility should be able to make possible in a few minutes, the response evoked without delay, and the abilities and tools put to use [13].

2.1 Decision-Making Levels

Organizations operate in very complex, interrelated cyber environments, in which the new and long used information technical system entities (e.g. system of systems) are utilized. Organizations are depended on these systems and their apparatus in order to accomplish their missions. The management must recognize that clear, rational and risk-based decisions are necessary from the point of view of business continuity. The risk management at best combines the best collective risk assessments of the organization's individuals and different groups related to the strategic planning, and also the operative and daily business management. The understanding and dealing of risks are an organization's strategic capabilities and key tasks when organizing the operations. This requires for example the continuous recognition and understanding of the security risks on the different levels of the management. The security risks may be targeted not only at the organization's own operation but also at individuals, other organizations and the whole society [7].

Joint Task Force Transformation Initiative [7] recommends implementing the organization's cyber risk management as a comprehensive operation, in which the risks are dealt with from the strategic to tactical level. That way, the risk-based decision-making is integrated into all parts of an organization. In Joint Task Force Transformation Initiative's research, the follow-up operations of the risks are emphasized in every decision-making level. For example, in the tactical level, the follow-up operations may include constant threat evaluations about how the changes in an area can affect the strategic and operational levels. The operational level's follow-up operations, in turn, may contain for example the analysis of the new or present technologies in order to recognize the risks to the business continuity. The follow-up operations of the strategic level can often concentrate on the organization's information system entities, the standardization of the operation and for example on the continuous monitoring of the security operation [7].

From the necessity of the organization's risk follow-up operations can be drawn the necessity of the whole organization's SA. As mentioned, the formation of the organizations' and decision-makers' SA is supported by the situation awareness arrangements. Thus, an appropriate SA supports the cyber risk management and more extensively the evaluation of the organization's whole cyber capability.

2.2 Situational Awareness Models

Ensley [4] has developed a SA model when working on several different research assignments in the service of United States Air Force. Figure 2 describes the general structure of the model. The core of SA consists of the three basic elements: detection (Level 1), situational understanding (Level 2) and its impact assessment towards the future (Level 3). This SA provides the foundation for conclusions and the following decision-making. Depending on the situation, the assignment- and system-specific

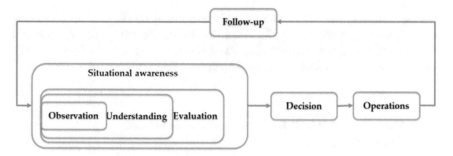

Fig. 2 Situational awareness and dynamic decision-making (adapted from Endsley [4])

features and the decision-maker's experiences and evaluation ability bring their own impacts on the table. Decision-making, in turn, guides the operation that reflects back to the observed operational environment.

Faber [5] regards the SA development operations, concerning both public and private businesses, as one of the most significant near future goals aiming to improve cyber security. He recommends applying Endsley's model to the follow-up needs of a cyber operational environment.

The general structure of Endsley's SA model is applied when solving this study's research questions. The framework for forming of critical infrastructure SA is introduced in Fig. 3. The detection part (Level 1) of Endsley's structure is presented as the organization-specific detection needs of the strategic (S), operational (O) and technical/tactical (T) decision-making levels. The goal is to gain perception that serves each decision-making level. The SA that is formed of observations is a prerequisite for understanding the observations (Level 2). After that, the impact analysis

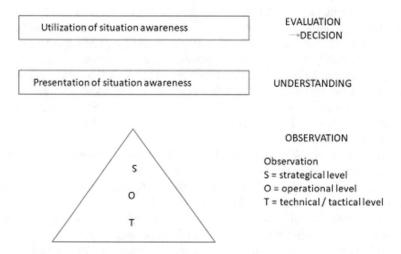

Fig. 3 Framework for forming situational awareness

and assessment of the observations is made possible by utilizing the understanding about situation awareness (Level 3). There, the analysis capability plays an important role. The final goal is to make appropriate and situation-specific decisions on each decision-making level, and conduct the operations followed by the decisions.

3   Formation of SA in a Critical Infrastructure Organization

This section explores the formation of SA in Finnish critical infrastructure organizations. Examinations are made at the technical/tactical, operational and strategic decision-making levels.

3.1   *Situational Awareness on a Tactical Level*

Both technical, networked and management situation awareness are emphasized when building SA. During the last years, Finland has formed its cyber situation awareness through the information sharing mechanisms of different operators. It is about national and international collaboration. The improvement of information sharing and perception is still a matter of development when it comes to Finland's cyber security [10].

The critical infrastructure operators use such protection techniques in their ICT systems that extend from the interface of Internet and the organization's internal network right up to the protection of a single workstation or apparatus. These technical solutions make possible to verify different harmful or anomalous observations. The typical technologies are related to security products such as network traffic analysis and log management by Security Information and Event Management (SIEM) systems, firewall protection, intrusion prevention and detection systems (IPS and IDS) and antivirus software. SA is built up in centralized monitoring rooms (Security Operations Center, SOC) whose technical solutions can be under the organization's own control, or the service can be outsourced to the information security operator. A crucial goal is the real-time SA that is an absolute prerequisite for protection of the business processes.

In addition, the critical companies for security of supply have the Finnish cyber threat prevention mechanisms *HAVARO* system in the external interface of their network. The system follows the network traffic and detects harmful and anomalous traffic. Then, the warnings come from the NCSC-FI.

The observation ability relates also to so called advance warning that can be received from the organization's international or national operation networks. In a center of operation, there is always the organization's capability to pay attention to the abnormal operation that possibly occurs in the system. The overall observation ability is developed for example by benchmarking and practicing.

The organizations implement the analysis of incidents and anomalies from their own starting points and at the hands of their own or carried out by the service provider. Securing of organizations' business process operations requires more and more ability of analyzing. The intensification of protection operations or for example the introduction of alternative operational models are the most important goals of the operation. The analyzing capability determines the choice of needed operations and, that way, plays an important role in the organization's decision-making process. The analyzing ability must enable a severity classification and a cyber-physical view.

The analyzing is based on SA and usually happens in SOCs, where the information coming from different censors is aggregated and a situation-specific analysis is formed. Based on the analysis, the needed operations are launched. The organization's possibilities to utilize the information gotten from the international or national operational networks relate to the analyzing ability. Personnel's capability to interpret available observations correctly has a significant meaning in composing situation-specific analyses.

A typical reaction to an incident or anomalous operation comes from an incident response manager based on SA and its analyzing. The magnitude and severity of an incident have an impact on the operations. Besides fast reacting, the organization's management can be congregated to decide on the extension of the operations, and the allocation of the needed resources. Depending on the magnitude of an incident, the organization's whole management all the way to the supervising board can be informed. Regarding the publicly traded companies, the organization's external informing is guided by the informing obligations based on the law.

In case of a nationally extensive incident, the CI organization in question keeps in touch to the NCSC-FI, and utilizes not only the authority network but also the industry's own network and its business networks. In this communication, the organization's SA and its situation-specific analyzing are combined.

The companies that are critical for emergency supply have a requirement to inform authorities, such as NCSC-FI, in case of an incident. Based on the Directive on security of network and information systems (NIS Directive), an authority can expand this demand to the critical infrastructure organizations whom the duty to notify does not yet apply.

3.2 Situational Awareness on an Operational Level

The operational level operations are used to advance the strategic goals. The comprehensive security- and trust-adding operations require a comprehensive cyber security management. Its starting point must be the target's risk assessment, and the operation analyses carried out based on it. The operational level's concrete hands-on operations must be targeted at the confirmation of information security solutions and the composition of the organization's continuity and disaster recovery plans. The goal must be continuous monitoring of the operational processes' usability, and the decision-making support in case of incidents that require analyzing and decisions.

NCSC-FI and NESA has been identified Finland's administrative point of contacts on the business level. NESA with different pools, especially the digital pool of Finland, support companies in developing and maintaining CSA. Because of the operation goals, NESA brings together authorities and IT businesses. The private sector should recognize its own tasks to advance national cyber security. Finnish collaboration models between authorities and private businesses have been created, and they are internationally comparable.

The KRIVAT—information sharing network between critical infrastructure companies—service of the State Security Networks Group Finland is an example of an information sharing and cooperation framework, which is specifically designed for the management of disturbances and continuity of CI operations. It, thus, exists to specifically enhance the preparedness of CI. KRIVAT is a framework for action, and its main purpose is to supplement the existing preparedness and disturbance-management activities of critical infrastructure operators during major disturbances. It responds to a recognized need for clearer communications structures and better situation awareness between organizations for disturbance management. Finland is one of a few countries where CI companies are, in case of disruptions, required to cooperate with one another, and facilitating open information sharing is key to the KRIVAT community [15].

The technical protection ability of most significant CI organizations and the observation ability based on that are on a good level in Finland. Different collaboration networks are widely used, and organizations keep in touch with NCSC-FI regularly. The analyzing ability of anomalous operation and the incident management ability are base on capable personnel in CI companies and functional collaboration networks.

3.3 Situational Awareness on an Uppermost Management Level

CI organizations' uppermost management should continuous develop and maintain the reliability of their piece of the national CI. The strategic choices relate to the reputation of an organization. The management must make concrete strategic choices and support and guide the performance of the chosen operations through the whole organization. An important task of the management is to take care of the adequate resourcing of operations. About the chosen operations must be communicated extensively with the organization's personnel and other interest groups.

It is important to create a cybersecurity assessment model for the needs of the uppermost management. By the model, other organizations may evaluate the company's cybersecurity level, and management can become aware of the company's weaknesses and possible insufficiencies in contingency planning, and take care at least of the basics. The operations require strategic level decisions from the organization's uppermost management.

Finland's national cyber security execution program 2017–2020 aggregates the pervasive and significant information and cyber security improving projects and operations of the state administration, business and associations, and their responsibilities. The progress of the execution program can be followed by following the development of the different organization's capabilities during the concerned inspection period. The execution program includes extensively effective operations that are developed by other administrative-specific operations, and by the work related to the development of cyber and information security and business continuity management. At the same time, the follow-up results in the formation of the national cyber SA [16, 16].

The implementation programme for Finland's cybersecurity strategy for 2017–2020 expresses the need for a light cybersecurity evaluation tool by which organizations are able to take care of reaching the minimum level of security. The National Cyber Security Index (NCSI) was developed for the follow-up of the national cyber security related capability. It is based on twelve sectors that are sorted into four groups as follows: (1) General Cybersecurity indicators, (2) Cybersecurity basic indicators, (3) Event and crisis management indicators, and (4) International event indicators [19]. The NCSI index has four cybersecurity viewpoints per each twelve sections: (1) effective legislation, (2) functioning individuals, (3) collaboration arrangements, and (4) the results from different processes. Table 1 introduces the measure tool that is based on the NCSI index. It measures the organization's cybersecurity capability. The tool is developed for the use of businesses and other organizations. The evaluation is based on the requirements, business, interest group collaboration and results.

The widespread commissioning of the measure tool in CI organizations would make it possible to follow the cyber security development of the whole area in the same way as it serves the strategic level needs of a single organization.

4 Situational Awareness Challenges

The most significant challenges of an organization's CSA deal with to the observation of the vulnerabilities and operational deviation of the complex technical system wholeness [8]. Main challenges at the technical level consist of two aspects: CSA of enterprise ICT assets and CSA of industrial automation.

4.1 Enterprise ICT Systems

The vulnerabilities of the ICT assets of the organization have to be identified continuously against the emerging threats. The organization considers real-time forming of the CSA of ICT systems and assets which is in accordance with the Information

Table 1 Structure of an organization-specific measure tool

	Requirements	Business	Interest group collaboration	Results
General Indicators				
Ability to develop the organization's cyber security culture				
Ability to analyze its cyber environment				
Magnitude of cyber security training				
Basic level indicators				
Confirmation of operational resources				
Risk assessments				
Quality requirements of the information systems' operation				
Operation follow-up and measures				
Event and incident management indicators				
Quality of contingency planning for incidents				
Situational awareness 24/7				
Ability to manage incidents				
Ability to recover from incidents				
National impact indicators				
Operation in cyber operational networks				
Points				

Technology Infrastructure Library (ITIL) service model. Figure 4 represents an organization's framework for forming the CSA of its own ICT systems. ICT assets and the ticketing system as well as CERT messages and SOC log analysis that arrive via Security Information and Even Management (SIEM) (see Fig. 4) represent Level 1 elements of Endslay's model (see Fig. 2). The objective is to achieve from them a real-time expression which describes information security. The CSA which is formed of the observations is a precondition for the understanding of observations (Level 2). After this the preconditions to estimate effects which are in accordance with the observations form good using the CSA and utilizing the knowledge of the structure of the ICT system of the interpreter of the CSA and technical know-how (Level 3). The final objective is naturally the making of the situation-specific right decisions

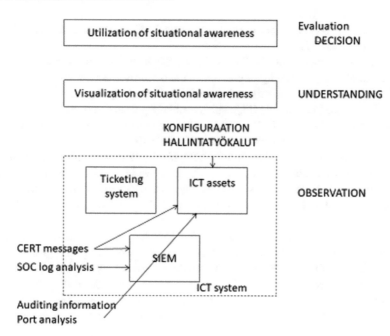

Fig. 4 Forming of the CSA of enterprise ICT systems

and the control of the measures which maintain the capacity of the ICT systems and information reserves which are in accordance with the decisions.

ICT systems and assets can be controlled with the tools of the control of the configuration (Discovery tools). These tools help to analyze the properties of the firewalls and to clarify the services which function in the ICT systems. The real-time situation information of the user experience of information processing systems and of information reserves which is in accordance with the ITIL service model consists of the support requests of the service which are directed to them (Ticketing system). On the other hand, the SIEM system helps to form the situation consciousness of known threatening factors (CERT messages) and of logs of ICT systems (SOC log analysis) as a subject of the examination. So every one of the above mentioned arrangements produces the CSA for its part. The final objective is to accomplish the automatic connecting of this information which is directed to the structure of the ICT system which is in accordance with the ITIL services of the target organization.

4.2 Industrial Automation

The Controller Area Network (CAN) is an automation bus that was originally designed for real-time data transfer of distributed control systems to cars. Later, the CAN bus was developed as a universal automation system for many automation solutions. CAN bus is widely used also in critical infrastructure and is used in this section as an example.

One of the characteristics of the CAN bus is that its traffic is not supervised in any way due to the lack of timing of control. In other words there are no authentication mechanism. The Cybersecurity and Infrastructure Security Agency [3] warned the users of the bus of the vulnerability, which enables a physical entrance to the automation system, and makes the DoS attack possible. The seriousness of the attack depends on how the CAN bus has been carried out in a destination system and how easily the potential attacker can use the input port (typically OBD-II). This differs from the earlier frame based attacks that were usually perceived by IDS or IPS. The utilizing of the vulnerability concentrates on the message bits which govern the bus and causes functional disorders in the CAN nodes in the sending of the right message frames [3].

The CAN communications protocol, ISO-11898:2003, describes how information is passed between devices on a network and conforms to the Open Systems Interconnection (OSI) model, which is defined in terms of layers. Actual communication between devices connected by the physical medium is defined by the physical layer of the model. The ISO 11898 architecture defines the lowest two layers of the seven-layer OSI/ISO model as the data-link layer and the physical layer, shown in Fig. 5.

The application layer establishes the communication link to an upper-level application specific protocol such as the vendor-independent CANopen™ protocol. This protocol is supported by CAN in Automation (CiA), the international users and manufacturers group. Many protocols are dedicated to particular applications, such as industrial automation, diesel engines, or aviation [2].

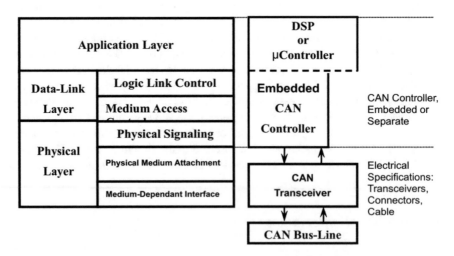

Fig. 5 CAN bus in the OSI/ISO model (adapted from Corrigan [2])

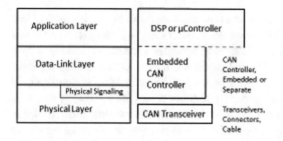

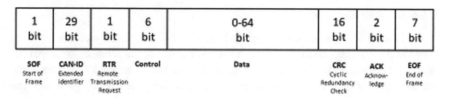

Fig. 6 CAN bus message

CAN message and frames

The four different message types, or frames (see Fig. 6), that can be transmitted on a CAN bus are the data frame, the remote frame, the error frame, and the overload frame.

The data frame is the most common message type, and comprises the arbitration field, the data field, the CRC field, and the acknowledgment field. In Fig. 6 the arbitration field contains a 29-bit identifier and the RTR bit, which is dominant for data frames. Next is the data field, which contains zero to eight bytes of data, and the CRC field, which contains the 16-bit checksum used for error detection. The acknowledgment field is last.

The intended purpose of *the remote frame* is to solicit the transmission of data from another node. The remote frame is similar to the data frame, with two important differences. First, this type of message is explicitly marked as a remote frame by a RTR bit in the arbitration field, and second, there is no data.

The error frame is a special message that violates the formatting rules of a CAN message. It is transmitted when a node detects an error in a message and causes all other nodes in the network to send an error frame as well. The original transmitter then automatically retransmits the message. An error mechanism in the CAN controller ensures that a node cannot tie up a bus by repeatedly transmitting error frames.

The overload frame is mentioned for completeness. It is similar to the error frame with regard to the format, and it is transmitted by a node that becomes too busy. It is primarily used to provide for an extra delay between messages.

Arbitration is a mechanism for conflict resolution between network nodes. When the network path is free, any of the nodes in the network can start the message send process. If another node also wishes to send at the same time, the order of the transmissions is decided using a bitwise arbitration mechanism. During arbitration, both nodes start their transmission. The transmission starts with a start bit, followed by an id field (identifier, CAN-ID). The sending order decision is made based on the value of the id field and the other node or nodes discontinue their transmissions. The messages are sent ordered by priority, where the zero value is dominant. In practice this means that if a node currently sending a bit with a value of one sees that another node is sending a zero bit, it backs off. In other words, it discontinues its own transmission, forfeiting its turn to the node sending the dominating bit. In practice the message with the smallest decimal id value has the highest priority [6].

CAN bus pros and cons

According to Voss and Comprehensible [18], CAN bus was designed for maximal speed and reliability. At the technical level this mean, among other aspects, that the network communication uses a provider–consumer model instead of the common sender–receiver model. The second feature aiming for performance gains was the lossless bus arbitration described above. Improving the reliability of the data transmitted between the nodes was achieved with a mechanism that insures the integrity and timeliness of the messages. These mechanisms are based on bus arbitration, using checksums checking the payload and resending failed messages [18].

Based on these design decisions, CAN bus is effectively a broadcast network, where any node can send a message and all nodes are listening to the network and reacting to the messages they are interested in. The only thing the recipients check is the protocol correctness of the received message [18].

CAN bus speed is 1 Mbit/s, which these days does not seem fast. Yet for transmitting short messages and having an effective collision avoidance mechanism, CAN bus is more suitable to be used in real-time applications than connected protocols such as TCP/IP, even if those would be using greater transmission speeds [18].

With further development the CAN bus has became a dominant technology for the data transmission of vehicle basic functions. During the last two decades the number of electronic systems in vehicles has increased and at the same time they have become more complex. CAN bus vulnerabilities can be traced back to design decisions described above, the most significant of these being the lack of authentication mechanism. The receiving entity does not have any mechanism to verify the origins of the received message or the validity of the data received. In other words, the control unit does not have a mechanism to detect message forgery. This characteristic makes vehicle CAN busses vulnerable to attacks, such as message forgery, unauthorized data use and denial of service. The DoS vulnerability can be exploited by sending a large number of high priority messages. These attacks can affect the vehicles systems in such a way as to cause loss of control, incorrect functionality, premature wear or rendering the vehicle unable to function at all [1].

Attack surfaces

The taxonomy of CAN bus attack surfaces is usually divided into two parts: remote exploits and exploits requiring physical access to the CAN bus. In addition to this, some researchers have expanded the use of physical connections by constructing experiments that enable man-in-the-middle type of attacks on the CAN bus [9].

Physical connection to a CAN bus is not technically complex to achieve. The simplest physical connection can be implemented through the vehicle's diagnostics port. This approach does not require any alterations to the vehicle in question. The limitation of this approach is the amount of network data observable at this point of entry, depending heavily on the make and model of the vehicle. CAN bus traffic seen through the diagnostic port is restricted by segmenting the network. These limitations can be avoided by choosing another point of entry from the desired segment. In most cases this approach requires alterations to the vehicle's wiring harnesses, because segment-specific connectors are rarely implemented in production vehicles.

Remotely exploitable attack surfaces that would have a direct effect on the vehicle's physical functionalities are usually more challenging to exploit. In practice, this normally means a multistage attack where the attacker first has to find a vulnerable and remotely accessible service to gain a foothold. As an example, this kind of service can be found from the vehicle telemetry or infotainment systems. After gaining a foothold on one of the connected systems, the attacker needs to find a way to gain access to another system that has connectivity to the more critical segments of the vehicle's CAN bus. This type of attack has been successfully conducted by some vehicle security researchers.

Cyber security of CAN bus (vehicle) study has been conducted at the University of Jyväskylä (AaTi study). The focus of the AaTi study was to survey anomaly detection methods applicable to CAN-networks. This research complements previous research and patents by understanding network-traffic characteristics using recordings obtained from a test vehicle. The study shows that attacks against vehicle networks can be categorized into three groups. The network can be injected (a) with special messages such as diagnostics messages; (b) with normal messages that disturb vehicle functionality or (c) by sending normal messages after the real sender has been rendered unfunctional. The most common situation is probably when the real sender is still functional, and the attacker sends normal CAN messages. These kinds of attacks can be detected by observing message send intervals, since in a normal situation the intervals should remain regular.

In the first phase of research a neural network implementation was tested for its ability to detect abnormalities in message data payloads. The aim of this implementation was to provide technical means to learn different payload possibilities and predict the data incoming in the following messages. This would have created the possibility to detect abnormal data payloads. The problem with using neural networks arose from its resource intensiveness and lack of prediction accuracy. The next experiments focused on anomality detection methods based on message timing.

The first time-based method we tested was One-Class Support Vector Machine (OCSVM), which is a variation of the popular machine learning method. This method defines boundaries around normal behavior and classifies all other traffic as abnormal. In the implementation, a moving window with a set number of messages was used as a data-entity. The characteristics of the messages are then calculated using OCSVM and, based on the results, the whole window is declared normal or abnormal. The characteristics used in this implementation were average interval and standard deviation.

After this fist experiment, other methods based on message interval were surveyed. Kernel density estimation models interval deviation for each message identifier. This value can then be compared to incoming messages in order to detect abnormalities. Modeled deviation provides a density function for the interval that can be used for likelihood value calculation for incoming messages. A drop in the calculated likelihood that exceeds a predetermined threshold can be detected as an anomaly and an alarm can be triggered. Because kernel density estimation is also a resource-intensive method and the observed test data did not show multipeak properties, a simplified version using the same principles of this method was implemented. This method aims to model message identifier deviation using key values. This implementation of absolute deviation achieves substantial gains in resource efficiency and without decline in the performance of the detection properties. The modeling was done using standard deviation in order to use the two key values: average and standard deviation. In the practical implementation training phase average, lower and upper bound values were calculated for each message identifier for classification purposes. A moving window was used as a data entity. If the values within the window went below the lower bound or exceed the upper bound, the whole window is declared an anomaly in the network traffic.

All of the above mentioned methods have their own challenges in either resource intensiveness, accompanied in some cases with inaccuracy of predictions. Based on the experience described in the method comparison chapter, a novel method for detecting CAN bus anomalies based on message arrival intervals was developed and a patent application for this method has been filed.

As different digital platforms become ever more common in automated processes, the protection of different processes and the cyber security of the infrastructure is going to play a significant role in the overall safety of these platforms. For future researchers in this field, the group would like to recommend the usage of outcomes found in the AaTi study as well as the utilization of the patented method as a part of future CAN bus implementations in order to improve cyber security and situation awareness of automation networks.

5   Conclusions

For the first research question (how the cyber situational awareness of a CI organization can be developed) it is stated that as the target state of the organization's cyber SA and its interest groups' information sharing can be set the operation where the recognition of threatening incidents and reacting to them happens in an efficient process. It must include all the organization's decision-making levels (strategic, operational and technical/tactical) and utilize the national and international strengths of information sharing.

Based on the research the following basic requirements apply to the development of the organization's incident management:

- Strategic goals: (a) Cyber security management in all circumstances; (b) Strategic choices for operational continuity management
- Critical success factors: (a) Good SA on all the organizational levels; (b) Fast reaction ability and executive guidance; (c) Clear operational models and their sufficient resourcing; (d) Good information sharing between the different interest groups; (e) Crisis communication
- Evaluation criteria and target levels: (a) Effectivity of the operation; (b) Optimal resourcing.

In May 2017, the WannaCry ransomware campaign was affecting various organizations with reports of tens of thousands of infections in over 150 countries. On December 23, 2015, Ukrainian power companies experienced unscheduled power outages impacting a large number of customers in Ukraine. These examples show necessity of the technical/tactical level CSA. However, as this article expresses, there are significant challenges with regard to the technical/tactical level CSA. The WannaCry ransomware utilized a Windows vulnerability and because CSA was insufficient in many organizations, it prevented quick countermeasure. Also in the Ukreine case, the attack was not perceived from the technical system because of inadequate cyber situational awareness.

References

1. Carsten, P., Yampolskiy, M., Andel, T., McDonald, J.: In-vehicle networks: attacks, vulnerabilities, and proposed solutions. In: CISR '15 proceedings of the 10th annual cyber and information security research conference, pp 477–482 (2015)
2. Corrigan, S.: Introduction to the Controller Area Network (CAN). s.l.: Texas Instruments (2016)
3. Cybersecurity and Infrastructure Security Agency: ICS Alert (ICS-ALERT-17–209–01), CAN Bus Standard Vulnerability, s.l.: s.n. (2017)
4. Endsley, M.R.: Toward a theory of situation awareness in dynamic systems. Hum Factors Ergon Soc 37(1), 32–64 (1995)
5. Faber, S.: Flow Analysis for Cyber Situational Awareness (2015). [Online] Available at: https://insights.sei.cmu.edu/sei_blog/2015/12/flow-analytics-for-cyber-situational-awareness.html. Accessed 8 June 2019
6. Johansson, K.H., Törngren, M., Nielsen, L.: Vehicle applications of controller area network. In: William, D.H.B., Levine, S. (eds.) Handbook of Networked and Embedded Control Systems. s.l.:s.n. (2005)
7. Joint Task Force Transformation Initiative: NIST Special Publication 800–39: Managing Information Security Risk—Organization, Mission, and Information System View. National Institute of Standards and Technology, Gaithersburg (2011)
8. Kokkonen, T.: Anomaly-Based Online Intrusion Detection System as a Senor for Cyber Security Situational Awareness System. s.l.: Jyväskylä studies in computing 251. University of Jyväskylä (2016)
9. Lebrun, A., Demay, J.C.: Canspy: A Platform for Auditing Can. s.l.: s.n. (2016)

10. Lehto, M., et al.: Suomen kyberturvallisuuden nykytila, tavoitetila ja tarvittavat toimenpiteet tavoitetilan saavuttamiseksi (Finland's cyber security: the present state, vision and the actions needed to achieve the vision). Prime Minister's Office, Helsinki (2017)
11. Lehto, M., et al.: Kyberturvallisuuden strateginen johtaminen Suomessa (Strategic management of cyber security in Finland). Prime Minister's Office, Helsinki (2018)
12. Ministry of Defence: Yhteiskunnan turvallisuusstrategia (The Security Strategy for Society). Ministry of Defence, Helsinki (2010)
13. National Audit Office of Finland: Kybersuojauksen järjestäminen. National Audit Office of Finland, Helsinki (2017)
14. Pöyhönen, J., Lehto, M.: Cyber security creation as part of the management of an energy company. In: 16th European Conference on Cyber Warfare and Security, Dublin, pp. 332–340 (2017)
15. Ruoslahti, H., Rajamäki, J., Koski, E.: Educational competences with regard to resilience of critical infrastructure. J Inf Warfare **17**(3), 1–16 (2018)
16. The Security Committee: Yhteiskunnan turvallisuusstrategia (The Security Strategy for Society). The Security Committee, Helsinki (2017)
17. Virrantaus, K., Seppänen, H.: Yhteiskunnan Kriittisen Infran Dynaaminen Haavoittuvuusmalli. Matine, Helsinki (2013)
18. Voss, W., Comprehensible, A.: Guide to Controller Area Network. Massachusetts. Copperhill Media Corporation, Massachusetts (2005)
19. e-Governance Academy: National Cyber Security Index (NCSI) (2017). [Online] Available at: https://ncsi.ega.ee/. Accessed 8 June 2019

Possible Instant Messaging Malware Attack Using Unicode Right-To-Left Override

Veneta K. Yosifova and Vesselin V. Bontchev

Abstract The right-to-left special Unicode character has a legitimate use for languages that are transcribed in a right-to-left direction or in an environment that combines both right-to-left and left-to-right languages, like web pages, emails, desktop documents and text messages. These writing systems include right-to-left languages such as Persian, Arabic and Hebrew. The "right-to-left" attacks have been used for many years for malicious purposes, mostly in email communications. Early in 2018, Kaspersky Lab published an article describing a vulnerability in the Windows client of the popular instant messenger Telegram. This vulnerability uses the Unicode "right-to-left" character to obfuscate the name of the malware file. In our paper, we shall describe a possible attack that we discovered and that uses a combination of the "right-to-left" override attack and instant messaging malware attack and presents a realistic threat for another widely used messenger—Microsoft's Skype for Linux. Our purpose for conducting this research was to describe an exploit that we discovered and to warn the people who use this communication application about it, as well as to appeal to the producer for fixing it. Additionally, it is important to emphasize that the attack scenario developed by us also impacts other applications that allow file transfer (e.g. e-mail clients) and run on Linux systems with Wine installed.

Keywords Instant messaging malware attack · Right-to-left Unicode override · File name obfuscation · Microsoft Skype for Linux · Wine

V. K. Yosifova · V. V. Bontchev (✉)
National Laboratory of Computer Virology—BAS, "Acad. George Bonchev" Str., Block 8, Sofia, Bulgaria
e-mail: vesselin.bontchev@nlcv.bas.bg

V. K. Yosifova
e-mail: veneta.yosifova@nlcv.bas.bg

T. Tagarev et al. (eds.), *Digital Transformation, Cyber Security and Resilience of Modern Societies*, Studies in Big Data 84,
https://doi.org/10.1007/978-3-030-65722-2_11

1 Introduction

1.1 Right-to-Left Override Used for Spoofing

This type of attack is not new. It is not limited to a specific operation system. We shall describe the danger it represents and how it is possible to exploit it for sending malware on Linux. In the past, this method has been used in Mac OS X and Windows OS, too.

First, let us explain the meaning of the "right-to-left" override. As shown in the official Unicode documentation [1] "right-to-left" is a special non-printing character, called "RIGHT-TO-LEFT OVERRIDE" (commonly abbreviated as RLO). Its Unicode number is U+202E. As shown in Table 1, there are several Unicode format characters related to language writing systems that serve as invisible operators used to control the text appearance.

Their legitimate usage refers to languages that use a right-to-left writing system, e.g., Persian, Arabic, or Hebrew. When these languages are used by a web page or desktop application, together with a text written in a left-to-right language, they need to be displayed correctly.

The Unicode character U+202E instructs the rendering software to display the text after that character in right-to-left direction. This can be used for malicious purposes if the file name is modified to include a right-to-left overriding character and is then attached to a carrier like email. For instance, if the file name is "myxtxt.exe" and

Table 1 Unicode writing systems format characters [1]

Unicode number	Name	Abbr	Usage [2]
200C	ZERO WIDTH NON-JOINER	ZWNJ	Breaks leading and trailing characters
200D	ZERO WIDTH JOINER	ZWJ	Joins leading and trailing characters
200E	LEFT-TO-RIGHT MARK	LRM	Acts as a Latin character
200F	RIGHT-TO-LEFT MARK	RLM	Acts as an Arabic character
202A	LEFT-TO-RIGHT EMBEDDING	LRE	Treats following text as embedded left-to-right
202B	RIGHT-TO-LEFT EMBEDDING	RLE	Treats following text as embedded right-to-left
202C	POP DIRECTIONAL FORMATTING	PDF	Restores direction previous to LRE, RLE, RLO, LRO
202D	LEFT-TO-RIGHT OVERRIDE	LRO	Forces following characters left-to-right
202E	RIGHT-TO-LEFT OVERRIDE	RLO	Forces following characters right-to-left

if RLO character is placed after the first "x" (i.e., "myx*U+202E*txt.exe"), the file name will be displayed as "myxexe.txt".

A section on the official Unicode site, named "Unicode Security Considerations", describes that it is possible to use bidirectional text spoofing [3] in IRI (Internationalized Resource Identifier) scheme, as well.

Usually, the "right-to-left" override attack is conducted via email attachments or links for downloading files and is used on phishing websites or for sending malware.

1.2 Instant Messaging Spoofing

When talking about instant messaging spoofing, people usually mean an instant messaging phishing attack. For more than ten years, instant messaging clients have been targeted by this type of phishing [4]. It relies on the fact that if the victim receives an instant message from someone on their contact list, they are likely to click on the link contained in the message and leading to a phishing or an otherwise malicious website. The site would then request that the victim enters their credentials associated with their instant messenger account. The account of the sender of the phishing message could have been compromised by the attacker because, e.g., it had been protected with a weak password [5], or the password had been phished using a similar attack, or the password had been stolen by malware.

This way instant messengers become a vector for a type of attack called "social phishing" with the victim "purportedly getting a phishing message from someone they know" [4].

2 Related "Right-to-Left" Override Based Threats that Have Inspired Our Research

2.1 Review of Past "Right-to-Left" Override Attacks, Conducted via Emails or File Downloads

As mentioned above, file name obfuscation with RLO characters is not anything new. For instance, back in 2005 [6, 7] phishing emails were using the "right-to-left" attack, in order to sneak past spam filters. In 2009 a Mozilla Foundation Security Advisory warned that the names of the downloaded files are spoofed with RTL override and when "downloading a file containing a right-to-left override character (RTL) in the filename, the name displayed in the dialog title bar conflicts with the name of the file shown in the dialog body" [8]. This vulnerability can be used to obfuscate the filename extensions. When such a file is downloaded and opened, it can result in the user running an executable file instead of opening the non-executable file that it pretends to be. In 2011 Brian Krebs in [9] described how the "right-to-left" override attack can "disguise malicious executable files (.exe) as relatively harmless documents, such as text or Microsoft Word files" delivered by emails. In 2013 F-Secure published an article about "signed Mac malware using right-to-left override trick" [10] to obfuscate application files as PDF files. In 2015 Talos Security Intelligence and Research Group wrote about a Remote Access Trojan (RAT) named DarkKomet (DarkComet), delivered by a spam campaign, attempting to obfuscate its own name again by using the RLO method [11].

It is worth mentioning that the same "right-to-left" override attack has been used against "activists, opposition members, and nongovernmental organizations in the Middle East" [12-1] for years. As described in [12], these are files with the DarkComet RAT payload, spyware, and C&C (Command and Control) servers affiliated with the government. The attack used some ideologically or movement-relevant files like a "document or a PE [executable file] as download or attachment with accompanying encouragement to open or act on the material, often masquerading as legitimate PDF documents or inadvertently leaked regime programs. Frequent use of RLO [is] to disguise [the] true [file] extension (such as .exe or .scr)" [12-7].

As mentioned in [13] and [14], some software applications black-list this Unicode character in the name of the downloaded file. Also, some email applications block executable files and they cannot be attached to messages, including files with names, obfuscated with this technique. This protection can be bypassed, however, if the obfuscated files are placed inside an archive file like .zip, .rar or similar. Anti-virus and anti-malware programs also still detect these obfuscated files if their content is known as being malicious. But many Linux users do not have an anti-virus program installed on their machines because the number of malware threats to Linux-based operating systems is still much lower than the corresponding number of malware threats for Windows.

All these examples of attacks are conducted via emails or downloading files—not via instant messenger.

2.2 "Right-to-Left" Override Attacks Conducted via Instant Messenger

Early in 2018, Kaspersky Lab published a research article, in which they described a vulnerability in the Windows client of the popular instant messenger Telegram. They called it a "zero-day vulnerability in Telegram [where] cybercriminals exploited [a] Telegram flaw to launch multipurpose attacks" [15]. This newest example, which uses a "right-to-left" override attack and an instant messaging malware attack, shows that this combination represents a realistic threat.

In this attack, the attackers used the Unicode "right-to-left" character to obfuscate the names of the malicious files before sending them. The researchers discovered several cases where it was actually exploited. This included taking remote control of the victim's machine and installing a cryptocurrency miner that used the system resources of the infected computer. As described in the previous section, the attackers were using similar approaches, obfuscating the file extensions like .exe, .src, and .rar, and pretending to be picture files like.jpeg and.jpg (name*U+202E*gpj.rar, name*U+202E*gpj.exe, photoadr*U+202E*gepj.scr, address*U+202E*gpj.scr).

The researchers have "informed the Telegram developers of the problem, and the vulnerability no longer occurs in Telegram's products" [15].

3 Description of the Discovered Problem

3.1 Proof of Concept

In this subsection we shall describe a possible attack that we discovered in another widely used messenger service—Microsoft's Skype messenger for Linux.

Skype was founded by Nordic entrepreneurs in 2003 offering free instant messaging and cheap overseas calls online. It reached tens of millions of users then and became a really popular messenger so that it even became a verb [16]. It was acquired by Microsoft in 2011. The published number of Skype users has not been updated since 2016 when they were about 300 million [17]. But the "downloads of Skype's Android app reached 1 billion in October" 2017 [16]. So Skype remains popular despite criticism from longstanding consumers and attempts to be more focused on the corporate market. Thanks to Microsoft's work in AI and integrating Cortana in the chat interface, Skype can now translate calls into a dozen languages [16, 17]. Therefore it is important to draw attention to security problems that can exploit the Skype instant messenger.

The vulnerability exists only in the Linux versions of Skype. The purpose of this paper is to describe a real possible to exploit problem and to warn people who use this communication service about. Below we describe how a possible attack could be performed. To explain the exploit of the vulnerability, we have to prepare a harmless executable file (as a proof of concept) and then send it in an instant message.

3.1.1 Preparing Obfuscated File

Unlike Windows, Linux presents a particular challenge to the attacker. In Linux, the "executability" of a file is not determined by the file extension but by one of the file attributes in the file system. Even if we succeed to trick the victim into downloading our malicious file, Skype will not set the executable attribute when saving it to the disk, so if the file has executable contents (e.g., if it is an ELF file), it will not be executed by opening it.

Therefore, we need to send a file, the contents of which will be treated as executable, even if the executable bit is not set in the file system. For instance, we could send a document containing macros, but in Linux there is a better alternative—Wine.

Wine was created 25 years ago and "is a compatibility layer capable of running Windows applications on several POSIX-compliant operating systems such as Linux, macOS, & BSD. Instead of simulating internal Windows logic like a virtual machine or emulator, Wine translates Windows API calls into POSIX calls on-the-fly, eliminating the performance and memory penalties of other methods" [18]. Because Wine is a free and open-source and allows computer programs such as application software and computer games developed for Microsoft Windows to run on POSIX-compliant operating systems, it is very popular and is widely used. We "can treat things like web traffic to WineHQ as a rough estimate of overall interest in the project" [19]. According to the statistics of WineHQ webpage usage for 2017, the hits on the page are more than 100 million [20]. Therefore, it is very likely that Wine is already present on the machine of the victim.

But what is even more important for our purposes, if Wine is installed, it will be used to launch any file with an EXE extension, without the file needing to have the executable attribute set. Although this attack is applicable also to other forms of file transfer like e-mail attachments, to the best of our knowledge, this attack against Linux users has not been discussed before.

As our proof-of-concept, we shall use the following Python script, named hexgpj.py (1). Line 1. changes the current directory to the user's home desktop. Line 2. creates a text file named he.txt there. Line 3. puts the string "some code here" in the file and, finally, line 4. closes the file.

```
(1)     os.chdir(expanduser("~")+'/Desktop')

        f = open('he.txt', 'a+')

        f.write('some code here')

        f.close()
```

We shall use a Windows machine to create a Windows executable (EXE) file from this script. (It can be done directly on a Linux machine, but in that case we would have to install Python for Windows there and to run it via Wine.) First, we need to make sure that Python is installed on the Windows machine and, if not, to install it there—e.g., from [21].

We also need the utility pyinstaller. It can be installed on Windows machine like that:

```
pip install pyinstaller
```

Fig. 1 Executable file
hexpgj.exe

hexpgj.exe

Fig. 2 Executable file
hexpgj.exe with changed
icon

hexpgj.exe

Next, we shall create an executable file (shown in Fig. 1) from the hexpgj.py script
with the pyinstaller utility:

```
pyinstaller -F -s --distpath . hexpgj.py
```

For improved obfuscation, we could also replace the default icon resource of the
generated EXE file with the icon for JPG files. Assuming that we have such an icon in
a file named icon.ico, the following command will generate a file named hexgpj.exe
(shown in Fig. 2) with this icon:

```
pyinstaller -F -s --distpath . -i icon.ico hexpgj.py
```

The next step is to obfuscate the file extension using a RLO character. Placing
this character after the "hex" string in the filename will make Skype display the
remaining string "pgj.exe" in reverse. On Ubuntu OS entering a Unicode character
is easy by typing Ctr+shift+U (code number) at the Console and pressing Enter or
Space. This operation does not change the actual file type. The real file extension
is still.exe. So in our example changing the file representation with "right-to-left"
override character should be:

```
cp hexpgj.exe hex*Ctr+Shift+U202e*gpj.exe
```

The obfuscated file is shown in Fig. 3. If we highlight the file in the file manager
(Fig. 4), we shall see that the status information is reversed, too. That is because

Fig. 3 Prepared executable
file hexpgj.exe looking like
an image file named
hexexe.jpg

hexexe.jpg

Fig. 4 Reversed status
information

hexexe.jpg

"hexexe.jpg" selected (3,5 MB)

the RLO character causes everything after "hex" to be reversed. To restore the original text display direction, we can use the U+202C code, which is the PDF (POP DIRECTIONAL FORMATTING) Unicode character:

```
cp hexpgj.exe hex*Ctr+Shift+U202e*gpj.exe*Ctr+Shift+U202c*
```

Now the status information is correctly displayed as shown in Fig. 5.

With these steps the obfuscated file is ready to be sent to the victim via a Skype message.

3.1.2   Launching an Attack by Sending an Obfuscated File

Now the attacker can send a message and the victim will see an incoming image file named hexexe.jpg, instead of a suspicious executable file named hexgpj.exe. This is shown in Figs. 6 and 7.

If the recipient clicks on "Download", they will see one of the following messages. The first one, shown in Fig. 8 is displayed when U +202C is not used in the file name. The one shown in Fig. 9 is displayed if U+202C character is used and in this case the process of misleading the victim is compete.

Fig. 5 Corrected status
information

hexexe.jpg

"hex" selectedexe.jpg (3,5 MB)

Fig. 6 Obfuscated file sent
via Skype

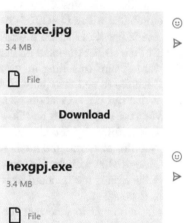

hexexe.jpg
3.4 MB

File

Download

Fig. 7 Unobfuscated file
sent via Skype

hexgpj.exe
3.4 MB

File

Download

Fig. 8 The obfuscation is
broken in the message

Warning!

This file may be unsafe for your computer/device. Accept files
only from people you trust. Do you want to accept the file
"hex?   morf "exe.jpg

Accept

Cancel

Fig. 9 The obfuscation is
corrected in the message

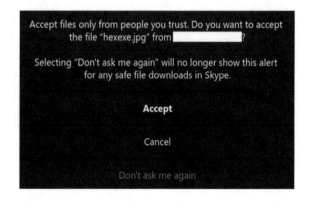

If the filename obfuscation is incomplete (the right-to-left text display direction is not terminated with a U +202C character), then the message is shown as "Warning" because Skype sees that the file has an executable extension (.exe). Also, the Skype name of the sender (blanked on the picture for privacy reasons) is reversed just like the word "from" (resulting in "morf"). But in the Fig. 9 where the character U + 202C is used to end the right-to-left displaying of the text in the file name, Skype thinks that this is a normal image file and offers to not show this message again.

When the victim clicks on "Download", the file is downloaded and saved to the ~ /Download folder by default. When examining this folder with a GUI file manager like Nautilus or Nemo, the file is shown as in Fig. 10. In some Linux distributions (e.g., Linux Mint with the Cinnamon desktop), the file icon is not handled correctly. There the icon is the standard icon for Wine executables as shown in Fig. 11. This seems to be a bug.

It is important to note that the downloaded executable file does not have the executable bit set (Fig. 12). However, if you double-click on its icon from the GUI file manager, Wine will launch it. Even better, after the downloading is complete, Skype offers to "Open" the file (Fig. 13). Clicking "Open" will result in Wine launching the file without any warring messages.

Fig. 10 Downloaded
executable file hexpgj.exe
looking like an image file
named hexexe.jpg

hexexe.jpg

Fig. 11 Downloaded
executable file hexpgj.exe
looking like an image file
named hexexe.jpg but with a
wrong icon

hexexe.jpg

`-rw-rw-r-- 1        3495985      17:44 hexgpj.exe`

Fig. 12 Executable bit is not set

Fig. 13 Skype offers to open the obfuscated file

Fig. 14 Downloaded file creates this file with some text in it

When the file is launched by Wine, a Terminal window will flash for a short moment, the file will run, and the code in (1) will be executed. This creates a text file named "he.txt" on the desktop of the user with the string 'some code here' in the file (Fig. 14). At this point the security of the victim is compromised.

There are two mitigating factors. First, the correct file extension is displayed in the Terminal (Fig. 12). It seems that Terminal ignores the "right-to-left" override character. But it is not common to open picture files from the Terminal. The second factor is that in the Cinnamon desktop environment, the icon file cannot be changed by the attacker (Fig. 11).

3.1.3 Possible Consequences

Since this attack results in code, written by the attacker, being executed on the machine of the victim, the possible consequences from this attack are limited only by the imagination of the attacker who is exploiting this vulnerability.

Some realistic scenarios are, for instance, installing malware like Ransomware, Viruses, Remote Access Trojans (RATs), Cryptocurrency miners; destroying user data, stealing sensitive information and so on. Also, it is possible to install inappropriate files (like child pornography or stolen classified information) on the system of the victim and to accuse the latter of possession or to try to blackmail them. Alternatively, a trivial change to the script (1) could make it append the SSH public key of the attacker to the file~/.ssh/authorized_keys of the victim, giving the attacker a backdoor (remote access) to the machine of the victim.

These are only some examples that are by no means an exhaustive list of the possible attacks exploiting this security problem.

3.2 Reporting the Problem

Given the serious implications following from this vulnerability, we decided to contact Microsoft's security service (Microsoft is the owner of Skype) and to report the problem to them. We presented them with a detailed description of the vulnerability, the steps needed to reproduce it, the affected platforms, the mitigating factors, the security implications, and a description of what had to be done, in order to fix it. Microsoft's response was that: "Unfortunately your report appears to rely on social engineering to accomplish, which would not meet the bar for security servicing".

The latest version of Skype for Linux at the time of the report (March 3, 2018) was 8.16.0.4 (from February 21, 2018) and all tests were conducted with it, too. But a three days after our report and their response, Microsoft rolled out a new release, 8.17.0.2. The release notes of version 8.17.0.2 specify as changes only the following: "Call international numbers with ease: International calls to mobiles and landlines is now quicker with the ability to select a country/region. From Calls, select the dial pad, and then choose the country/region you want to call from the drop-down" [22]. There is nothing mentioned there about the "right-to-left" override character vulnerability but our further experiments demonstrated that Microsoft has fixed it and the vulnerability no longer exists in Skype for Linux after this version.

4 Conclusion

The paper presents only those cases that we were able to test and experiment with. We do not have information which other versions of the Skype for Linux have been affected by this vulnerability. The contribution of our research is that it helped solving this security issue. Also note that this attack against Linux users is applicable to other forms of file transfer. For instance, it could be used to send malicious files as e-mail attachments and if the user tries to open the saved attachment, the malicious program will run. To the best of our knowledge, this attack against Linux users has not been discussed before.

Acknowledgements The preparation of this paper is supported by the National Scientific Program "Information and Communication Technologies for a Single Digital Market in Science, Education and Security (ICT in SES)", financed by the Ministry of Education and Science of the Republic of Bulgaria.

References

1. Unicode 10.0 Character Code Charts. General Punctuation Range: 2000–206F. Accessed on May 2018. [Online]. Available: https://bit.ly/2IKJ9zT
2. Woodvine, S.: RLO spoofing is another Internet security reverse, December 2015. Accessed on May 2018. [Online]. Available: https://bit.ly/2ImPjma
3. Davis, M., Suignard, M.: Unicode Technical Report #36. Unicode Security Considerations. Bidirectional Text Spoofing. Accessed on May 2018. [Online]. Available: https://bit.ly/2L2 T6Hd
4. Stavroulakis, P., Stamp, M. (eds.): Handbook of Information and Communication Security, p. 438. Springer Science & Business Media (2010). https://doi.org/10.1007/978-3-642-041 17-4
5. Boyd, C.: Steer Clear of this Skype Spam, September 2015. Accessed on May 2018. [Online]. Available: https://bit.ly/2KnKCcf
6. Batchelder, N.: Phishing fun with Unicode, April 2005. Accessed on May 2018. [Online]. Available: https://bit.ly/2IhpkAK
7. Katz, D.: Worst Phisher Ever, April 2005. Accessed on May 2018. [Online]. Available: https:// bit.ly/2GfbkBy
8. Mozilla Foundation Security Advisory 2009–62. Download filename spoofing with RTL override. Accessed on May 2018. [Online]. Available: https://bitly.com/2rL3haj
9. Krebs, B.: 'Right-to-Left Override' Aids Email Attacks, September 2011. Accessed on May 2018. [Online]. Available: https://bit.ly/2L1w2bJ
10. Brod: Signed Mac Malware Using Right-to-Left Override Trick. F-secure Labs, July 2013. Accessed on May 2018. [Online]. Available: https://bit.ly/2GhXr5u
11. Biasini. N.: Threat Research. Ding! Your RAT has been delivered, July 2015. Accessed on May 2018. [Online]. Available: https://bit.ly/2KpFzIi
12. Marczak, W.R., Scott-Railton, J., Marquis-Boire, M., Paxson, V.: When Governments Hack Opponents: A Look at Actors and Technology.In USENIX Security Symposium, pp. 511–525 (2014). Available: https://www.icir.org/vern/papers/govhack.usesec14.pdf
13. Pieter, A.: The RTLO method, January 2014. Accessed on May 2018. [Online]. Available: https://bit.ly/2rKcakz
14. Chancel, J.: Right to Left Override Unicode Can be Used for Multiple Spoofing Cases. Accessed on May 2018. [Online]. Available: https://bit.ly/2Ill66U
15. Firsh, A.: Zero-day vulnerability in Telegram, February 2018. Accessed on May 2018. [Online]. Available: https://bit.ly/2r0RrbB
16. Bass, D., Lanxon, N.: Don't Skype Me: How Microsoft Turned Consumers Against a Beloved Brand, May 2018. Accessed on May 2018. [Online]. Available: https://bitly.com/2jQgQSa
17. Skype has more than 300 million monthly active users, will get bots. Accessed on May 2018. [Online]. Available: https://bit.ly/2lxPA9W
18. WineHQ. Accessed on May 2018. [Online]. Available: https://www.winehq.org
19. WineHQ Usage Statistics. Accessed on May 2018. [Online]. Available: https://bit.ly/2ImRPgs
20. Usage Statistics for WineHQ winehq.org. Accessed on May 2018. [Online]. Available: https:// bit.ly/2Kqcb4R
21. ActiveState—The Open Source Languages Company. ActivePython. Accessed on May 2018. [Online]. Available: https://www.activestate.com/activepython
22. Release Notes for Skype for Desktop, Mac and Linux. Accessed on May 2018. [Online]. Available: https://bit.ly/2L1Aqr9

A VBA P-Code Disassembler

Vesselin V. Bontchev

Abstract Recently, we have observed a significant increase in the frequency with which Microsoft Office macros are being used as an attack vector. Microsoft Office uses a macro programming language called Visual Basic for Applications (VBA), which is powerful enough to do whatever the attacker needs. Usually, the malicious VBA macros are used to download the second stage of the malware (ransomware, banking Trojan, backdoor, etc.). They can be relatively small and are easy to modify or even to completely rewrite them each time, thus making them difficult to detect at the perimeter defenses (e.g., with an e-mail scanner) with known-malware detection tools. So far, we have seen malicious VBA macros being distributed with Microsoft Word, Excel, PowerPoint, Access, Visio, Project and Publisher documents. Microsoft Office has built-in protections against execution of foreign macros, but unless properly administered, they are easy for the user to disable and the malicious documents usually use some form of social engineering to convince the user to do so. Therefore, we need proper tools for inspecting the macro content of the received documents, in order to decide whether it contains any malicious code. During our research we have discovered that the publicly available tools lack the capability to discover all forms in which a malicious macro can exist. We have applied our findings from reverse-engineering the formats of Microsoft Office documents and have created a tool, which allows disassembling of the p-code into which VBA is compiled.

Keywords Macros · Malware · Microsoft office · VBA · Visual basic for applications

V. V. Bontchev (✉)
National Laboratory of Computer Virology, Bulgarian Academy of Sciences, Sofia, Bulgaria
e-mail: vesselin.bontchev@nlcv.bas.bg

1   Introduction

Recently we have observed a significant increase in the frequency at which Microsoft Office macros are being used as an attack vector. Microsoft Office uses a macro programming language called Visual Basic for Applications (VBA), which is powerful enough to do whatever the attacker needs. Usually, the malicious VBA macros are used to download the second stage of the malware (ransomware, banking Trojan, backdoor, etc.). They can be relatively small and are easy to modify or even to completely rewrite them each time, thus making them difficult to detect at the perimeter defenses (e.g., with an e-mail scanner) with known-malware detection tools.

So far we have seen malicious VBA macros being distributed with Microsoft Word, Excel, PowerPoint, Access, Visio, Project and Publisher documents.

Microsoft Office has built-in protections against execution of foreign macros, but unless properly administered, they are easy for the user to disable and the malicious documents usually use some form of social engineering to convince the user to do so.

Therefore, we need proper tools for inspecting the macro content of the received documents, in order to decide whether it contains any malicious code. During our research we have discovered that the publicly available tools lack the capability to discover all forms in which a malicious macro can exist. We have applied our findings are our knowledge from reverse-engineering the formats of Microsoft Office documents and have created a tool, which allows the disassembling of the p-code into which VBA is compiled.

2   Problem Description

VBA macros are stored in OLE2 files. OLE2 was the native document format used by the older versions of Microsoft Office. The contemporary versions use a different, mostly text-based format. The information is stored into multiple files, most of which are text files, and these files are gathered together and compressed into a ZIP archive, which is the document. But even in this case, the VBA macros reside in a file named vbaProject.bin, which is an OLE2 file.

The format of the OLE2 files is extremely complex. It is basically a "file system in a file"—with its own clusters, file allocation table, "files" (called "streams") and "directories" (called "storages") [1]. Each VBA module is stored in a separate OLE2 stream, called a "code module". The exact position of this stream inside the OLE2 stream and storage structure varies among the different Office applications but, generally, all VBA-related streams (the code module streams as well as some other streams) reside under a storage named VBA.

It is not widely known but a VBA macro can exist in up to three different executable forms. Which one of them is actually executed when the document, containing the macro, is opened, depends on various circumstances.

These three forms are:

- *Source code.* When a VBA module is created with the VBA Editor, its source code is compressed and stored at the end of the corresponding code module stream.
- *P-code.* As soon as a single line of VBA code is entered (or modified), it is compiled into an assembler-like language, called "p-code" (from "pseudo code") for a virtual stack machine. The p-code for each VBA line is also stored in the code module stream, before the compressed source code. The p-code of each line is stored in the order in which the lines have been created – which is not necessarily the order in which the VBA lines appear in the source. The p-code is preceded by a table, the entries in which point to each line of p-code. These entries appear in the same order as the lines in the VBA source.
- *Execodes.* After any of the VBA modules has been executed at least once (even if just debugging has been started on a module), the p-code is compiled further and stored in some form of internal representation in OLE2 streams, the names of which begin with __SRP_, followed by a number.

The most "universal" form of internal VBA representation is the source code. It is one and the same in every version of VBA higher than 3. (Version 3, used only in Excel 95, has neither source code, nor execodes—only p-code.) That is, you can take a document containing VBA macros that was created with a version of Office using VBA5 (i.e., Office 97) and open it with a version of Office that uses a different VBA version (e.g., VBA7, used by Office 2013) and the VBA program will work just fine. This is despite the fact that each different version of VBA uses a different set of p-code instructions. (More exactly, the opcodes of many of the instructions change, because new instructions are inserted in the middle of the instruction set in every new VBA version.)

This is why Microsoft advised the anti-virus industry to locate the compressed macro source, decompress it, and scan it for known macro malware. Microsoft has documented the format of the module streams sufficiently to locate the compressed source code area [2]. Therefore, many of the publicly available open-source tools for VBA inspection (and even many anti-virus products) follow this advice and show the contents of the decompressed source. For the anti-virus products this has the additional benefit that a single detection entry will match all VBA forms (VBA5, VBA6, VBA7) of the same macro malware, since the source code remains the same.

Paradoxically, however, the source code is only rarely used by VBA. In particular, it is used only when a document created by a version of Office that uses one version of VBA is opened by a version of Office that uses a different VBA version. In such cases the source code is retrieved and it is compiled into p-code understandable by the VBA version of the Office application that has opened the document.

In practice, this means that if the VBA version does not change (Office 97 uses VBA5, Office 2000, 2002, 2003, 2005, 2007 and 2009 use VBA6, Office 2010, 2013 and 2016 use VBA7), the contents of the source code will be ignored and the contents of the p-code will dictate what actions will be actually performed.

This situation creates a window of opportunity for the macro malware author. He could modify the contents of the document so that the compressed source indicates some benign VBA program or is missing completely, while the p-code performs the desired malicious action. As long as such a doctored document is not opened with a version of Office that uses a different version of VBA, the malicious p-code will be executed while all source code inspecting tools will be fooled by the benign (or missing) source code.

We have created a proof-of-concept set of documents for each version of Microsoft Word that uses a different version of VBA (`Word97.doc`, `Word2000-2009.doc` and `Word2010+ .doc`) [3]. Most of the free macro inspection tools will report that these documents contain no macros, because the source code in the code modules has been overwritten with zeroes. However, if opened with the corresponding version of Microsoft Word, these documents will display a short message and will start the Windows calculator.

The execodes pose a similar problem, but on a much smaller scale. While they, too, could contain a program that is different from what both the source code and the p-code suggest, the execodes are specific to the particular version of Office that has created them. They will not run on anything but on that particular version. Execodes created by Word 2000 will execute only on Word 2000 and not, for instance, on Word 2002, despite that both the same VBA version (VBA6). Furthermore, the format of the execodes is extremely complex and is outside the scope of this work.

There are some very real examples of the discrepancy between source code, p-code and execodes being used in macro malware:

- A Word macro virus (`W97M/Class.EZ`) destroys the p-code area in the infected documents and sets the VBA version to an impossibly large number, forcing re-compilation from source, when the document is opened.
- A variant of the `X97M/Jini` virus has been misdisinfected by some inferior anti-virus product. The code module streams have been removed (thus the source code and the p-code are absent) but the execodes remain. With a normal, parasitic virus, this would have prevented the virus from working, since the execodes wouldn't be able to copy a module that is not present. However, Jini is a mass-mailer. It sends the infected document as an Outlook e-mail attachment, so the inability to infect other documents does not prevent it from replicating.

In fact, given that the possibility of discrepancy between the source code, the p-code and the execodes has been known since the 90 s, it is surprising that malware authors are not exploiting it more often. However, the recent resurgence of macro malware suggests that they might start doing so in the future. Therefore, we need more powerful tools that let us inspect not only the source code but also the p-code of any VBA programs residing in suspicious documents.

3 Related Work

Our work would have been impossible without the work done by other authors who have created tools that allowed us to easily access the contents of the OLE2 files and the module streams.

Of particular relevance are the following:

- `oledump` by Didier Stevens [4] is a Python program for displaying the stream structure of OLE2 files and dumping the contents of any particular stream (including the VBA source code)
- `oletools` by Philippe Lagadec is a Python library for accessing programmatically the stream structure of an OLE2 file and in particular the VBA-related stream and storage subtree [5].

4 P-Code Disassembler

While working in the anti-virus industry (the author has designed the macro malware scanning engine in the anti-virus product F-PROT), we did some extensive reverse-engineering of the VBA code module formats. We refused to accept Microsoft's advice of looking at the source code, since it only rarely determines what would be actually executed and is effectively absent in VBA3. Instead, we concentrated on understanding how the p-code is stored and how to locate, extract and analyze it.

We used our knowledge and the results of our research to construct a Python program, which is able to locate and disassemble the p-code instructions of any VBA module. It is built upon Philippe Lagadec's excellent package `oletools`, which allows to obtain access to the VBA code module streams without having to handle ourselves such things as the OLE2 file formats. We have released the resulting program, a VBA p-code disassembler, as an open-source project on GitHub [6].

The program takes as a command-line argument a list of one or more names of files or directories. If the name is an OLE2 document, it will be inspected for VBA code and the p-code of each code module will be disassembled. If the name is a directory, all the files in this directory and its subdirectories will be similarly processed. In addition to the disassembled p-code, by default the script also displays the parsed records of the dir stream, as well as the identifiers (variable and function names) used in the VBA modules and stored in the _VBA_PROJECT stream.

The script supports VBA5 (Office 97, MacOffice 98), VBA6 (Office 2000 to Office 2009 inclusive) and VBA7 (Office 2010 and higher).

The script also accepts the following command-line options:

`-h, -help`	Displays a short explanation how to use the script and what the command-line options are.
`-v, -version`	Displays the version of the script.

-n, -norecurse   If a name specified on the command line is a directory, process only the files in this directory; do not process the files in its subdirectories.

-d, -disasmonly   Only the p-code will be disassembled, without the parsed contents of the dir stream, the contents of the PROJECT stream, or the identifiers in the _VBA_PROJECT stream.

-verbose   The contents of the dir and _VBA_PROJECT streams is dumped in hex and ASCII form. In addition, the raw bytes of each compiled into p-code VBA line is also dumped in hex and ASCII.

For instance, using the script on the proof-of-concept document [3] produces the following results:

```
python pcodedmp.py -d poc2b.doc

Processing file: poc2b.doc
=========================================================================
Module streams:
Macros/VBA/ThisDocument - 1949 bytes
Line #0:
        FuncDefn (Sub / Property Set) func_00000078
Line #1:
        LitStr 0x001D "This could have been a virus!"
        Ld vbInformation
        Ld vbOKOnly
        Add
        LitStr 0x0006 "Virus!"
        ArgsCall MsgBox 0x0003
Line #2:
        LitStr 0x0008 "calc.exe"
        Paren
        ArgsCall Shell 0x0001
Line #3:
        EndSub
```

For reference, it is the result of compiling the following VBA code:

```
Private Sub Document_Open()
    MsgBox "This could have been a virus!", vbInformation+vbOKOnly, "Virus!"
    Shell("calc.exe")
End Sub
```

We have also created a plug-in for Didier Stevens' tool oledump [7]. Unfortunately, this plug-in works only for documents that use VBA6 and only dumps the p-code in hex and ASCII, without doing any disassembling. This is because, in order to determine the VBA version (and thus the location of the p-code and the meaning of each p-code opcode), we need access to the _VBA_PROJECT and dir streams and oledump simply does not provide access to that to its plug-ins via the plug-in interface.

5 Future Work

Our tool requires further testing, in order to make sure that it is compatible with some marginal cases (like MacOffice, which uses big-endian format for storing the various structures in the VBA-related streams, as opposed to the little-endian format used by the PC version of Microsoft Office).

We also plan on extending the tool to support VBA3. Although VBA3 uses p-code too, the OLE2 stream format structure is quite different. While VBA3 was used only in a now-obsolete version of Excel (Excel 95), Excel documents produced by later versions of Excel can be saved in a "compatible" format (which contains both the VBA3 and VBA5 forms of the same VBA macros), so VBA3 macros pose a non-negligible threat.

Acknowledgements The preparation of this paper is supported by the National Scientific Program "Information and Communication Technologies for a Single Digital Market in Science, Education and Security (ICT in SES)", financed by the Ministry of Education and Science of the Republic of Bulgaria.

References

1. Microsoft: Microsoft Compound Document File Format. https://www.openoffice.org/sc/com pdocfileformat.pdf
2. Microsoft: Office VBA File Format Structure. https://msdn.microsoft.com/en-us/library/cc3 13094%28v=office.12%29.aspx
3. Bontchev, V.: https://bontchev.my.contact.bg/poc2.zip
4. Stevens, D.: Oledump. https://blog.didierstevens.com/programs/oledump-py/
5. Lagadec, P.: Oletools—Python tools to analyze OLE and MS Office files. https://www.decalage.info/python/oletools
6. Bontchev, V.: A VBA p-code disassembler. https://github.com/bontchev/pcodedmp
7. Bontchev, V.: A p-code dumper plugin for oledump. https://github.com/DidierStevens/Didier StevensSuite/blob/master/plugin_pcode_dumper.py

Cybersecurity of Critical Infrastructures and Industrial Systems

Scoping the Scenario Space for Multi-sector Cybersecurity Analysis

Todor Tagarev and Nikolai Stoianov

Abstract The paper presents results from the Horizon 2020 ECHO project, supporting the identification and development of cyberattack scenarios. It explores the scenario space along four dimensions: (1) critical infrastructures and essential services, critically dependent on the ICT infrastructure; (2) types of malicious actors and their capabilities; (3) exploited vulnerabilities; and (4) short- versus longer term horizon. The exploration serves to span comprehensively the scenario space. The authors present a partial list of selected scenarios, storylines, and use cases, that are used in follow-up research to define key components of the capability requirements: technology roadmaps, cyber skills framework, information exchange and certification requirements.

Keywords Cybersecurity · Cyberattacks · Scenario space · Capability requirements · Technology roadmap · Cyber skills

1 Introduction

In various research and application fields, scenarios are increasingly used to handle uncertainty in complex systems. The purposes encompass analysis of interdependencies [1], vulnerabilities and hazards [2], estimation and mitigation of risk with account of local context [3], analysis of human reliability [4], analysis of the dynamics of the responses and actions in plausible futures [5], systems design in the pursuit of multiple objectives with consideration of uncertainty [6, 7], the exploration of options in the decision space [8], prioritising resource allocation [9], elaboration of investment strategies [10], etc.

T. Tagarev (✉)
Institute of Information and Communicatoin Technologies, Acad. G. Bonchev Str., Bl.2, Sofia 1113, Bulgaria
e-mail: tagarev@bas.bg

N. Stoianov
Bulgarian Defence Institute, 2, Prof. Tzvetan Lazarov Blvd., Sofia 1152, Bulgaria
e-mail: n.stoianov@di.mod.bg

T. Tagarev et al. (eds.), *Digital Transformation, Cyber Security and Resilience of Modern Societies*, Studies in Big Data 84,
https://doi.org/10.1007/978-3-030-65722-2_13

Scenario-based analysis is of foundational importance in the Horizon 2020 project ECHO—European network of Cybersecurity centres and competence Hub for innovation and Operations—aiming to pilot novel cybersecurity organisational and technological solutions. Along with three other pilot projects, ECHO was launched in the beginning of 2019 to support the establishment of a European Cybersecurity Competence Network, to develop and implement a common cybersecurity research & innovation roadmap, to support the European industry with technologies and skills for developing innovative security products and services and protect vital assets against cyberattacks [11].

Scenarios are used in ECHO to reflect diverse requirements to the provision of cybersecurity in pursuing a number of objectives:

- Comprehensive definition of requirements (where each of the four pilot projects focuses on selected sectors);
- Rigorous analysis of interdependencies and opportunities for exchange of methods, research results and good practice;
- Facilitating the allocation of research tasks among project partners and projects;
- Assessing cybersecurity risks and, on that basis, prioritising proposals for development of new tools;
- Selecting scenarios and use cases for demonstration of novel tools and techniques; etc.

As a rule, in any research or practical initiative, scenarios and their main parameters are defined by subject matter experts in communication with other stakeholders, e.g. policy makers. The required effort for analysis afterwards is proportional to the number of scenarios designed and processed. Further, a large number of scenarios challenges the information-processing capacity of decision makers [12]. Therefore, a consistent effort is needed to guarantee that the selected set of scenarios represents well the problem at hand.

This paper describes the rationale for selecting scenarios and use cases in the ECHO project. The next section outlines the underlying methodological approach. It is followed by two sections presenting respectively an elaboration of the four selected dimensions of the problem space (sectors and sub-sectors; actors; exploited vulnerabilities; time horizon) and a sample of the initial list of developed scenarios and use cases. The concluding section outlines the current and the future use of the scenarios and the ways of identifying and prioritising identified needs and proposals to add further scenarios and use cases to the set already elaborated.

2 Methodological Approach

The rigorous application of scenario-based analysis requires proper structuring of the problem space [13]. A variety of methods can be utilised for that purpose, including morphological analysis, system dynamics modelling, Bayesian networks and influence diagrams [14].

The approach undertaken by the authors draws primarily from the morphological analysis. The advantages of this approach are that it offers a structured process for developing of scenarios; establishes a consistent typology; leaves a clear audit trail; and enables better communication of results [15].

The first step is to define the main parameters, or 'factors', characterising the problem, described by the scenario (in our case that is a cyberattack and the direct and indirect consequences it will potentially bring). For example, in an application of the approach to the field of defence planning, Johansen defines four main parameters: Actor, Goal, Method and Means [15]. In the discussion on exploring the scenario space, we prefer to designate such main parameters as 'dimensions' of this space.

On the second step, along each dimension one defines distinct parameter values, or 'states', so that each parameter is represented by an exhaustive set of possible values [15, 16]. In the following steps of implementation (not examined in detail here), the study team utilises various methods and techniques to make sure that the candidate scenarios are plausible, consistent, relevant to and representative for the purpose of the study [15, 16].

This approach allows to capture all potential requirements of interest, while at the same time assuring that requirements stemming from different scenarios are sufficiently distinct. That allows to guarantee comprehensive scoping of the scenario space, as well as economy of effort.

3   Spanning the Scenario Space

This section presents the initial application of the approach to the field of cybersecurity. Partially echoing the structure used by Johansen [15], the following questions were subject of consideration:

- What is the target of the attack (both direct and indirect)?
- Who are the attackers and what are their skills, intentions, and dedication, i.e. readiness to invest money and effort into preparing, conducting and exploiting the results of the attack?
- What vulnerabilities are exploited in the attack?
- To what extent the attacker makes use of currently existing vulnerabilities, tactics and techniques or, alternatively, plausible future exploits?

Based on the analysis of these questions, the authors structured the scenario space in the following four dimensions:

- Sectors/sub-sectors of critical infrastructures or essential services;
- Malign actors;
- Exploited vulnerabilities; and
- Time horizon.

Each dimension is presented in a sub-section below, along with the distinct 'values' that can be used to define a scenario.

3.1   Sectors and Sub-sectors

Not only the information and communication technology (ICT) infrastructure, but an increasing number of other critical infrastructures and essential services are dependent on advanced communications, networks and data, and can be affected though cyberspace. The definition of sectors and sub-sectors of such infrastructures in this paper is based exclusively on documents of the European Union (EU). Nevertheless, the authors consider the proposed structuring applicable beyond the EU.

In 2005, the European Commission proposed the Green Paper on Critical Infrastructure Protection (CIP) [17], and since then the number of relevant official documents is consistently growing, each one adding additional sectors of concern.

The direct heir of the Green Paper, Directive 2008/114 [18], defines two sectors of critical infrastructure, energy and transport, each with sub-sectors, as follows:

- Energy:
    - Electricity (generation, transmission, supply);
    - Oil (production, refining, treatment, storage and transmission by pipelines);
    - Gas (production, refining, treatment, storage and transmission by pipelines, LNG terminals);

- Transport:
    - Road transport;
    - Rail transport;
    - Air transport;
    - Inland waterways transport;
    - Ocean and short-sea shipping and ports.

This represented significant reduction of the coverage of the Green Paper, which envisioned 11 sectors of critical infrastructures:

I   Energy
II   ICT
III   Water
IV   Food
V   Health
VI   Financial
VII   Public & Legal Order and Safety
VIII   Civil administration
IX   Transport
X   Chemical and nuclear industry
XI   Space and Research

with 37 "products or services" (sub-sectors), including:

- Telecommunication services: Provision of fixed telecommunications, Provision of mobile telecommunications, Radio communication and navigation, Satellite communication, Broadcasting;

- Interned as digital infrastructure;
- Financial infrastructure such as payment services/payment structures (private) and government financial assignment,

and structuring the energy sector somewhat differently.

The ICT sector was emphasised in Directive 2016/1148 on network and information security (NIS Directive) [19]. Further, in its main text and Annex II, the NIS Directive referred to the following sectors:

- Energy: Electricity (generation, transmission, supply); Oil (production, refining, treatment, storage and transmission by pipelines); Gas (production, refining, treatment, storage and transmission by pipelines, LNG terminals);
- Transport: Road transport; Rail transport; Air transport (including airports and air carriers), Water transport (including maritime ports);
- Banking, e.g. credit institutions and investment firms;
- Financial market infrastructures;
- Health sector;
- Drinking water supply and distribution;
- Digital Infrastructure: IXPs; DNS service providers; TLD name registries;
- Cloud computing services;
- Hardware manufacturers and software developers;
- Personal data.

Importantly, the NIS Directive introduced obligations for the operators of "essential services" and defined criteria for designating such operators, leaving however the actual designation to national authorities.

The 2019 Cybersecurity Act [20] provided detail to the provision of digital services, referring to search engines, cloud computing services and online marketplaces.

The final official document analysed by the authors was the Joint Communication "Resilience, Deterrence and Defence: Building strong cybersecurity for the EU" [21], which added explicitly Internet of Things stating that "A failure to protect the devices which will control our power grids, cars and transport networks, factories, finances, hospitals and homes could have devastating consequences and cause huge damage to consumer trust in emerging technologies." It further examined the risk of politically-motivated attacks on civilian targets, and of shortcomings in military cyber defence.

Accounting for these official documents, the call for project proposals [11] added to the more "traditional" sectors of critical infrastructures the manufacturing and the defence sectors.

Analysing all these documents the authors developed a structure of critical sectors and sub-sectors, presented in Table 1.

Table 1 Structuring the cybersecurity relevant sectors and sub-sectors of critical infrastructures and essential services

	Call H2020-SU-ICT-2018–2020	Dir 2008/114– CIP	Dir 2016/1148 – NIS	2005 Green Paper CIP	JOIN(2017) 450 final	ECHO
ICT/CIS/cyber infrastructure						
• Telecom	✓			✓		
• Digital Infrastructure: IXPs; DNS service providers; TLD name registries			✓	✓	✓	
• Digital service providers			✓			
• Cloud computing services			✓			
• Hardware manufacturers and software developers			✓			
Personal data			✓			
Energy	✓					
• Electricity (generation, transmission, supply)		✓	✓	✓	✓	✓
• Oil (production, refining, treatment, storage and transmission by pipelines)		✓	✓	✓		
• Gas (production, refining, treatment, storage and transmission by pipelines, LNG terminals)		✓	✓	✓		
Transport	✓				✓	
• Road transport		✓	✓	✓		

(continued)

Table 1 (continued)

	Call H2020-SU-ICT-2018–2020	Dir 2008/114– CIP	Dir 2016/1148 – NIS	2005 Green Paper CIP	JOIN(2017) 450 final	ECHO
• Rail transport		✓	✓	✓		
• Air transport		✓	✓	✓		
• Inland waterways transport		✓	✓	✓		
• Ocean and short-sea shipping and ports		✓		✓		✓
Banking			✓	✓	✓	
Financial market infrastructures	✓		✓	✓	✓	
Health	✓		✓		✓	
• Medical and hospital care				✓		✓
• Medicines, serums, vaccines and pharmaceuticals				✓		
• Bio-laboratories and bio-agents				✓		
• Networked devices, Implantable medical devises						✓
Drinking water supply and distribution			✓	✓	✓	
Food (provision of food and safeguarding food safety and security)				✓		

(continued)

Table 1 (continued)

	Call H2020-SU-ICT-2018–2020	Dir 2008/114– CIP	Dir 2016/1148 – NIS	2005 Green Paper CIP	JOIN(2017) 450 final	ECHO
eGovernment	✓			✓		
(cyber) Defence	✓			✓	✓	
Public & Legal Order and Safety				✓		
Chemical and nuclear industry				✓		
Space	✓			✓		✓
Research				✓		
Manufacturing	✓			✓	✓	
Homes					✓	

3.2 Actors

Actors with malicious intent may have differing skills, intent, access to resources and level of dedication. For the design of scenarios, the authors consider as key the skills and the overall capabilities of the actors to perform cyberattacks. There is a variety of actors reflected in the so-called hackers' classifications. 'The Hats' approach is used to classify actors as 'White Hats,' 'Black Hats,' and 'Gray Hats' [22]. Thirteen additional types of actors are proposed based on a combination between 'The Hats' approach and an expertise-based classification [23]. In our approach, the 'Hats' classification is accompanied by eight types of actors.

To achieve the goal of the project and to cover all required types of malicious activities in the scenario development, the following types of actors were defined:

- *Script kiddie*—A person lacking programming skills and network knowledge, who is using tools existing on the net to try to exploit computer systems;
- *Lamer*—A person who does not know what he is doing. In different definitions that can be a 'cracker' or a 'phreaker,' but also an employee in the organization doing wrong things in wrong ways;
- *Experienced/Elite hacker*—A person with knowledge and experience in the field of computer systems and networks, carefully planning and executing attacks using tools developed by himself/herself or other well-studied hacking tools. Actors of this type usually do not destroy systems or information to which they have gained access but just use a successful penetration to access the next target (a system or a person);
- *Hacking group*—A group of hackers, usually experienced, sharing a common understanding, tools and services. There are two major subtypes—groups dedicated to very specific types of attack and groups combining actors specialising on different types of attack, tools or services;
- *Hacktivist*—A person who is trying to gain access to computer systems or information with the intent to use this access (information) for social or political purposes;
- *Organized criminal gang*—A group of criminals using internet not only to organize themselves and to communicate but also to launch specific attacks in order to steal money or extort ransom;
- *State or state-affiliated hacker group*—A hacker or group of hackers "employed" by a country or an agency to attack another country or another governmental organization;
- *Cyber weapons dealer*—An actor who is developing "tools" for specific vulnerabilities and sells them on the dark net, or an actor who is developing full consequence of steps of a dedicated attack against a designated system or service and is hired by another person, group or an organization. Usually, actors of this type do not launch attacks against targets, but focus on the development of tools and techniques.

3.3   Exploited Vulnerabilities and Attacks

Various definitions of vulnerability are used in the context of cybersecurity. Some are focused on weaknesses in computer systems, others deal with security procedures and policies, a third group focusses on implementation modes that can be exploited. For the purposes of the ECHO project, and aiming to reflect broad combinations of all of the above, user/personnel related vulnerabilities were also considered. Classifying cybersecurity vulnerabilities requires addressing the so-called horizontal and vertical aspects. In the approach proposed by Jang-Jaccard and Nepal vulnerabilities are classified on the basis of systems' components, e.g. software, hardware, and networks [24]. Additionally, the need to account for 'social' types of vulnerabilities and cloud-based vulnerabilities was identified. As early as 2005, Hansman and Hunt proposed to classify vulnerabilities in four categories [25]. In order to address cross-sectoral dependencies, the following types of vulnerabilities are used in the ECHO project:

- *Hardware*—this type of vulnerabilities exists in a computer or network system and, more specifically, in their hardware components. These vulnerabilities can give access, locally or remotely, to the system to be exploited;
- *Software*—weaknesses in program code or running services that can be exploited by attackers;
- *Networking*—this type of vulnerabilities relates mainly to network protocols or data/information exchange. The information process flow also can be affected, e.g. by an attack exploiting existing weaknesses;
- *Personnel-related*—this type of vulnerabilities is closely related with gaps in awareness and training and education of personnel. Humans are often seen as the "weakest" security point in every computer system; therefore, this type of vulnerabilities was also addressed in the scenario development process;
- *Organisational*—these types of vulnerabilities address capacity and maturity readiness of an organization to identify, prevent, stop and mitigate existing or foreseen cyberattacks.

3.4   Time Horizon

Theoretically, cybersecurity scenarios may span the spectrum from well-established attack patterns to futuristic, long term hypotheses on vulnerabilities, complex interdependencies and attackers' skills and intent.

Since the study presented here aims to define requirements to future systems, develop and demonstrate novel tools, frameworks and certification schemes, the study team decided to discard the low end of that spectrum.

Scenarios find extensive application in foresight studies as representation of uncertainty and supporting the search for robust, long-term solution. Of particular interest are scenario- and foresight-based approaches to defence and security planning, looking 15 or more years into the future (e.g. [15, 16]). However, the filed of cybersecurity, and more generally—the development and implementation of new ICT and their impact, is more fluid and any attempt to take a long-term view is fraught with risk to increase, rather than reduce uncertainty. Therefore, cybersecurity related foresight studies (e.g. [26]) prefer to look into drivers for future threats (and not the actual instantiation of a threat, i.e. a scenario) and have found increasing disagreement among experts asked to estimate the maturity of certain technologies anticipated more than five years into the future [26]. Based on this rationale, the study team decided to discard also the high end of the time spectrum.

As a result, this dimension of the scenario space is represented by two values:

- Most recent attack patterns, that still lack effective counteraction in terms of technology, organisation or competences;
- Short-term forecasts (building on a parallel study within the ECHO project of foreseen technologies, vulnerabilities and threats, looking three to five years into the future and used to generate a research roadmap).

In addition, the plan is to conduct the forecast exercises roughly every two years and to update the roadmap accordingly, thus allowing to consider new scenarios based on the updated forecasts.

4 The ECHO Approach and Results

At the end of the first year of the ECHO project, the study team delivered the first set of scenarios and use cases.

The decisions along the first dimension were taken based on the competences represented in the ECHO consortium with the understanding that sufficiently broad coverage of the sectors of interest will be provided by all four pilot projects launched in 2019. The ECHO project covers:

- Healthcare;
- Maritime transport, with focus on shipping;
- Energy, with focus of electrical energy transmission and supply;
- Defence.

In terms of actors, dedicated social engineering, lamer, script-kiddies, hacktivists and state affiliated actors were all addressed in the scenarios. Taking into account the complexity of inter-, multi- and transversal cyber sector dependencies, reflecting the capacity of these types of actors allowed to cover most of the activities of initial interest and to develop sufficiently detailed scenarios.

Regarding vulnerabilities and related types of attack such as malware, ransomware, fishing, etc., the goal in scenario development was to address not only one type of attack but to combine several attack types. For example, if the goal of the attacker is to grant access to a specific system, the scenario included social engineering through a phishing attack, combined with password attacks. This approach allows to examine different attack paths and to study different kill chains.

As for the time horizon, the first set of scenarios was developed on the basis of recent cyberattacks as well as attacks forecasted by other researchers in their published work.

The team accepted the following working definition for scenario:

> A set of coordinated operations and activities of one or a group of malicious actors aiming to accomplish a common goal, normally within a given time and space, and the corresponding defence activities aiming to prevent, stop or mitigate attacks.

and developed one scenario per sector/sub-sector.

Then, each scenario includes several storylines providing operational and functional descriptions of the system under attack, including technical details on devices and services of specific interest, and the effect of the attack on the technical system and the business processes.

Each storyline is accompanied by two or more use cases, each giving technical description of the consequence of activities of the attacker and the defending side of specific interest.

For example, one of the storylines in the healthcare scenario is named "Tampering with Medical Devices" and is accompanied by three use cases:

- Theft of electronic health records through hacking a connected medical device;
- Hospital havoc through hacking a connected medical device;
- Cyber assassination through an implanted medical device.

5 Way Ahead

Currently, the scenarios and use cases are used to identify user needs, security issues in different sectors, and existing gaps (from technical, operational and human point of view), and to develop a dependency matrix between sectors. To ensure that the developed scenarios and use cases cover the main elements for specific sectors and also address sector interdependencies, the scenarios with their associated use cases are modelled and simulated with the CAIRIS tool (https://cairis.org/).

The following enhancements are planned for the second cycle of scenario design:

- Studying if similar or the same attacks, described for one sector, can be launched in other sectors and cause effects of interest;
- Adding a scenario for the space sector, as well as scenarios representing cross-sector interdependencies and transversal effects;
- Adding specific use cases, covering two or more sectors to represent inter-, multi-, and transversal sector dependencies;
- Addressing additional types of actors, including dedicated groups of hackers, hackers' weapons and social/political hacktivists is considered through extension of existing scenarios or newly developed ones;
- Accounting for cognitive and behavioural aspects for attackers and defenders;
- Adding more sophisticated attacks, including APT-based ones.

- Reflecting forecasts from the first version of the ECHO research and technology roadmap.

Of particular interest will be the implementation for several sectors of critical infrastructure, allowing to identify interconnections and dependencies between sectors, and exploring effects in various dimensions, i.e. technical, transversal, and operational. Designing the respective scenarios, modelling and simulating attackers, defenders and systems' activities will be in the focus for the follow-on scenario improvement and use case development.

The methodology for scoping the scenario space will be refined by linking it to several taxonomies currently developed within the project. Further, scenarios should allow to address all layers in the ECHO Multi-sector Assessment Framework (E-MAF) and will feed into the following studies under the umbrella of the ECHO project:

- Development of technology roadmaps taking into account current user needs and forthcoming cybersecurity issues incorporated into ECHO tools and services;
- Transversal aspect of scenarios will be further elaborated by developing the ECHO Cyber Security Skills Framework;
- The ECHO Cyber Security Certification Scheme is another element that will be developed with support of scenarios and use cases;
- The study of sector relations/dependencies is a critical aspect of the ECHO project. By using this scope of dedicated scenarios, "hidden" inter-, multi- and transversal connections between five identified sectors will be identified and further explored.

To validate the approach and achieved results in the ECHO project, including tools and services, five demonstration cases will be developed. All of them will be based on already defined and specified scenarios. Thus, the systematic exploration of the scenario space will facilitate the common understanding for processes, tools and services, generation and dissemination of best practices in scenario-based analysis of cybersecurity.

Acknowledgements This work was supported by the ECHO project, which has received funding from the European Union's Horizon 2020 research and innovation programme under the grant agreement no. 830943.

References

1. Tagarev, T., Stoianov, N., Sharkov, G.: Integrative approach to understand vulnerabilities and enhance the security of cyber-bio-cognitive-physical systems. In: Cruz, T., Simoes, P. (eds) 18th European Conference on Cyberwarfare and Security (ECCWS19), University of Coimbra, Portugal, 4–5 July 2019, pp. 492–500. Academic Conferences & Publishing International, Reading, UK (2019)
2. Liu, Y., So, E., Li, Z., Su, G., Gross, L., Li, X., Qi, W., Yang, F., Fu, B., Yalikun, A., Wu, L.: Scenario-based seismic vulnerability and hazard analyses to help direct disaster risk reduction

in rural Weinan, China. Int. J. Disaster Risk Reduction **48**, 101577 (2020). https://doi.org/10. 1016/j.ijdrr.2020.101577

3. Menoni, S., Schwarze, R.: Recovery during a crisis: facing the challenges of risk assessment and resilience management of COVID-19. Environ. Syst. Decis. **40**, 189–198 (2020). https:// doi.org/10.1007/s10669-020-09775-y

4. Golestani, N., Abbassi, R., Garaniya, V., Asadnia, M., Khan, F.: Human reliability assessment for complex physical operations in harsh operating conditions. Process Saf. Environ. Prot. **140**, 1–13 (2020). https://doi.org/10.1016/j.psep.2020.04.026

5. Wan, S., Radhakrishnan, M., Zevenbergen, C., Pathirana, A.: Capturing the changing dynamics between governmental actions across plausible future scenarios in urban water systems. Sustain. Cities Soc. **62**, 102318 (2020). https://doi.org/10.1016/j.scs.2020.102318

6. Pourshahabi, S., Rakhshandehroo, G., Talebbeydokhti, N., Nikoo, M.R., Masoumi, F.: Handling uncertainty in optimal design of reservoir water quality monitoring systems. Environ. Pollut. **226**, 115211 (2020). https://doi.org/10.1016/j.envpol.2020.115211

7. Hassler, M.L., Andrews, D.J., Ezell, B.C., Polmateer, T.L., Lambert, J.H.: Multi-perspective scenario-based preferences in enterprise risk analysis of public safety wireless broadband network. Reliab. Eng. Syst. Safety **197**, 106775 (2020). https://doi.org/10.1016/j.ress.2019. 106775

8. Mena, C., Melnyk, S.A., Baghersad, M., Zobel, C.W.: Sourcing decisions under conditions of risk and resilience: a behavioral study. Decis. Sci. **51**(4), 985–1014 (2020). https://doi.org/10. 1111/deci.12403

9. Tagarev, T., Stankov, G.: A gaming approach to enhancing defense resource allocation. Connections Q. J. **8**(2), 7–16 (2009). https://doi.org/10.11610/Connections.08.2.02

10. Yu, X., Wu, Z., Wang, Q., Sang, X., Zhou, D.: Exploring the investment strategy of power enterprises under the nationwide carbon emissions trading mechanism: a scenario-based system dynamics approach. Energy Policy **140**, 111409 (2020). https://doi.org/10.1016/j.enpol.2020. 111409

11. "Establishing and operating a pilot for a Cybersecurity Competence Network to develop and implement a common Cybersecurity Research & Innovation Roadmap," Cybersecurity Call H2020-SU-ICT-2018-2020. https://ec.europa.eu/info/funding-tenders/opportunities/portal/scr een/opportunities/topic-details/su-ict-03-2018

12. Comes, T., Hiete, M., Schultmann, F.: An approach to multi-criteria decision problems under severe uncertainty. J. Multi-Criteria Decis. Anal. **20**(1–2), 29–48 (2013). https://doi.org/10. 1002/mcda.1487

13. Carlsen, H., Eriksson, E.A., Dreborg, K.H., Johansson, B., Bodin, Ö.: Systematic exploration of scenario spaces. Foresight **18**(1), 59–75 (2016). https://doi.org/10.1108/FS-02-2015-0011

14. Ritchey, T.: General morphological analysis as a basic scientific modelling method. Technol. Forecast. Soc. Chang. **126**, 81–91 (2018). https://doi.org/10.1016/j.techfore.2017.05.027

15. Johansen, I.: Scenario modelling with morphological analysis. Technol. Forecast. Soc. Chang. **126**, 116–125 (2018). https://doi.org/10.1016/j.techfore.2017.05.016

16. Tagarev, T., Ivanova, P.: Analytical support to foresighting EU roles as a Global Security Actor. Inf. Secur. Int. Journal 29, no. 1 21–33 (2013), https://doi.org/10.11610/isij.2902.

17. Commission of the European Union: Green Paper on a European programme for critical infrastructure protection. COM (2005) 576 Final. Brussels, 17 November 2005

18. Council Directive 2008/114/EC of 8 December 2008 on the identification and designation of European critical infrastructures and the assessment of the need to improve their protection. Official J. L **345**, 75–82, 23 December 2008. https://data.europa.eu/eli/dir/2008/114/oj

19. Directive (EU) 2016/1148 of the European Parliament and of the Council of 6 July 2016 concerning measures for a high common level of security of network and information systems across the Union. Official J. L **194**, 1–30, 19 July 2016. https://data.europa.eu/eli/dir/2016/114 8/oj

20. "Regulation (EU) 2019/881 of the European Parliament and of the Council of 17 April 2019 on ENISA (the European Union Agency for Cybersecurity) and on information and commu- nications technology cybersecurity certification and repealing Regulation (EU) No 526/2013, Official J. L **151**, 15–69, 7 June 2019, https://data.europa.eu/eli/reg/2019/881/oj.

21. Resilience, Deterrence and Defence: Building strong cybersecurity for the EU. JOIN(2017) 450 final, Brussels, 13 September 2017, https://eur-lex.europa.eu/legal-content/en/TXT/?uri= CELEX%3A52017JC0450
22. Walker, M.: CEH Certified Ethical Hacker All-in-One Exam Guide, 4th edn. McGraw-Hill Education, New York, NY (2019)
23. Shah, M.: Different types of hackers: why they hack. Tech Funnel, 13 March 2020. https:// www.techfunnel.com/information-technology/different-types-of-hackers/
24. Jang-Jaccard, J., Nepal, S.: A survey of emerging threats in cybersecurity. J. Comput. Syst. Sci. **80**(5), 973–993 (2014). https://doi.org/10.1016/j.jcss.2014.02.005
25. Hansman, S., Hunt, R.: A taxonomy of network and computer attacks. Comput. Secur. **24**(1), 31–43 (2005). https://doi.org/10.1016/j.cose.2004.06.011
26. Raban, Y., Hauptman, A.: Foresight of cyber security threat drivers and affecting technologies. Foresight **20**(4), 353–363 (2018). https://doi.org/10.1108/FS-02-2018-0020

Automotive Cybersecurity Testing: Survey of Testbeds and Methods

Shahid Mahmood, Hoang Nga Nguyen, and Siraj A. Shaikh

Abstract Computing and connectivity capabilities in modern cars have introduced new cybersecurity challenges that can potentially affect the safety of an automobile and its occupants. Effective cybersecurity testing of vehicles can play a crucial role in discovering and addressing security flaws; however testing a real vehicle (involving cyber-physical components) carries safety and economic risks. Therefore, many researchers and practitioners rely on testing environments (commonly known as testbeds) for uncovering cybersecurity vulnerabilities. Effective and efficient security testing needs the application of appropriate and systematic testing methods. This study presents a survey of seven different automotive cybersecurity testbeds proposed over the last ten years (between 2012 and 2019) as well as four different types of cybersecurity testing methods employed by cybersecurity researchers. This survey will help students, researchers, and professionals with designing and building their testbeds and refining their testing techniques and methodologies.

Keywords Automotive · Cybersecurity · Testbed · Testing · Approaches · Methods

1 Introduction

Contemporary vehicles are increasingly vulnerable to cybersecurity attacks due to the embedded computing and internet connectivity capabilities they are equipped with. As cyberattacks have the potential to seriously undermine the safety of an automobile

S. Mahmood (✉) · H. N. Nguyen · S. A. Shaikh
Systems Security Group, Institute for Future Transport and Cities,
Coventry University, Coventry CV1 5FB, UK
e-mail: mahmo136@coventry.ac.uk

H. N. Nguyen
e-mail: hoang.nguyen@coventry.ac.uk

S. A. Shaikh
e-mail: siraj.shaikh@coventry.ac.uk

© The Author(s), under exclusive license to Springer Nature Switzerland AG 2021   219
T. Tagarev et al. (eds.), *Digital Transformation, Cyber Security and Resilience
of Modern Societies*, Studies in Big Data 84,
https://doi.org/10.1007/978-3-030-65722-2_14

and its occupants, effective testing for detecting software flaws and weaknesses is crucial. However, cybersecurity testing of automobiles is not always feasible due to their technical complexity, physical size, safety risks, and high financial costs. In order to overcome these challenges, cybersecurity researchers and professionals often rely on testing environments (commonly referred to as testbeds) by mainly using virtual devices and sometimes real components as well.

Although, using real devices/components in the cybersecurity testing can provide very high degree of fidelity; there are safety and financial ramifications to consider carefully. While testbeds provide a conducive and safe testing environment, appropriate security testing techniques help tremendously in identifying cybersecurity threats in a systematic way. We present a review of seven major automotive cybersecurity testbeds and four security testing approaches that are widely used in the field of automotive cybersecurity.

1.1 Motivation, Objective and Scope of This Study

To the best of our knowledge, there are no prior studies that survey automotive cybersecurity testbeds and testing methods. Therefore, this study aims at closing this gap by providing a comprehensive survey of the recent developments in automotive cybersecurity testbeds and testing approaches. This study does not explore or consider testbeds designed for cybersecurity testing of autonomous vehicles. Moreover, we have deliberately not included works involving real vehicles (e.g., [35]) as a testing platform. The reason for excluding such setups is because they target a very specific make and model of a certain real vehicle. The main objective of this study is to find answers to the following questions: Over the last ten years, What testbeds have been proposed for automotive cybersecurity testing? What are the key characteristics, strengths, and weaknesses of each testbed? What methods have been proposed and widely used for automotive cybersecurity testing?

1.2 Outline

This paper is organized as follows: In Sect. 2, background of the automotive cybersecurity is presented, providing an overview of different cybersecurity threats that modern vehicles are vulnerable to. Section 3 provides an overview of the related work, highlighting similar studies that survey cybersecurity testbeds and methods in other domains. Section 4 discusses seven different testbeds proposed over the last ten years for automotive cybersecurity testing. This is followed by Sect. 5, which discusses and compares various attributes of the testbeds including adaptability, portability, fidelity, safety and cost. Four different types of cybersecurity testing approaches for automotive security evaluation are presented in Sect. 6 followed by the conclusion in Sect. 7.

2 Automotive Cybersecurity

Software flaws or vulnerabilities in the vehicle can lead to serious consequences, ranging from incidents of information theft to life-threatening situations. Numerous previous studies show how a vehicle can be maliciously controlled by exploiting one or more weaknesses in its software systems. Koscher et al. [35] demonstrate that it is possible for an adversary to maliciously influence a car's behaviour (e.g., engaging or disengaging its brakes) if they are able to access the car's internal network.

A connected vehicle may have several internal and external connections for accomplishing various important tasks. While these connections support correct functioning of different applications in the car, they can be exploited by cybercriminals to launch cyberattacks targeting various digital systems in the vehicle. Some of the external connections to the vehicle include cellular network, WiFi, Bluetooth, Keyless Entry System, KES, and Tyre Pressure Monitoring System (TPMS). Whereas, Onboard Diagnostic II (OBD II) port, USB, and in-vehicle infotainment are some of the internal connections [43]. This section provides an overview of some of the common cybersecurity threats and risks faced by modern automobiles.

2.1 CAN Bus

Most cars today come equipped with a variety of computing devices, known as Electronic Control Units (ECUs). A typical modern vehicle may contain a number of different ECUs, each of which has unique responsibilities for performing one or more functions of the vehicle. For example, one ECU may be responsible for detecting whether there is a passenger present in the vehicle, whereas another one may be monitoring the tyre pressure. In order to perform their duties correctly, these ECUs often need to communicate with each other as well as external world [28].

For local communication, ECUs rely on various automotive networking technologies including Controller Area Network (CAN), Local Interconnect Network (LIN), Media Oriented Systems Transport (MOST), and FlexRay. Each of these technologies has been designed to meet specific needs of a particular automotive application. For instance, MOST is a high-speed network technology to support audio, video, and voice data communications. Whereas, LIN is used in automotive applications requiring low network bandwidth and speed, such as mirror control and door lock/unlock features [20].

Although CAN is a robust and fault-tolerant network technology, it lacks any security mechanisms because it has not been designed with security in mind [5]. Therefore, it is vulnerable to numerous cybersecurity threats which have been reported in many existing studies, such as [22, 37, 38]. For example, an adversary could gain access to the internal network of the vehicle remotely followed by injecting messages onto it in order to compromise and control a target ECU. Once an ECU is compromised, the attacker can potentially control the safety-critical functions of the car, such as braking, acceleration, and steering. It is however important to note that safety-critical components usually reside on a network that is separate from other non-critical components. Nevertheless, hackers may still be able to break into these networks by leveraging a gateway ECU [30].

Other examples of CAN bus exploitation include installation of a malicious diagnostic device to send packets to the CAN bus, using CAN bus to start a vehicle without a key, leveraging the CAN bus to upload malware, installing a malicious diagnostic device in order to track the vehicle and enable remote communications directly to the CAN bus [25].

By compromising one of the connections listed above, cybercriminals can potentially attack in-vehicle systems in order to take over a vehicle remotely, shut down it, unlock it, track it, thwart its safety systems, install malware on it, or spy on its occupants. For example, an adversary can access the vehicle's internal network or the remote diagnostic system remotely by means of cellular connection. Similarly, an attacker can exploit the WiFi connection for gaining access to the vehicle network (from up to 300 yards), intercepting data traffic of the WiFi network, breaking the WiFi password and more [43].

2.2   In-Vehicle Infotainment

An In-Vehicle Infotainment (IVI) or an automotive infotainment system is an integrated unit, providing information services and entertainment functionality to the driver and other vehicle occupants for an enhanced in-vehicle experience.

Infotainment systems, one of the major attack vectors in connected cars, are growing both in terms of their capabilities and popularity. As more and more features are being added to infotainment systems, this will likely to increase the number of new vulnerabilities, attack vectors, and threats that can undermine the privacy and safety of the vehicle and its occupants. Typically, an IVI is interconnected with the CAN bus for communicating with other devices. From cybersecurity perspective, this connectivity may have serious implications. Prior studies have evidenced that cybercriminals can target automotive infotainment systems for mounting sophisticated attacks on automobiles [24].

An attacker can exploit weaknesses in the infotainment system or can use it as an entry point to gain access to in-vehicle network, thus to safety-critical features of the vehicle. Some possible use cases include utilising a remote connection to the infotainment system for exploiting the application in the IVI responsible for handling incoming calls, accessing the subscriber identity module (SIM) through the IVI, installing malicious code on the infotainment system, putting the infotainment console into debug mode, using a malicious application to access the internal CAN bus network, using a malicious application to eavesdrop on actions taken by vehicle occupants.

2.3 Onboard Diagnostic (OBD) Port

Modern vehicles have Onboard Diagnostic (OBD) ports inside them that are used for ECU firmware updates, vehicle repairing and inspections. Implementation of these ports is obligatory since 1998 in the USA and since 2001 for gasoline-powered vehicles and since 2003 for diesel-powered vehicles in the EU respectively [44]. Onboard Diagnostic is mainly used for reporting the data gathered by various sensors in the car to the outside world, providing information on the health status of the vehicle. This information is often used by service providers for fixing any reported problems [12]. Since inexpensive OBD dongles are readily available in the market, attackers can leverage them as an entry point for breaking into in-vehicle networks.

Nilsson and Larsan in [33] demonstrate how a virus can be injected on to the CAN bus through the OBD port that issues some messages for controlling some aspects of the vehicle behaviour (e.g., locks, brakes, etc.) if certain conditions are found to be true.

Unlike the attack mentioned above in [33], which requires physical access to the vehicle, many modern automobiles allow remote access to these dongles via WiFi connections from a computer, allowing adversaries to launch cyberattacks remotely. As reported in a survey [47], more than 50% of the surveyed dongles, were found to be containing vulnerabilities (e.g., exposed keys, weak encryption), which can be exploited by cybercriminals to compromise the security of a vehicle.

3 Related Work

To the best of our knowledge, there is no prior published study that surveys automotive cybersecurity testbeds and testing methods. A few previous studies, such as [34, 45] describe exiting testbeds for automotive cybersecurity testing, but they are limited to very brief, high level descriptions only. For example, Toyama et al. [45] compare their proposed testbed with some existing testing environments [15, 29, 31] and briefly outline their strengths and limitations. Likewise, brief descriptions of some existing testbeds have been presented by Oruganti et al. in [34]. Similarly, there is no prior known work that presents a survey of the testing methods in automotive cybersecurity.

A number of studies surveying testbeds, and some others reviewing security testing techniques in other domains do exist. In this section, we briefly describe some of those surveys.

Holm et al. [21] present a survey of 30 Industrial Control System (ICS) testbeds primarily focusing on facilitating vulnerability analysis, test and education of defense mechanisms. The study aimed at investigating what ICS testbeds exist, what specific ICS objectives they propose, how ICS components are implemented into these, and how they manage testbed requirements. Cintulu et al. [9] provide a survey on cyber-physical smart grid testbeds along with a taxonomy based on smart grid domains as well as a set of guidelines for developing the testbed. Their survey include a detailed discussion and evaluation of existing smart grid testbeds. Furthermore, they also outline future trends and possible developments in cyber-physical smart grid testbeds.

A comprehensive survey of cybersecurity testing approaches in SCADA systems has been presented by Nazir et al. [32] They provide an overview of various common vulnerabilities that may exist in many SCADA systems. Additionally, among other approaches, authors describe and discuss model-based and simulation-based methods and frameworks for cybersecurity testing of SCADA systems. A number of SCADA testbeds have also been discussed in the study. Approaches, such as machine learning and penetration testing have also been discussed in the context of SCADA systems testing. Finally, the authors highlight some recent developments and future trends in the domain, specifically referring to cloud computing, virtualization, software defined networks (SDN), as well as open standards such as OpenSCADA.

4 Overview of Automotive Cybersecurity Testbeds

In this section we present an overview of seven different automotive cybersecurity testbeds that have been proposed in the last ten years.

Testbeds can generally be categorised in three different types: simulation based, hardware based, and hybrid. Simulation-based testbeds rely solely or substantially on software to simulate the behaviour of ECUs and in-vehicle networks. Since they do not include real cyber-physical components, simulation-based testbeds are generally cheaper to build, and provide a safer environment for the testers. Hardware-based testbeds, on the other hand, include real or emulated hardware components. As opposed to software-based testbeds, hardware-based testbeds enable testers to study interactions between components through physical inputs and outputs. Hybrid testbeds include both software and hardware components, offering strengths of simulation-based and hardware-based testbeds.

Table 1 presents an overview of the surveyed testbeds, indicating whether they are simulation-based, hardware-based, or hybrid. Mobile testing platform from [29] is the only testbed that uses real physical components and a vehicle (go-cart) for investigating cybersecurity threats. OCTANE and the Testbed for Security Analysis of Modern Vehicle Systems are hybrid testing environments. All other testbeds rely on virtual/software components only.

4.1 OCTANE: Open Car Testbed and Network Experiments

Open Car Testbed and Network Experiments (OCTANE) [4] includes a hardware framework and a software package providing capabilities to reverse engineer and test automotive networks. In particular, the tool can be used for fuzz testing various proprietary vehicular network protocols.

Table 1 Types of automotive cybersecurity testbeds

Name of testbed	Test platform	Year	References
Open car testbed and network experiments (OCTANE)	Hybrid	2013	[4]
Mobile testing platform	Hardware	2015	[29]
Cyber assurance testbed for heavy vehicle electronic controls	Simulator	2016	[10]
Testbed for automotive cybersecurity	Simulator	2017	[17]
Testbed for security analysis of modern vehicle systems	Hybrid	2017	[48]
Portable automotive security testbed with adaptability (PASTA)	Simulator	2018	[45]
Hardware-in-loop based automotive embedded systems cybersecurity evaluation testbed	Simulator	2019	[34]

The software package allows transmission and monitoring of CAN messages for general purpose network diagnostic and debugging as well as automated replay testing of Electronic Control Units (ECUs), whereas the hardware framework assists in setting up hardware components of the automotive networks for two main different configurations: lab setup and real-world setup. The hardware framework outlines a structured step-by-step approach to set up a particular environment without prescribing any specific type of hardware components.

The testbed has been designed to enable entry into automotive cybersecurity testing research and teaching in a safe and cost-effective way. In order to maintain clear separation of concerns between software and hardware components, the hardware middle layer plays a pivotal role. This makes it easy to add a new hardware adapter to replace the existing one without affecting other layers. In order to enable adaptability and flexibility, the software package has been designed using a layered architecture consisting of a presentation layer, a business layer composed of a processing layer and a thread layer, a hardware middle layer, and a hardware layer.

This testbed allows security testing of various vehicular network protocols including CAN, LIN, MOST and FlexRay. An appropriate adapter needs to be used when working on a specific network technology. One of the main limitations of the testbed is that it only uses the OBD port as an attack surface. The testbed is not capable of testing vulnerabilities related to wireless connectivity. We observe that although the source code is available to download on the Google Code platform, we were unable to find any related documentation.

4.2 A Mobile Testing Platform

Miller and Valasek [29] implemented a mobile testbed by modifying a go-cart to emulate a real vehicle. They equipped it with various ECUs and sensors, which help study the behaviour of actual devices in an economical way. As compared to a real vehicle, there is low financial risk involved, because the go-cart is much cheaper than a real vehicle. However, risk of physical injuries is still present as it is a moving vehicle.

One of the major capabilities of the real vehicle they included was a power steering control module (PSCM) on the go-cart. Additionally, they integrated different sensors including proximity and speed sensors. A real pre-collision system was also Incorporated for actual distance readings while the vehicle is in motion. While using a moving vehicle instead of a bench setup certainly enabled the testers to study the behaviour of the moving vehicle which a testbed set up on a bench is incapable of, there are some shortcomings as well that the original developers of the environment identified.

While real components were used in the go-cart, they may not represent the complete functionality and behaviour of an actual car. For example, the developers note that PSCM does not work properly after some right and left turns, as the it enters its final state. Another limitation reported by authors is steering wheel radius that does not allow steering to be controlled by the CAN bus. This limitation was also hurdle for auto-park capability, which otherwise could have been realized. Finally, remodelling of the go-cart vehicle is another key challenge for the researchers who may be interested in using this setup. To summarize, while this mobile testing platform enables the tester to evaluate the impact of cyberattacks on the moving vehicle cost-effectively, it has some considerable limitations as well.

4.3 A Cyber Assurance Testbed for Heavy Vehicle Electronic Controls

This testbed [10] has mainly been proposed for cybersecurity testing of heavy vehicles remotely. It primarily supports J1939 networks that are found in heavy vehicles including buses and trucks. Authors used real ECUs, Linux-based, simulated node controllers for their testing setup.

The testbed allows the researcher to study and manipulate the network traffic by providing various features. For example, one of the distinctive characteristics of this testbed is the capability of remote experimentation, which allows researchers to access the data remotely without physically interaction with the vehicle. For this purpose, the authors of the testbed introduced a custom-built five-layer application. The five layers are web interface, experiment processing, experiment logic, CAN data processor and a database for experiment and J1939 data. The web interface layer allows the tester/researcher to interact with ECUs for monitoring and modifying network traffic. Experiment processing layer is responsible for converting the CAN messages into a human readable format. The database layer is mainly used for storing the experiment data.

4.4 Testbed for Automotive Cybersecurity

Fowler et al. [17] built a testbed for automotive cybersecurity testing consisting of an established industry, real-time CAN simulator from Vector Informatik.

The simulator along with its associated software CANoe is widely used in the automotive industry primarily by automakers for the development and testing of ECUs. The simulator provides CAN data traffic monitoring, capturing, and analysis capabilities, which help in reverse engineering of vehicles. To validate the testbed, CAN message-injection was performed by using a Bluetooth-enabled dongle connected to an OBD port on the simulator. The messages were successfully injected validating the correct functioning of the testbed. The description presented in the paper is limited to some high level information only without going into details about the architecture and other characteristics of the testbed.

4.5 Testbed for Security Analysis of Modern Vehicles

Zheng et al. [48] developed a prototype of their proposed testbed, which was built around a real-time CAN bus simulator using dedicated hardware from National Instruments and a simulated vehicular infotainment system (using LabVIEW software). The testbed is able to capture CAN messages for security analysis and can inject malicious messages through simulated infotainment system.

It is argued by the authors that while use of real vehicles or vehicle components for testing is more effective and produce accurate results, such test environments provide little or no flexibility in terms of their configurations. The proposed testbed by Zheng et al. is reconfigurable, enabling the testbed to replicate many test configurations. Furthermore, the testbed is able to reproduce the complexity of interconnected ECUs in the in-vehicle network.

The authors performed a denial-of-service attack targeting the CAN bus by leveraging the emulated infotainment system as an entry point into the in-vehicle network. A dump containing a large number of previously captured CAN messages was injected causing the CAN bus to fail to operate properly by rejecting legitimate CAN messages. This testbed is reconfigurable, inexpensive to reproduce, and provides a safe environment for automotive cybersecurity testing. However, being largely a simulated environment its obvious limitation is the lack of physical input and output ports which seriously affects security evaluation requiring these ports.

4.6 PASTA: Portable Automotive Security Testbed with Adaptability

Portable Automotive Security Testbed with Adaptability (PASTA) [45] is another automotive security testbed with a special focus on white-box ECUs, high adaptability and portability. Authors explain why white-box ECUs can be more effective when it comes to automotive cybersecurity testing. First of all, white-box ECUs provide the ability to observe their inputs and outputs as well as disassembly of the ECU programs without involving any suppliers or OEMs. Secondly, ECUs can be reprogrammed and rearranged in a number of different configurations in the automotive networks allowing evaluation of the security technology against cyberattacks. Finally, the ability to modify different parameters, such as CAN ID, payload, or transmission cycle enables the reproducibility of a commercial vehicle.

The authors outline the requirements that they considered while designing their proposed testbed. The first factor is the cost of the testbed, which is typically very high when involving a real vehicle containing a variety of ECUs and in-vehicle networks. High financial cost is one of the barriers to automotive cybersecurity research. To minimize the cost of their testbed, they eliminated expensive sensors and other similar components including simulators such as speed, angle of tyres, and status of headlights. Authors argue that such expensive components are not essential for cybersecurity testing.

Portability of the testbed is another key consideration; PASTA has been designed with portability in mind, its compact size allows it to be easily carried to different places for demonstrating research experiments and results. Another aspect of the testbed is the generalizability of the vehicle to ensure the testing is not restricted to a specific make and model.

Safety is also critical aspect, especially when the testing involves physical subsystems, such as actuators, which can behave in an unexpected way causing injuries to the researchers or any other parties involved. Such safety risks can be addressed by using an emulated actuator instead of a real one, for example. Finally, the target testbed must be designed in a way that it supports the learning of all the stakeholders, especially software developers, because software bugs and flaws in the software design due to human error often result in catastrophic consequences.

PASTA, according to its designers, meets all the requirements outlined above. The testbed can be customized to fulfill specific needs of a researcher, as it has been designed using non-proprietary technologies. The testbed allows testers to use custom security technology and provides the flexibility to design the in-vehicle network as per their particular requirements.

While PASTA is safe, flexible, portable and adaptable, it has some shortcomings too. Its software vehicle simulator is not able to replicate a vehicle's behaviour accurately. As the designers of PASTA have noted that speed of the vehicle reaches 199 km/h in a very short time when the acceleration is applied, which is obviously not reflective of true behaviour of a real vehicle. The software needs tweaking to resolve this issue. Another limitation of PASTA is that it currently supports CAN protocol only. Other protocols such as LIN, FlexRay, and MOST are not supported. OBD-II port and a tapped CAN cable are the only physical intrusion points that PASTA provides for launching attacks on the CAN bus. Moreover, it currently lacks attack surfaces such as Bluetooth, WiFi, and cellular networks. Finally, the software architecture for the implementation of ECUs environment is not Automotive Open System Architecture compliant.

4.7 Hardware-In-Loop Based Automotive Embedded Systems Cybersecurity Evaluation Testbed

Oruganti et al. [34] propose a testbed for automotive cybersecurity testing. Authors report their current progress towards the development of their testbed which will include hardware-in-loop components. The current testbed is completely a virtual setup, thus limited to a software simulation only. Using this virtual testbed, the authors demonstrate a GPS location spoofing attack on a virtual vehicle. The authors list essential elements of the testbed that should be present, which include connectivity, vehicular networks, controller modeling and algorithm implementation, hardware-in-loop and telematics. Each of these subsystems allows the cybersecurity evaluation and validation of a connected car for a range of attack surfaces and attack vectors.

5 Evaluation of Testbeds

This section compares the reviewed testbeds based on their various characteristics, such as adaptability, portability, fidelity and cost. An overview of other capabilities (e.g., types of attacks, attack surfaces, attack targets, and communication protocols supported) of the testbeds have also been discussed.

5.1 An Overview of Supported Network Protocols

Modern cars have multiple network types for facilitating various applications. Not all testbeds that have been surveyed offer support for testing all types of communication standards. Table 2 gives an overview of the protocols supported by each testbed. As can be noticed, OCTANE is the only testbed that claims to support testing for all major vehicular network protocols. All other testbeds do not cover any protocol other than CAN. This means they are unable to support study of threats/attacks related to other network standards found in modern automobiles.

5.2 Testbeds and Supported Attack Surfaces, Types of Attacks, and Attack Goal

Table 3 highlights types of attack surfaces exposed by each testbed, types of attack supported or demonstrated, and attack target or goal. OBDII port is the most popular choice as an entry point into the in-vehicle network. This is probably due to the fact that all cars do have an OBD port, (since it is legal requirement to have one) OBD scanners are cheap and easily available in the market.

Table 2 Overview of what types of in-vehicle network protocols are supported by each testbed for cybersecurity testing

Testbed name	CAN	LIN	FLR[a]	MOST	References
Open car testbed and network experiments (OCTANE)	✓	✓	✓	✓	[4]
Mobile testing platform	✓	N/A	N/A	N/A	[29]
Cyber assurance testbed for heavy vehicle electronic Controls	✓	N/A	N/A	N/A	[10]
Testbed for automotive cybersecurity	✓	N/A	N/A	N/A	[17]
Testbed for security analysis of modern vehicle Systems	✓	N/A	N/A	N/A	[48]
Portable automotive security testbed with adaptability (PASTA)	✓	N/A	N/A	N/A	[45]
Hardware-in-loop based automotive embedded systems cybersecurity evaluation testbed	✓	N/A	N/A	N/A	[34]

[a]FLR stands for FlexRay

Similarly, most popular type of attack is message/code injection. This is obviously because CAN does not have an authentication or other security mechanism capable of identifying and rejecting malicious contents. Since many testbeds lack support for wireless/remote attack surfaces, they are only confined to testing attack scenarios assuming physical access to the vehicle.

Each testbed type has its own strengths and limitations. Ideally, a testbed should be able to reproduce the behaviour of a real vehicle as accurately and faithfully as pos-

Table 3 An overview of the types of exposed attack surfaces, types of attacks, target and/or goal of the attacks supported by each testbed

Testbed name	Attack surface	Attack type	Attack target/goal	References
Open car testbed and network experiments (OCTANE)	OBDII Port	Message sniffing denial of service (Dos) replay	N\A	[4]
Mobile testing platform	OBDII port	CAN message injection	Take over vehicle control	[29]
Cyber assurance testbed for heavy vehicle electronic controls	ECU, Ethernet, USB	Brute force, DoS	Evaluation of seed\key exchange Strength and intrusion detection system	[10]
Testbed for automotive cybersecurity	OBDII port	CAN message injection	Comfort subsystem manipulation (e.g., headlamp ON/OFF)	[17]
Testbed for security analysis of modern vehicle systems	Infotainment gateway	Can sniffing, code injection, DoS	Control vehicle maneuver	[48]
Portable automotive security testbed with adaptability (PASTA)	OBDII Port/clipping area	Code injection/execution	N\A	[45]
Hardware-in-loop based automotive embedded systems cybersecurity evaluation testbed	Navigation system	GPS spoofing	Spoof GPS location	[34]

sible (fidelity), adaptable, portable, safe, and inexpensive to construct, as explained in [45]. In this section, we provide an overview of how and to what extent each of the testbeds meets the requirements of adaptability, portability, fidelity, safety, and cost-effectiveness.

Adaptability is a measure of a testbed's ability to support different testing configurations, i.e. how well a testbed can adapt to different configurations. Portability refers to how easy it is for the testbed to carry around. Fidelity of testbed can be measured by evaluating how accurately it can imitate the behaviour of a real vehicle in response to a an cyberattack. Testing environments involving cyber physical components have safety implications for the testers and any hardware/equipment involved;

therefore it is vital to determine how a given testbed ensures the safety of testers and the equipment involved in the testing. Finally, cost of the entire setup can be compared based on whether it contains real physical devices or virtual components. Table 4 provides a comparison of how adaptable, portable, accurate, safe and costly each testbed is.

5.3 Adaptability

In-vehicle networks and ECUs are the major targets of cyber attacks, their security testing is important to identify and fix any security issues. Unfortunately, due to copyright restrictions and closed-source proprietary ECU technologies, it is very difficult to perform security testing on commercial ECUs. In addition, testing specific ECUs does not provide insights and results that can be helpful when testing the ECUs from different manufactures. Similarly, it is also important that the testbed is adaptable to a variety of testing configurations (e.g., with different vehicular network types) and not confined to a particular technology.

For instance, PASTA [45] includes white-box or programmable ECUs that allow the researcher to program an ECU to replicate the behaviour of a specific ECU, with the knowledge of internal implementation, which is usually not possible with real proprietary ECUs. Also, because it is largely software-based setup, it can be used for different testing configurations.

The layered-based design of the OCTANE [4] allows it to be adapted to different testing setups by replacing hardware components in the hardware middle layer without affecting other layers. Furthermore, its bespoke software package can be modified to extend its capabilities according to specific testing scenarios.

The prototype testbed proposed by Zheng et al. [48] has a flexible architecture allowing additional ECUs to be added easily. The setup can also be used for testing and investigating attacks launched via remote connections. Since it is entirely software-based, the testbed presented by Oruganti et al. [34] is adaptable to various testing configurations, as virtual components can be easily added or removed in the software environment.

Daily et al. have relied on actual ECUs and sensor simulations primarily focusing on J1939 based networks, which are specifically designed for heavy vehicles, such as trucks and buses. Although, the authors do not explicitly consider or discuss adaptability of the testbed, based on the information provided, *it seems probable for the testbed to be adapted to various testing configurations.*

5.4 Portability

A key factor to consider while designing or using a testbed is the portability. Sometimes it may be necessary to carry the testbed to a different location for demonstration

Table 4 A comparative overview of the reviewed testbeds based on adaptability, portability, fidelity, safety and cost

Testbed name	Adaptability	Portability	Fidelity	Safety	Cost
Open car testbed and network experiments (OCTANE) (real-world setup)	●	●	●●●	●	●●●
Open car testbed and network experiments (OCTANE) (lab setup)	●●	●●	●●	●●	●
Mobile testing platform	●	●●	●●	●	●●
Cyber assurance testbed for heavy vehicle electronic controls	●	●●	●●	●●●	●
Testbed for automotive cybersecurity	●●	●●	●	●	●●
Testbed for security analysis of modern vehicle systems	●●	●●●	●	●●●	●
Portable automotive security testbed with adaptability (PASTA)	●●●	●●●	●●	●●●	●●
Hardware-in-loop based automotive embedded Systems cybersecurity evaluation testbed	●●	●●●	●	●●●	●

(●●● = High, ●● = Medium, ● = Low)

purposes (e.g., in a conference or workshop). A testbed with compact or virtual components is obviously easy to carry around as opposed to the ones that include large actual components. Below we describe how each of the testbeds reviewed supports portability.

OCTANE has two types of main setups: lab based and real world. The lab-based testing environment typically relies on small components and does not involve real vehicle. So, it is possible to carry the lab setup as necessary with ease. However, in the case of a real-world testing setup, the portability depends on the actual components involved. The portability will be affected if, for example, a real car or heavy components are used. PASTA has been designed to be portable, so all its components are able to fit in a briefcase allowing high degree of mobility. Instead of using a real vehicle, a simulated or scale model of a real vehicle is a key factor in allowing this testbed to be more portable. The testbed from Zheng et al. contains simulated and emulated components so it should be easy to relocate if required.

Similar to the Zheng et al., the testbed from Oruganti et al. is purely a software-based environment which allows it to be moved around easily. The cyber assurance testbed by Daily et al. does not use a real vehicle, can be accessed remotely and uses simulated components with real ECUs, hence it satisfies the portability requirements.

5.5 Fidelity

Fidelity of a testbed refers to its ability to accurately reproduce the behaviour of a real vehicle or components in response to a specific event. To achieve high degree of fidelity, real vehicle and/or real hardware components must be included in the test. Software-based testing environments cannot faithfully reflect the conditions of a real car. Thus, fidelity of the test results is directly linked to the type of systems/components involved in the testing. Most importantly, complex interactions among various ECUs and other cyber-physical components inside the vehicle cannot be simply reproduce with high accuracy in a virtual environment.

The software vehicle simulator used in PASTA, for example, reaches 199 km/h in a very short time which does not mimic the actual behaviour of the vehicle. While virtual, software-based testbeds have their own merits, they do not generally replicate actual behaviour of a real vehicle.

5.6 Cost

A virtual or software-based testbed is generally cheaper than a testbed which includes cyber physical components. OCTANE, and mobile testing platform (involving go-cart) rely on physical components, they are therefore more expensive. On the other hand, Zheng et al. Fowler et al. Oruganti et al. are software-based testbeds their cost is lower. PASTA and Daily et al. both contain ECUs, their cost will be higher than the pure software-based testbeds.

5.7 Safety Implications

Testing real vehicles help study the actual impact and behaviour of a vehicle as a result of a cyber attack. However, this has serious safety implications for the researcher and the vehicle under test. Physical safety of all stakeholders as well as of all the components/equipment involved must be the top priority. We look what safety implications each of the reviewed testbeds may have. In general, safety risks are high when a real vehicle or large cyber-physical components are used in the testing. The risk is even higher when the testing involves a moving vehicle on the road. While designing a new or using an existing testbed, it is a good idea to carefully consider any safety issues that can potentially surface.

OCTANE has two testing environments - laboratory-based and real world. In the lab-based setup, there are virtually no concerns related to human safety as it is a controlled environment with no real vehicle involved. The real-world testing setup potentially can lead to situations that can affect safety of both the vehicle and the testers.

Since [34, 45, 48] are primarily simulation based, these testbeds do not raise any safety concerns for the testers/researchers. Similarly, because the testbed from Daily et al. [10] is remotely accessible, it is safe to use.

It can be noticed that while software-based testbeds are generally more adaptable, portable, inexpensive, and safe, they however lack physical inputs and outputs (I/O) which may not be useful in the scenarios where evaluation of physical I/O is essential. Moreover, software-only testbeds do not provide accurate results and are often unable to reproduce the behaviour of actual systems.

6 Automotive Cybersecurity Testing Methods

While testbeds play a key role in security assessment of in-vehicle computing systems, effective testing methods are equally crucial for successful security evaluation of these systems. Knowledge of different testing approaches can be useful in choosing and applying the best possible technique for optimal results. We present a survey of four different automotive cybersecurity testing approaches here, as at the time of this writing, there is no existing work presenting such a survey.

Interconnected computing components (i.e. ECUs) in a modern vehicle control various features including safety-critical functions, such as airbags, braking, acceleration etc. Attackers can exploit security loopholes in these systems to take over control, steal information, or cause damage to the vehicle and/or its occupants. Prior studies [8, 19, 28, 35] discuss different attack scenarios that are possible and practical. Therefore, thorough and systematic testing of automotive components is paramount.

There are effective approaches employed by cybersecurity testers, professionals and researchers, which help detect potential security weaknesses in automotive systems. Following subsections discuss some major cybersecurity testing approaches.

6.1 Automotive Penetration Testing

Penetration testing, in general, is a security assessment approach which is usually adopted by security testing professionals to carry out security testing from the perspective of an attacker to discover security weaknesses in a system. While there are different variants of the approach, it generally has the following key stages as outlined in the NIST Guide to Information Security Testing and Assessment [40]:

1. Planning—this phase is concerned with collecting as much information as possible about the target system as well as the boundaries and relevant components involved in the testing.
2. Discovery—in this phase, all the available public external interfaces of the system are systematically discovered and enumerated.
3. Attack—in order to test the identified interfaces, a series of attacks are launched on the system by exploiting the found vulnerabilities.
4. Reporting—the reporting takes place simultaneously with other three steps. Documentation of the findings is done in this phase. Figure 1 illustrates the four stages of the penetration testing.

When the tester has no or limited knowledge of the system under test, they largely depend on publicly available information of the target system. In this case, the target system is treated as a black box, as such the specification of the system is not accessible. In contrast, when the tester has detailed knowledge of the system, the system can be referred to as white box, as the internal details of the system are known to the tester. Whereas, the system may be considered a grey box when the tester has

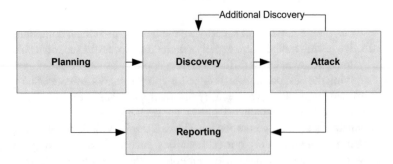

Fig. 1 The four stages of the penetration testing

partial information about it [2]. Black-box approach is the most appropriate choice for automotive cybersecurity assessment due to the unavailability of the functional specifications of in-vehicle systems.

Durrwang et al. [11] propose an improved penetration testing methodology for testing automotive systems by combining safety and threat analysis for deriving test cases in a systematic manner. The authors integrate attack trees [41] as a threat modelling technique for deriving quality test cases. Their proposed technique is based on the Penetration Testing Execution Standard (PTES) [36], a technical guide for penetration testing, which recommends threat modeling and integrates it as a key step. The authors perform an experiment attack involving an airbag ECU to demonstrate the application of their proposed technique.

PTES [36] defines the key stages or phases of the penetration testing as follows:

1. Pre-engagement Interactions
2. Intelligence Gathering
3. Threat Modelling
4. Vulnerability Analysis
5. Exploitation
6. Post Exploitation
7. Reporting.

Cheah et al. [7] propose a similar penetration testing framework for security evaluation of automotive interfaces. Very similar to the work described above, attack trees are an essential part of the approach for threat modeling. The authors apply the technique to the automotive Bluetooth interfaces for uncovering potential vulnerabilities. They also introduce a proof-of-concept tool to perform testing on vehicles using the proposed framework. The proposed proof-of-concept tool follows an attack tree for carrying out testing on the Bluetooth interface of the automobile. The tool is semi-automated i.e., some manual interventions are required for the tool to complete security assessment of the target system.

All the testbeds surveyed in this study support the penetration testing.

6.2 Automotive Fuzz Testing (Fuzzing)

Fuzz testing or fuzzing is used to discover new vulnerabilities by exposing the system to invalid, malformed, or unexpected inputs and the target system is monitored for any unusual behaviour, which may cause the system to crash. Fuzzing involves three main steps [26, 27]:

1. Preparing the input
2. Delivering the input to the target
3. Observing the behaviour of the target.

Fuzzer, a software application specifically designed for performing fuzz testing, is used for bombarding the system under test with a huge number of automatically generated data values. The software then observes system's behaviour to see any reactions to the input data. The input values are crafted either from existing valid input dataets or from a prescribed set of values.

While fuzzing has been around since 90s, and widely used as an effective testing technique in other domains for vulnerability discovery, it is not very popular in automotive security testing yet. This is probably due to the presence of specific challenges that require some adjustments for successful application of the technique to automotive security assessment. For example, monitoring of the system for unusual behaviour is crucial, but since the same interface is usually used for both the fuzz massage-injection and monitoring purposes, this means the internal reactions of the system might not be visible to the observers. A virtual testing environment with adequate support for observing the reaction of the target ECU can be an effective solution to this challenge as Bayer et al. report in [3].

In their study, Fowler et al. [16] describe a basic experiment attack they performed on a virtual vehicle using fuzz testing with a custom-built fuzzer. OBD port was used to interface the fuzzer with the CAN bus. The attack involved locking/unlocking the door lock of the virtual vehicle by injecting messages onto CAN bus. This was achieved by injecting random CAN messages for a short period of time. Based on their experience by executing the attack successfully and influencing the behaviour of the vehicle, the authors conclude that the fuzzing can be useful in reverse engineering of CAN messages as well as causing disruption to the vehicular networks. Most importantly, they note that the fuzzing can be detrimental for the vehicle under test.

In a more recent work [18], Fowler et al. emphasize the importance and usefulness of fuzzing (and other security testing methods), especially, when it is performed prior to production for allowing the discovery and fixing of bugs, which can lead to serious security issues, in the early phases of the system development.

6.3 Model-Based Security Testing

Model-based security testing is concerned with specifying, documenting and generating security test objectives, test cases, and test suites in a systematic and efficient manner [42]. It primarily uses models to verify if the target system meets its security requirements [14].

Santos et al. [39] propose their automotive cybersecurity testing framework, which uses Communication Sequential Processes (CSP) for representing the models of the vehicle's bus systems as well as a set of attacks against these systems. CSP—a language with its own syntax and semantics—is a process-algebraic formalism used to model and analyze concurrent systems. Using CSP, they create architectures of the vehicle's network and bus systems along with the attack models. One of the key challenges that authors claim to address in their work is the scalability of the testing in distributed environments.

Their system model is comprised of networks, bus systems connected to each network, and the gateways. Additionally, network parameters, such as latency can also be modelled. An attack model is also created, defining the attackers' capabilities as channels. An attacker's capabilities may include command spoofing, communication disruption, eavesdropping and influencing behavior of the system. According to the authors, the ability for a detailed definition of the scope of the attack and test cases is a key advantage of using these models for security testing.

Wasicek et al. [46] present aspect-oriented modelling (AOM) as a powerful technique for security evaluation of Cyber-Physical Systems (CPS), especially focusing on safety-critical elements in automotive control systems. AOM is based on the ideas inspired by aspect-oriented programming, which is concerned with crosscutting aspects being expressed as concerns (e.g., security, quality of service, caching etc.) [13]. Aspect-oriented modelling is used to express crosscutting concerns at a higher level of abstraction by means of modelling elements [6].

The technique presented by Wasicek et al. [46] models attacks as aspects, and aims at discovering and fixing potential security flaws and vulnerabilities at design time, because it becomes highly costly to find and fix the bugs if they are discovered later in the development life-cycle stages for automotive systems. Some of the main benefits that can be achieved by using AOM for security assessment of automotive systems include: separation of functional and attack models into aspects allows domain experts to work on different aspects without any interference; real-world attack scenarios involving high degree of risks can be modelled easily; general models can be reused in other systems.

An automotive case study is presented by the authors, involving adaptive cruise control system as an example. They use a special modeling and simulation framework, called Ptolemy II, for developing their models. The authors intended to explore effects of attacks on the communication between two vehicles. A discussion of four different attacks (i.e., man-in-the-middle, fuzzing, interruption, and replay) is presented.

6.4 Automotive Vulnerability Scanning

Automotive vulnerability scanning focuses on testing the system for existing known weaknesses in the system to ensure that the system is protected against known threats. An automotive system is typically scanned for identifying known weaknesses in the source code, ICT infrastructure and networks by using a regularly updated database of known vulnerabilities.

Vulnerability scanning can be performed in several different ways, depending on the types of target weaknesses for which the system is being examined. For example, in order to verify whether certain software flaws (e.g., buffer/heap overflows) present in the software, static and dynamic analyses can be performed on the source code. Various interfaces including WiFi, cellular network, and Ethernet can be scanned for open ports and running services in automotive systems. In particular, in-vehicle networks, such as CAN and on-board diagnostic port should be scanned. Finally, analysis of the entire system specifically focusing on various configurations to verify if there are any loopholes that can be leveraged by adversaries to compromise the system [1].

Vulnerability scanning of an automotive infotainment system is presented in a recent study [23] by Josephlal and Adepu. The infotainment system used in the study has various connectivity interfaces including WiFi, Bluetooth, USB port, CAN and others. The authors used different tools (e.g., Nmap, Nessus) to support their experiment involving a attack vector analysis and vulnerability scanning of the infotainment system. The scan was able to detect various types of vulnerabilities of varying levels of risks. In particular, IP address of the infotainment system, an infotainment service running on a certain port, as well as a number of information leaking vulnerabilities were identified.

In addition to the vulnerability scan described above, they also report different attacks including a denial-of-service attack they conducted using a malicious smartphone app.

7 Conclusion

Modern cars are open to various cyberattacks due to in-vehicle ICT capabilities they are equipped with. Discovery of any security weaknesses that may potentially be present in the automotive systems is a first important step towards strengthening their security. Testing real automotive systems involves safety and economic risks. One effective solution to this issue is using testing environments instead of relying on real vehicles, as it offers several benefits including a safe and cost-effective testing setup. In order to ensure that maximum number of security flaws are revealed and fixed, a systematic and suitable testing approach must be employed. There are no known studies exist providing information on cybersecurity testbeds and security methods, which can be useful for students, researchers, and security professionals in the automotive cybersecurity domain for setting up their own testing environment and use established, systematic testing methods.

This study presents a survey of seven different automotive cybersecurity testbeds and four different types of testing approaches including automotive penetration testing, automotive fuzz testing, model-based security testing and automotive vulnerability scanning. Core features, merits, limitations and various characteristics of all testbeds and testing methods have been highlighted.

References

1. Bayer, S., Enderle, T., Oka, D.K., Wolf, M.: Automotive security testing—the digital crash test. In: Energy Consumption and Autonomous Driving, pp. 13–22. Springer, Cham (2016)
2. Bayer, S., Enderle, T., Oka, D.K., Wolf, M.: Security crash test-practical security evaluations of automotive onboard it components. Autom-Saf Secur **2014**, (2015)

3. Bayer, S., Kreuzinger, T., Oka, D., Wolf, M.: Successful security tests using fuzzing and HiL test systems, (2016, December) [Online]. Available: https://www.etas.com/download-center-files/products_LABCAR_Software_Products/Hanser-automotive_Successful-security-tests-hil-system_en.pdf

4. Borazjani, P., Everett, C., McCoy, D.: OCTANE: an extensible open source car security testbed. In: Proceedings of the Embedded Security in Cars Conference, p. 60 (2014)

5. Buttigieg, R., Farrugia, M., Meli, C.: Security issues in controller area networks in automobiles. In: 2017 18th International Conference on Sciences and Techniques of Automatic Control and Computer Engineering (STA), pp. 93–98 (2017). IEEE

6. Chavez, C., Lucena, C.: A metamodel for aspect-oriented modeling. In: Workshop on Aspect-Oriented Modeling with UML (AOSD-2002) (2002)

7. Cheah, M., Shaikh, S.A., Haas, O., Ruddle, A.: Towards a systematic security evaluation of the automotive Bluetooth interface. Veh. Commun. **9**, 8–18 (2017)

8. Checkoway, S., McCoy, D., Kantor, B., Anderson, D., Shacham, H., Savage, S., Koscher, K., Czeskis, A., Kohno, T.: Comprehensive experimental analyses of automotive attack surfaces. USENIX Security Symposium **4**, 447–462 (2011)

9. Cintuglu, M.H., Mohammed, O.A., Akkaya, K., Uluagac, A.S.: A survey on smart grid cyber-physical system testbeds. IEEE Commun. Surv. Tutor. **19**(1), 446–464 (2016)

10. Daily, J., Gamble, R., Moffitt, S., Raines, C., Harris, P., Miran, J., Johnson, J.: Towards a cyber assurance testbed for heavy vehicle electronic controls. SAE Int. J. Commer. Veh. **9**(2016-01-8142), 339–349 (2016)

11. Dürrwang, J., Braun, J., Rumez, M., Kriesten, R., Pretschner, A.: Enhancement of automotive penetration testing with threat analyses results. SAE Int. J. Trans. Cyber. Priv. **1**(11-01-02-0005), 91–112 (2018)

12. Eiza, M.H., Ni, Q.: Driving with sharks: rethinking connected vehicles with vehicle cyberse-curity. IEEE Veh. Technol. Mag. **12**(2), 45–51 (2017)

13. Elrad, T., Filman, R.E., Bader, A.: Aspect-oriented programming: introduction. Commun. ACM **44**(10), 29–32 (2001)

14. Felderer, M., Zech, P., Breu, R., Büchler, M., Pretschner, A.: Model-based security testing: a taxonomy and systematic classification. Software Test. Verification Reliab. **26**(2), 119–148 (2016)

15. Fisher, K.: HACMS: high assurance cyber military systems. In: ACM SIGAda ada letters, vol. 32, No. 3, pp. 51–52. ACM (2012)

16. Fowler, D.S., Bryans, J., Shaikh, S.A., Wooderson, P.: Fuzz testing for automotive cyber-security. In: 2018 48th Annual IEEE/IFIP International Conference on Dependable Systems and Networks Workshops (DSN-W), pp. 239–246. IEEE (2018, June)

17. Fowler, D.S., Cheah, M., Shaikh, S.A., Bryans, J.: Towards a testbed for automotive cybersecu-rity. In: 2017 IEEE International Conference on Software Testing, Verification and Validation (ICST), pp. 540–541. IEEE (2017, March)

18. Fowler, D.S., Bryans, J., Cheah, M., Wooderson, P., Shaikh, S.A.: A method for constructing automotive cybersecurity tests, a CAN fuzz testing example. In: 2019 IEEE 19th International Conference on Software Quality, Reliability and Security Companion (QRS-C), pp. 1–8. IEEE (2019, July)

19. Haas, R.E., Möller, D.P.: Automotive connectivity, cyberattack scenarios and automotive cyber security. In: 2017 IEEE International Conference on Electro Information Technology (EIT), pp. 635–639. IEEE (2017, May)

20. Hafeez, A., Malik, H., Avatefipour, O., Rongali, P.R., Zehra, S.: Comparative study of can-bus and flexray protocols for in-vehicle communication (No. 2017-01-0017). SAE Technical Paper (2017)

21. Holm, H., Karresand, M., Vidström, A., Westring, E.: A survey of industrial control system testbeds. In: Nordic Conference on Secure IT Systems, pp. 11–26. Springer, Cham (2015, October)

22. Hoppe, T., Kiltz, S., Dittmann, J.: Security threats to automotive CAN networks—practical examples and selected short-term countermeasures. In: International Conference on Computer Safety, Reliability, and Security, pp. 235–248. Springer, Berlin, Heidelberg (2008, September)

23. Josephlal, E.F.M., Adepu, S.: Vulnerability analysis of an automotive infotainment system's WIFI capability. In: 2019 IEEE 19th International Symposium on High Assurance Systems Engineering (HASE), pp. 241–246. IEEE (2019, January)
24. Kim, H.Y., Choi, Y.H., Chung, T.M.: Rees: malicious software detection framework for meego-in vehicle infotainment. In: 2012 14th International Conference on Advanced Communication Technology (ICACT), pp. 434–438. IEEE (2012, February)
25. Klinedinst, D., King, C.: On board diagnostics: risks and vulnerabilities of the connected vehicle. Softw. Eng. Inst.-Carnegie Mellon Univ. **10** (2016)
26. Li, J., Zhao, B., Zhang, C.: Fuzzing: a survey. Cybersecurity **1**(1), 6 (2018)
27. Manès, V.J.M., Han, H., Han, C., Cha, S.K., Egele, M., Schwartz, E.J., Woo, M.: The art, science, and engineering of fuzzing: a survey. IEEE Trans. Soft, Eng (2019)
28. Miller, C., Valasek, C.: A survey of remote automotive attack surfaces. Black hat USA **94**, (2014)
29. Miller, C., Valasek, C.: Car hacking: for poories. Technical report, IOActive Report (2015)
30. Miller, C., Valasek, C.: Remote exploitation of an unaltered passenger vehicle. Black Hat USA **2015**, 91 (2015)
31. Munera, J., Fuentes, J.M.D., González-Tablas, A.I.: Towards a comparable evaluation for VANET protocols: NS-2 experiments builder assistant and extensible test bed (2011)
32. Nazir, S., Patel, S., Patel, D.: Assessing and augmenting SCADA cyber security: a survey of techniques. Comput. Secur. **70**, 436–454 (2017)
33. Nilsson, D.K., Larson, U.E.: Simulated attacks on can buses: vehicle virus. In: IASTED International Conference on Communication Systems and Networks (AsiaCSN), pp. 66–72 (2008, August)
34. Oruganti, P.S., Appel, M., Ahmed, Q.: Hardware-In-Loop based automotive embedded systems cybersecurity evaluation testbed. In: Proceedings of the ACM Workshop on Automotive Cybersecurity, pp. 41–44. ACM (2019, March)
35. Patel, S., Kohno, T., Checkoway, S., McCoy, D., Kantor, B., Anderson, D., Shacham, H., et al.: Experimental Security Analysis of a Modern Automobile. In: 2010 IEEE Symposium on Security and Privacy
36. Penetration Testing Execution Standard, PTES Technical Guidelines (2014)
37. Riggs, C., Rigaud, C.E., Beard, R., Douglas, T., Elish, K.: A survey on connected vehicles vulnerabilities and countermeasures. J. Traff Logistics Eng. **6**(1), (2018)
38. Rizvi, S., Willet, J., Perino, D., Marasco, S., Condo, C.: A threat to vehicular cyber security and the urgency for correction. Proc. Comput. Sci. **114**, 100–105 (2017)
39. Santos, E.D., Simpson, A., Schoop, D.: A formal model to facilitate security testing in modern automotive systems (2018). arXiv preprint arXiv:1805.05520
40. Scarfone, K., Souppaya, M., Cody, A., Orebaugh, A.: Technical guide to information security testing and assessment. NIST Spec. Publ. **800**(115), 2–25 (2008)
41. Schneier, B.: Attack trees. Dr. Dobb's Journal **24**(12), 21–29 (1999)
42. Schieferdecker, I., Grossmann, J., Schneider, M.: Model-Based Security Testing (2012). arXiv preprint arXiv:1202.6118
43. Smith, C. (2016). The Car Hacker's Handbook: A Guide for the Penetration Tester. No Starch Press
44. Studnia, I., Nicomette, V., Alata, E., Deswarte, Y., Kaâniche, M., Laarouchi, Y.: Survey on security threats and protection mechanisms in embedded automotive networks. In: 2013 43rd Annual IEEE/IFIP Conference on Dependable Systems and Networks Workshop (DSN-W), pp. 1–12. IEEE (2013, June)
45. Toyama, T., Yoshida, T., Oguma, H., Matsumoto, T.: PASTA: portable automotive security testbed with adaptability, London, blackhat Europe 2018 (2018, December)
46. Wasicek, A., Derler, P., Lee, E.A.: Aspect-oriented modeling of attacks in automotive cyber-physical systems. In: 2014 51st ACM/EDAC/IEEE Design Automation Conference (DAC), pp. 1–6. IEEE (2014, June)
47. Yan, W.: A two-year survey on security challenges in automotive threat landscape. In: 2015 International Conference on Connected Vehicles and Expo (ICCVE), pp. 185–189. IEEE (2015, October)

48. Zheng, X., Pan, L., Chen, H., Pietro, R.D., Batten, L., Testbed, A.: Security Analysis of Modern Vehicle Systems: IEEE Trustcom/BigDataSE/ICESS. Sydney, NSW **2017**, 1090–1095 (2017). https://doi.org/10.1109/Trustcom/BigDataSE/ICESS.2017.357

Generalized Net Model of Cyber-Control of the Firm's Dumpers and Crushers

Dafina Zoteva, Peter Vassilev, Lyudmila Todorova, Krassimir Atanassov, Lyubka Doukovska, and Valery Tzanov

Abstract A generalized net model of a system for tracking and monitoring the movement and actions of a firm's dumpers and crushers is proposed. The presented model permits analysis regarding possible cyber manipulation of data, and highlights the crucial vulnerabilities of such systems.

Keywords Generalized net · Model · Data mining · Security

1 Introduction

The modelling of transport and other production activities in open area production allows to consider physical processes in the information space (digitalization) as a necessary step towards complete automation of the production management. The obvious advantage of such an approach is that the efficiency of production manage-

D. Zoteva · P. Vassilev (✉) · L. Todorova · K. Atanassov
Institute of Biophysics and Biomedical Engineering, Bulgarian Academy of Sciences,
Acad. G. Bonchev Str., Bl. 105, Sofia 1113, Bulgaria
e-mail: peter.vassilev@gmail.com

D. Zoteva
e-mail: dafy.zoteva@gmail.com

L. Todorova
e-mail: lpt@biomed.bas.bg

K. Atanassov
e-mail: krat@bas.bg

L. Doukovska
Institute of Information and Communication Technologies, Bulgarian Academy of Sciences,
Acad. G. Bonchev Str., Bl. 2, Sofia 1113, Bulgaria
e-mail: l.doukovska@mail.bg

V. Tzanov
Scortel LTD, 29 Deliiska Vodenitza Str., Sofia 1582, Bulgaria
e-mail: valery@scortel.com

ment increases significantly. But at the same time the information space sustains a number of specific hazards and risks in terms of both the management process and the production itself, which can be classified as cyber risks. The elimination of these risks is an element of the cybersecurity measures, an important component of the system security of an automated management and control systems for open area production.

The key to managing open processes is the modelling of loading and transportation of raw materials at open area construction sites, which is a part of project BG16RFOP002-1.005-0037-C01/25.5.2018 "Development and testing of a prototype of an innovative module for operational control and update of production schedule in open area production in changing production environment in real time."

Generalized Nets (GNs, [1–3]) are a suitable tool for modelling parallel processes. They are an extension of the ordinary Petri nets and all of their extensions and modifications. For this reason they are used here as a tool for modelling the process of loading and transporting raw materials, which are an essential element of the complex technological process in open production. In the present work, for simplicity, dumpers are considered the main work mechanisms. In general, different types of other mechanization may be considered: loaders, excavators and others.

The model of the production process considered in the present paper, which covers also the management of activities involving production machinery and mechanisms, makes it possible to identify various possibilities for illegal influence via data manipulation on elements of the control loop, including by organized cyber attacks. Manipulating the data can lead to the generation of incorrect commands, which in turn will cause confusion in the managed process, loss of time, resources and could even completely block the execution of production tasks.

Traditionally, the use of net models does not imply separation of material flows from the flows of data and commands. But such an approach does not permit the explicit description and modelling of the data flow and the changes caused by data manipulation. The manipulation process itself and the actions used to accomplish it are hard to spot. This presents difficulties in implementing measures to eradicate or minimize their consequences. Since, in our case, the net model serves as a basis for the management algorithms being developed, the differentiation of these flows allows a more adequate description of the object and the process of its management, and at the same time provides an opportunity to propose a mechanism for detecting irregularities in the information exchange, including in the control loop.

The main events that may cause problems in the functioning of the operational management system, using vulnerabilities in the information exchange, are the following:

- blocking or errors in data received by the sensors that control the operation of machines and vehicles;
- disturbing the integrity and indexed sequence of the (mobile) network by infiltrating it;
- replacement of data used in the management process with invalid or erroneous data.

It is very important to identify the risks and the potential problems which can be caused by cyber attacks from external or internal (for the system) sources. Without being exhaustive, we consider only these related to the specifics of the processes (production and management) in the automated management system for open production which are modelled.

2 A Generalized Net Model

The present paper illustrates the approach by considering the operating environment for functioning of the automated management system, presented as a simplified GN model (see Fig. 1).

Token ρ_i enters place l_0 with the initial characteristic

"current request for dumper(s) and/or crusher(s)"

for $i = 1, 2, \dots$.

Token Γ stays permanently in place l_4 with the initial and current characteristic

"manager of dumpers and crushers".

Token Λ stays permanently in place l_7 with the initial and current characteristic

"list of currently available dumpers".

Token T stays permanently in place l_{10} with the initial and current characteristic

"list of currently available crushers".

Token Ω stays permanently in place l_{11} with the initial and current characteristic

"current quantity of the material".

Token Σ stays permanently in place l_{21} with the initial and current characteristic

"information system that collects and processes the information for
the dumpers and crushers movement".

This information system collects the information from the sensors and evaluates the behaviour of the dumpers and crushers on the way between their garage and the particular open production site where they work as well as during their work on site. In this manner, an operational cyber-control of the firm's dumpers and crushers is realized. Using Data Mining tools, this information is processed and send to the

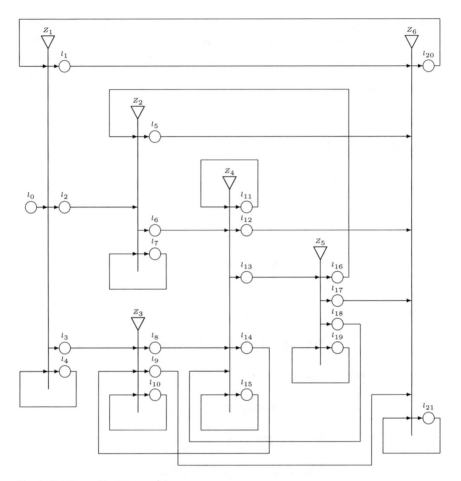

Fig. 1 The Generalized net model

garage manager. It is important to mention that all used Data Mining tools can be described by GNs, as discussed in [4, 5].

The GN-transitions have the following forms.

$$Z_1 = \langle \{l_0, l_4, l_{20}\}, \{l_1, l_2, l_3, l_4\},$$

	l_1	l_2	l_3	l_4
l_0	$false$	$false$	$false$	$true$
l_4	$W_{4,1}$	$W_{4,2}$	$W_{4,3}$	$true$
l_{20}	$false$	$false$	$false$	$true$

$\rangle,$

where
$W_{4,2}$ ="there is a request for dumper(s)",
$W_{4,3}$ ="there is a request for crusher(s)",
$W_{4,1} = W_{4,2} \lor W_{4,3}$.

When token ρ enters the GN through place l_0, on the next time-step it enters place l_4 and merges with token γ that obtains the characteristic

"new request for resources (dumpers or crusher)".

On the next time-steps token γ splits to three or four tokens—the original token γ that continue to stay in place l_4 with the above mentioned characteristic, tokens λ_0 and/or τ_0 and token σ_0. Token λ_0 represents the order for directing the dumpers from the garage to a quarry. Token τ_0 represents the order for directing the crushers from the garage to a quarry. Token σ_0 that represents a signal to the computer system with database gathering information for the dumpers and crushers, the visual feedback obtained via sensors, information regarding the type and weight of the load and the road travelled by the dumper. All these data are processed by different data mining tools in order to check the behaviour of the dumpers and crushers in any given day. These tokens obtain the following characteristics:

"list of dumpers that must go to a quarry"

in place l_2;

"list of crushers that must go to a quarry"

in place l_3;

"list of dumpers and/or crushers that must be observed"

in place l_1.

$$Z_2 = \langle \{l_2, l_7, l_{16}\}, \{l_5, l_6, l_7\},$$

	l_5	l_6	l_7
l_2	$false$	$false$	$true$
l_7	$W_{7,5}$	$W_{7,6}$	$true$
l_{16}	$false$	$false$	$true$

$\rangle$,

where
$W_{7,5} = W_{7,6}$ ="there is a dumper for direction to the construction site and to the quarry".

The λ-token from place l_2 enters place l_7 and merges with token Λ that obtains the characteristic

"order for direction of particular dumpers to the quarry".

On the next time-steps token Λ splits to three tokens—the original token Λ that continue to stay in place l_7 without a new characteristic, σ-token and λ-token. The number of the σ- and λ-tokens is given in the order received by the Λ-token. On the next time-moments, step by step, σ- and λ-tokens will enter places l_5 and l_6, respectively.

The current λ_i-token (for brevity, below we will mark each one of them as a λ-token) enters place l_6 with characteristic

"dumper's parameters (capacity); direction".

The current σ_i-token (for brevity, below we will mark each one of them as a σ-token) enters place l_5 with characteristic

"signal for dumper's images register; plate number; hour, minutes, loaded

material".

$$Z_3 = \langle \{l_3, l_{10}, l_{14}\}, \{l_8, l_9, l_{10}\},$$

$$
\begin{array}{c|ccc}
 & l_8 & l_9 & l_{10} \\
\hline
l_3 & false & false & true \\
l_{10} & W_{10,8} & W_{10,9} & true \\
l_{14} & false & false & true
\end{array}
\rangle,
$$

where
$W_{10,8} = W_{10,9} =$"there is a crusher to be directed to the quarry".

The τ-token from place l_3 enters place l_{10} and merges with token T that obtains the characteristic

"order for directing particular crushers to the quarry".

On the next time-steps token T splits to three tokens—the original token T that continue to stay in place l_{10} without a new characteristic, σ-token and τ-token. The number of the σ- and τ-tokens is given in the order received by the T-token. On the next time-moments, step by step, σ- and τ-tokens will enter places l_8 and l_9, respectively.

The current τ_j-token (for brevity, below we will mark each one of them as a τ-token) enters place l_8 with characteristic

"crusher's parameters (capacity); material; direction".

The current σ-token enters place l_9 with characteristic

"signal for crusher's image, ID number; hour, minutes".

$$Z_4 = \langle \{l_6, l_8, l_{11}, l_{15}, l_{18}\}, \{l_{11}, l_{12}, l_{13}, l_{14}, l_{15}\},$$

	l_{11}	l_{12}	l_{13}	l_{14}	l_{15}
l_6	true	true	false	false	false
l_8	false	true	false	false	true
l_{11}	true	true	true	false	false
l_{15}	false	$W_{15,12}$	false	$W_{15,14}$	$W_{15,15}$
l_{18}	true	true	false	false	false

$\rangle$,

where

$W_{15,12} = W_{15,14} =$ "the crusher has finished its work on site",

$W_{15,15} = \neg W_{15,14}$,

where $\neg P$ is the negation of predicate P.

The λ-token from place l_6 enters place l_{11} and merges with the Ω-token there. On the next time step, the token in place l_{11} splits into three tokens: the original token Ω, a σ-token with additional information for the database and a λ-token, which represents a dumper and its characteristics.

The original token Ω continues to stay in place l_{11} and obtains new characteristic:

"reduced quantity of the material by quantity unit".

In place l_{12} token σ obtains the characteristic

"dumper loaded image".

In place l_{13} token λ obtains the characteristic

"quantity unit".

In place l_{14} token τ obtains the characteristic

"total quantity units loaded/dug".

In place l_{15} token τ obtains the characteristic

"one more quantity unit loaded/dug".

$$Z_5 = \langle \{l_{13}, l_{19}\}, \{l_{16}, l_{17}, l_{18}, l_{19}\},$$

	l_{16}	l_{17}	l_{18}	l_{19}
l_{13}	false	false	false	true
l_{19}	$W_{19,16}$	true	$W_{19,18}$	false

$\rangle$,

where
$W_{19,16} =$"the dumper has finished its work in the quarry",
$W_{19,18} = \neg W_{19,16}$.

In place l_{19} the token does not obtain any characteristic.

When λ-token enters place l_{16} or place l_{18}, an identical copy of it (σ-token) enters l_{17} with a characteristic

"dumper image".

$$Z_6 = \langle \{l_1, l_5, l_9, l_{12}, l_{17}, l_{21}\}, \{l_{20}, l_{21}\},$$

	l_{20}	l_{21}
l_1	$false$	$true$
l_5	$false$	$true$
l_9	$false$	$true \rangle,$
l_{12}	$false$	$true$
l_{17}	$false$	$true$
l_{21}	$W_{21,20}$	$true$

where
$W_{21,20} =$"there is new information for the dumper or crusher available for next action".

All σ-tokens enter place l_{21}, where they unite with the token Σ, extending the information for the dumpers and crushers movements stored in it. This information is sent to the token Γ through place l_{20}, where token ι, generated by token Σ, enters with a characteristic

"information for the daily movement and status of the dumpers and crushers of the firm".

3   Discussion and Conclusion

The example under consideration illustrates the following features of the real management process in an automated management system in terms of information security.

The system data of the utilized vehicles comes from an on-board unit and can be: lost (blocked), damaged or manipulated (replaced) with incorrect ones. Embedded software in the devices eliminates incorrect data (out of range, abnormal measurement deviations, etc.). Lack of data is troublesome, but it does not lead to severe consequences directly in the management:

- Data exchange protocols ignore data that cannot be identified in an indexed sequence;
- The management procedures, implemented in the algorithms of transitions' functioning, ignore results from unacceptable values of the input variables (data).

These ways of protection work well enough in practice, but it is difficult to predict (and analyze) what happens when the manipulated data is still able to pass through these security levels. The proposed model provides a reliable enough tool to detect "breakthroughs" by registering structural-functional discrepancies (abnormalities in the net functioning) when infiltration of false data has occurred.

Acknowledgements The first, third, fifth and sixth author wish to thank for the partial support provided by Grant BG16RFOP002-1.005-0037-C01/25.5.2018 "Development and testing of a prototype of an innovative module for operational control and update of production schedule in open area production in changing production environment in real time" of the Operational Program "Innovation and Competitiveness", 2014–2020, and the second and fourth author for the support provided by the National Science Fund of Bulgaria under Grant DN02/10 "New Instruments for Knowledge Discovery from Data, and their Modelling".

References

1. Alexieva, J., Choy, E., Koycheva, E.: Review and bibliography on generalized nets theory and applications. In: Choy, E., Krawczak, M., Shannon, A., Szmidt, E. (eds.) A Survey of Generalized Nets, pp. 207–301. Raffles KvB Monograph No. 10 (2007)
2. Atanassov, K.: Generalized Nets. World Scientific, Singapore, New Jersey, London (1991)
3. Atanassov, K.: On Generalized Nets Theory, Prof. M. Drinov. Academic Publishing House, Sofia (2007)
4. Atanassov, K.: Generalized nets as a tool for the modelling of data mining processes. In: Sgurev, V., Yager, R., Kacprzyk, J., Jotov, V. (eds.) Innovative Issues in Intelligent Systems. Springer, Cham, pp. 161–215 (2016)
5. Zoteva, D., Krawczak, M.: Generalized nets as a tool for the modelling of data mining processes. A survey. In: Issues in Intuitionistic Fuzzy Sets and Generalized Nets, Vol. 13, pp. 1–60 (2017)

Knowledge Management Model Based Approach to Profiling of Requirements: Case for Information Technologies Security Standards

Sergiy Dotsenko, Oleg Illiashenko, Iegor Budnichenko, and Vyacheslav Kharchenko

Abstract The paper provides analysis of existing knowledge-management models. It justifies the need of integrated model of knowledge management for both industry and academia. It is proposed to build such a model using well-known standards of IT security—Common criteria and methodology for IT security evaluation. The formation of a model of knowledge management is carried out by analyzing the content of the relevant elements of standards and establishing the content of knowledge that determines the forms of relations between them. The architecture of four-factor models is proposed for application towards the formation of knowledge management models in the organization of the information security management system in accordance with the standards of the series ISO/IEC 27000.

Keywords Knowledge management · Information security · Information technologies · Security standards

S. Dotsenko · I. Budnichenko
Department of Specialized Computer Systems, Ukrainian State University of Railway Transport, Feijerbakha Square, 7, Kharkov 61050, Ukraine
e-mail: sirius_3k3@ukr.net

I. Budnichenko
e-mail: y.budnichenko@gmail.com

O. Illiashenko (✉) · V. Kharchenko
Department of Computer Systems, Networks and Cybersecurity, National Aerospace University "KhAI", 17 Chkalova Street, Kharkov 61070, Ukraine
e-mail: o.illiashenko@csn.khai.edu

V. Kharchenko
e-mail: v.kharchenko@csn.khai.edu

1 Introduction

It is essentially to form an integrated security system to bring into life the Industry 4.0 concept in full. The focus of this paper is on security of information technologies. In the field of information technologies security there are two approaches to the formation of security systems.

On the one hand, there are studies that are being carried out to develop business models for information security [1], as well as an integrated model of security awareness to evaluate its risks [2]. Studies of compliance of information security policies are also carried out, specifically, the integration of the theory of planned behavior and the protection motivation theory [3].

On the other hand [1], the relevant international and national standards for security of informational technologies are analyzed and implemented. In particular, provision of enterprise security systems is based on the standards of the ISO/IEC 15408 series, parts 1–3 [4–6], which define the methods and tools for providing and criteria for evaluating the security of information technologies. Together with the standards of ISO/IEC 18045 series, which defines methods and tools for security of information technologies, the security evaluation methodology [7] is used.

During the application process of the abovementioned standards the problem of knowledge management, which is formed in these standards, appears. By that, the process of gaining new knowledge through the establishment of appropriate links between the structural elements of these standards is meant.

However, the development of knowledge management models for these standards, which would have been the basis for the formation of appropriate information security management systems, has not yet been carried out. At the same time, standard ISO/IEC 15408-1 [4] states that there are relevant links among the elements of these standards, but the form of these links is not clearly defined and the method of their establishing is not proposed. Consequently, the task of defining the method of forming such connections arises, for example, in the form of an appropriate *knowledge management model*. Its development will enable the creation of a library (or a set) of standard forms of knowledge management models for appropriate information technology security systems. This is especially relevant when implementing an in-formation security management system in accordance with a series of standards ISO/IEC 27000 [8].

The goal of this paper is in the justification of the choice of the method of knowledge management model development for well-known set of information security standards ISO/IEC 15408 and standard ISO/IEC 18045. The structure of the paper is the following. Section 2 analyses the existed models of knowledge management and discusses possibilities of their application for development of integrated model. Section 3 provides theoretical foundation four-factor logical knowledge model of activity Sect. 4 describes the suggested integrated model of knowledge and algorithms of its development and application for the standards ISO/IEC 15408 and ISO/IEC 18045. Section 5 concludes the paper, discusses the scientific and implementation results as well as future research directions. The paper contains acknowledgments for support made by Horizon 2020 project ECHO and by colleagues of the authors of the paper.

2 Analysis of Models of Knowledge Management

2.1 Knowledge Management Perspectives

The overall understanding of the relationship of organizational creativity to the enablers of knowledge management, including culture, structure, people and information technology, through examining the role of knowledge management processes is given in [9]. An analysis of some well-known models in knowledge management performed in:

- Lawson's model (2003) for identifying one of the knowledge management tools [10];
- Organizational culture, Lee and Choi model (2003) for measuring other factors [11].

Several studies have also been carried out to measure the knowledge management process as *an indirect factor* [12].

The study identifies two perspectives on the knowledge management process: social and technical. For the social perspective and the organizational culture the structure and people are distinguished. For the technical perspective, the information technologies are distinguished. On the basis of expert assessments, interrelations between structure, culture, people and information technologies are established through knowledge management and organizational creativity.

2.2 Business Model

The business model for information security is presented in [13] (see Fig. 1). It contains four components:

- organization (design and strategy);
- people;
- technology;
- process.

For these components, the forms of connections are established as in the previous work. However, the task of forming a *security system* in the organization is not raised in it.

And so it could be truly said that there is a certain contradiction obviously introduced be the developers of the studied regulations. On the one hand, standards

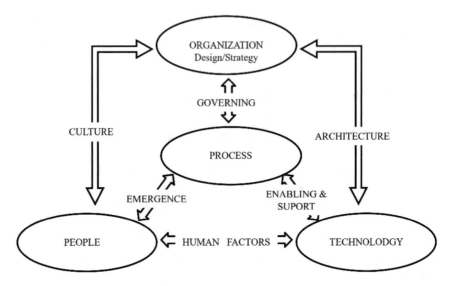

Fig. 1 The business model for information security

have been developed that will determine the requirements for security systems, but they do not have the theoretical justification. And on the other hand, there are theoretical works that establish appropriate recommendations for integrated knowledge management models for information technology security systems, but these recommendations do not have sufficient practical implementation.

It should also be noted that the concept of a culture of security is included as an important component of the business model for information security (see Table 1 [13]).

This table represents the content of the four elements of the *business model for information security*, which was investigated in [13], before the implementation of the security culture concept in this model and after its implementation.

2.3 Conceptual Model

The paper [14] proposed an improved conceptual model of knowledge management process (see Fig. 2).

The specific recommendations for the formation of the composition and content of elements of knowledge management models are not included in the considered models of knowledge management. This problem can be solved developing an appropriate model of knowledge management as an ex-pert system. However, its creation is connected to the necessity of pre-designing the database, output machine and other elements of the expert system.

Table 1 Shifting from functional to deliberate security culture

From	To
Technology	
• Uncertainty about the level of security the technology provides • Seeing security-related technology as disruptive and cumbersome to use	• Technology used is based on an assessment of the risk • Seeing new security technology as a means to enhance the sales process
Process	
• Security brought in when there is a suspected breach • Security is maintained by an expert knowledge	• Security involvement in the earliest planning phases of campaigns • Security shares its knowledge and expertise, developing broader security awareness across the enterprise
People	
• Security as an entity that enforces compliance • Security as a functional expert	• Security "as a partner" that creates awareness and commitment • Security "as a partner" that transfers security knowledge and expertise to its sales customers
Enterprise	
• Limited visibility or awareness of security issues • Security structure focused on technical expertise	• Receiving regular updates about potential risk • Security structure supports processes of its customers

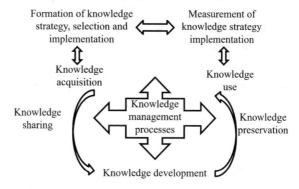

Fig. 2 Improved conceptual model of knowledge management process

Such systems, as a rule, are unique in design and require significant material, human and financial resources.

Therefore, the problem arises to find other methods of forming a knowledge management model. Requirements for such a model are as follows:

- knowledge management should be simple;
- the knowledge management model should have an open architecture;
- the model must be universal, without dependency of the content of knowledge.

2.4 DMT-Based Model

According to [15] for the solution of this problem, the didactic multidimensional technology is actively developing. It is based on visual didactic multidimensional instruments.

The concept of visual didactic multidimensional tools (DMT) consists in transforming the verbal, textual or other form of information representation into a visual, figurative-conceptual form, which is characterized by three parameters: *semantic* (meaningful), *logical* and *special graphic*.

The *multidimensionality* of the topic displayed by the tool is provided by three components (see Fig. 3 [15]), where $K1–K8$—Coordinates—direct measurements of the topic being studied:

- logical-semantic modelling;
- cognitive presentation of knowledge;
- radial-circular organization.

The content of the specified coordinates is determined for a specific subject area at the stage of development. Therefore, there is no universal definition of the content of these coordinates. When implementing DMT information is converted using the following principles:

- the principle of system-multidimensionality in the selection and consolidation of content;
- the principle of splitting and merging and the related principle of additionality in the formation and using of DMT;
- the principle of trinity in the formation of semantic groups that enhance psychological stability.

Fig. 3 Trinity basis of DMT

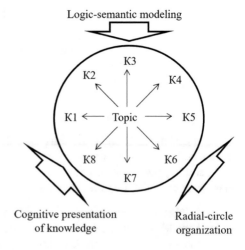

This concept does not have an unambiguous theoretical justification for each of these bases of multidimensionality of the subject being studied. In these models eight coordinate models of knowledge management are used as a rule.

2.5 Model of Strategical Thinking

In [16] fifty models of strategic thinking are presented. It should be noted that among them there are ten models that have a four-vector architecture, which is similar to the Cartesian coordinate system architecture. Figure 4 shows an example of such an architecture. Model analysis shows that the four-vector graphical representation of knowledge can be the basis for a knowledge management model for a particular subject area, since there are relations between pairs of factors that are characterized by a certain content of knowledge about these relationships, specifically: *Me—Thought*; *Me—Actions*; *They—Thought*; *They—Actions*. On this basis, a knowledge base about the subject area, which is characterized by these factors, is formed.

From the analysis of the methods of forming knowledge management models based on the concepts of a business model [13], a conceptual model [14], didactic multidimensional tools [15] and models of strategic thinking [16] it follows that these concepts correspond to specific subject areas and do not have a clear theoretical justification. At the same time, a special attention should be paid to the last two models. In these models the relations are established in an explicit form between the elements of adjacent factors.

To quantify the content of these relationships as a metric, it is recommended to use the metric in the form of the ratio of the number of non-zero elements for Cartesian products of each factor pair of adjacent model vectors (see Figs. 3 and 4) to the total number of elements for all pairs of adjacent vectors.

Revealing the content of these relationships provides the establishment of the content of knowledge that characterizes these relationships. However, in [15, 16] the recommendations on the formation of the appropriate knowledge base are not being given. This impedes the practical application of these models. Therefore, the task of developing proposals for the practical implementation of these models for the standards of ISO/IEC 15408, ISO/IEC 18045 is actual.

Fig. 4 Version of model of strategic thinking

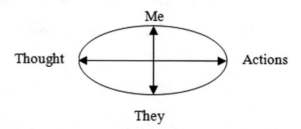

As shown in Sect. 2, four-factor logical models of activity knowledge representation are applied to different subject areas, but none of these applications offers a theoretical justification for these logical models. Therefore, the following solution to this problem is proposed.

3 Theoretical Foundation Four-Factor Logical Knowledge Model of Activity

3.1 Prerequisites

In fact that the common terms to describe organizational activities are "processes" and "resources". As well as term "factor" in economics. The correlations between these terms are defined in [17].

The following definitions of relations between "resource" and "factor" in terms of Hegel's dialectics [17]:

- resource factors of organizational activities (RFOA)—(common);
- resource factors of technological activities (RFTA)—(single).

So comparing those terms: *"factor"* is *an action* and *"resource"* is *a usage*. Thus, the expression *"resource factors"* literally means as *use in action*. And the following expressions have clear univocal meaning:

- resource factors of organizational activities—usage when organizational activity is performed;
- resource factors of technological activities—usage when technological activity is performed.

The relations between terms *"process"* and *"factor"* are defined in terms of Hegel's dialectics [17] as well:

- process factors of organizational activities (PFOA)—*common*;
- process factors of technological activities (PFTA)—*single*.

Thereby comparing the following terms: *"factor"* as *an action* and *"process"* as *an implementation*, the expression *"process factors"* literally means *"implementation in action"*.

And the following expressions have clear univocal meaning:

- process factors of organizational activities—*implementation during organizational activity*;
- process factors of technological activities—*implementation during technological activity*.

To summarize the results of investigated terms *"process"*, *"resource"* and *"factor"* it is important to provide mathematical relations of the established patterns.

To form the mathematical model, it is proposed to use sign "$\lhd$" which is absent in set theory as a sign of dialectical relations in category "general" and "partial".

In that case *the dialectical unity of resource factors* of organizational and technological activity possible to represent as follows:

$$RFTA \lhd RFOA. \tag{1}$$

And *the dialectical unity of process factors* of organizational and technological activity possible to represent as follows:

$$PFTA \lhd PFOA. \tag{2}$$

Because of both process and resource factors are implemented simultaneously it should be linked accordingly. It is clear that implementation of any activity must apply process factors as well as resource factors. In this case it has univocal representation on the plane and could be described as [17] Fig. 5.

On this model of knowledge for factors of activity possible to describe causal relations between process and resource factors in form of Cartesian multiplication of factors set. As follows from [17]:

$$P_I \subseteq PFOA \times RFTA \tag{3}$$

$$P_{II} \subseteq PFOA \times RFOA \tag{4}$$

$$P_{III} \subseteq PFTA \times RFOA \tag{5}$$

$$P_{IV} \subseteq PFTA \times RFTA. \tag{6}$$

Based on this logical model representation of knowledge of activity it is possible to define the mathematical model on the basis of formal theory of mathematical description of models.

For this purpose, it is proposed to interpret the elements of the model (factors) as alphabet B_a, and dialectics relations as "*common*" $\lhd$ "*single*" together with form of Cartesian multiplication ($\times$) consider as elements of final set of relations R_a. Mathematical model of such architecture could be described as follows [18]

$$S_a = \langle B_a, R_a \rangle. \tag{7}$$

The value of elements in a set of *alphabet* B_a as a form of factors determined according to Fig. 5.

So on the one hand there are four factors defined above which describe the *process of activity* to get valuable results. On the other hand there are factors that describe

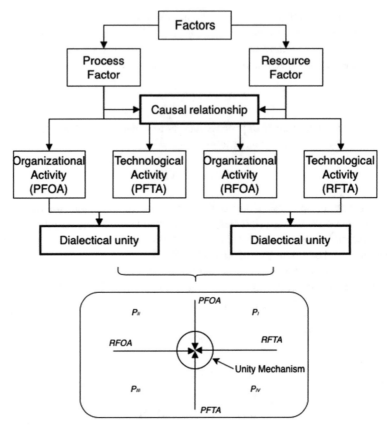

Fig. 5 Four-factor logical model of knowledge of activity representation

brain activity about generating the project of such activity. And the question is how those factors are related to each other?

The next hypothesis follows from the foregoing: since a holistic dialectically organized technological activity of the natural intellectual system is implemented based on mentioned form factors then the human brain should generate *the project of future activity results* for the particular moment of timeline based on processing these specific factors [17].

In [19] it was described that according to P. K. Anokhin the activity of the functional system <intellectual system> is always purposeful. And the final purpose is generated based on the form of future project results which became a standard. And all the acceptors of this results compare their parameters with this standard.

On the other hand, the proposition is to define exactly the model of this project of future results as *the model of knowledge* of the subject activities in subject area. It is always generated (predicted) for a certain point in the future and for certain results of activity.

This raises the question of how the factors defined above correlate with factors that are simultaneously processed according to central regularity of the integrative brain activity, namely: currently dominant motivation situational afferentation, triggering afferentation and memory according to P. K. Anokhin?

In [20] the models of knowledge of subject activities in subject area based on central regularity of integrative brain activity was developed. For the existing natural intellectual system all these three forms of measuring knowledge (motivation, situational and triggering afferentation) come from the environment in real time. And just the fourth dimension of knowledge—the memory of past experience, is stored in the brain permanently. Therefore, it is correct to formalize the problem of modelling knowledge of subject activities in subject area in the form of a project of the future result for a certain stage.

3.2 Four-Factor Model

From production system point of view the *situational afferentation* is a knowledge about *external* conditions of activity implementation. First of all, it is regulatory documents, licensing agreement of activity implementation. This category of knowledge is related to process factors of organizational activity (M_o, Fig. 6) [17].

For the production system the *dominant motivation* contains of external factors such as: consumers needs which generate internal factors of implementation these

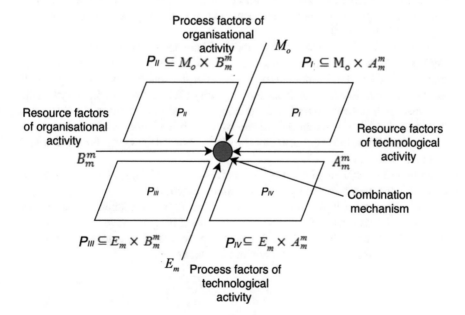

Fig. 6 Four-factor logical model of knowledge of energy saving activity representation

needs in activity mission form of production system, policy, strategy, goals, objectives and activity indicators. This category is related to resource factors of organizational activity ($B_m{}^m$, Fig. 6) [17].

As for triggering afferentation it is possible to present it as a knowledge of existing resources for activity implementation ($A_m{}^m$, Fig. 6). Upon information receiving and specific resources are available the action of synthesis mechanism of the purpose of activity as a project of future results is implemented. The production resources are material, financial, non-material, human, informational, intellectual etc. This category of knowledge is related to resource factors of technological activity [17].

The last question is what kind of knowledge is necessary to obtain from the memory to implement the generation process of the purpose of activity as a project of future results? Obviously, the knowledge about technological process (set of processes) is needed. So that process could be implemented to achieve the purpose of the activity using the resources we have in external conditions at this moment. This category of knowledge is related to process factors of technological activity E_m [17].

Thus, the forms of measuring knowledge defined by P. K. Anokhin for one neuron as well as for set of neurons have a concrete meaning for production activities. It is another question as well. How to implement the process of synthesizing the project of future results having these parameters.

For this we need to find out possible mechanisms to combine these parameters of knowledge.

It is also defined that the mechanism of *dialectical unity* is primary related to mechanism of causal relationship. Based on this patterns next logical model of knowledge of process-resource representation of thinking process *organization* is suggested (Fig. 7) [17].

This way, the four-factor logical model of knowledge representation about the thinking process based on the central regularity of the integrative activity of the brain proposed by P. K. Anokhin, on the basis of the simultaneous convergence of excitations <motivation, situational and triggering afferentation and memory> on the same neuron is basically based on the principles of dialectical unity of the categories "general" and "single" for each dimension of knowledge "process" and "resources". At the same time the causal relationship is established between the process and resource factors [17].

In Table 2 the content of described factors is defined and their mapping with forms of knowledge measurement by P. K. Anokhin is determined.

From the above it follows that between the formally established factors of processes and resources (Fig. 6) and factors ensuring the generation of the project of future activity results by P. K. Anokhin based on the central regularity of integrative brain activity (Fig. 7) the univocal relation is established.

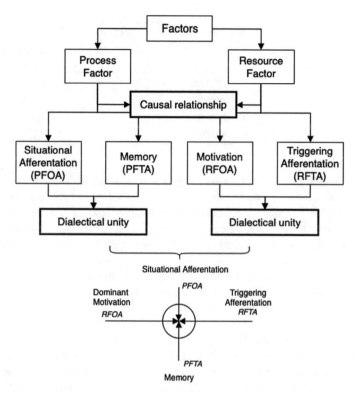

Fig. 7 Four-factor logical model of knowledge of thinking activity representation

Table 2 Determination the meaning of input factors that implement the process

Forms of input factors that implement the process	Contents of factors forms	
	For one neuron by P. K. Anokhin	For organization
PFOA (common)	Situational afferentation	Legislation, rules
PFTA (single)	Memory	Activity experience
RFOA (common)	Dominant motivation	Consumers needs
RFTA (single)	Triggering afferentation	Resources availability

3.3 Models Analysis

For further mathematical analysis of logical models development, it is proposed to apply formal theory by [21, p. 68]:

"The basis of logical models is the concept of formal theory which is given by the next four:

$$S = \langle B, F, A, R \rangle. \tag{8}$$

Table 3 The results of relations between factors of logical model of activities and interpretation of content of formal theory elements

Elements of formal theory (formula 8)	Logical models of	
	Knowledge (5.8)	Factor process representation
B—alphabet	B	PFOA
F—formula	F	PFTA
A—axiom	A	RFOA
R—relation	R	RFTA

where B—the set of *basic symbols (alphabet) of theory S*. The key point of the basic symbols called *the expressions of theory S. F*—subset of expressions of theory S called theory's formulas…. *A*—set of formulas called *axioms of theory S* i.e. set of a priori true formulas. *R—final set of relations* $\{r_1, …, r_n\}$ between formulas called the output rules. The most common formal system used to represent knowledge is predicate calculation."

The next question is how the elements of the formal theory S (formula 10) and the factors of the logical model for activity representation are related (Fig. 5). Alphabet of theory *B* can be *interpreted* as a process factor of organizational activity, since the process of its formation is a process of meaningful activity (process of thinking). It can be defined as a "general" concept. Single to it is the set of formulas *F* of theory *S*. **It is a process factor of technological activity because the formulas provide a dialectical connection between the axioms of theory *A* and the relations *R*.** At the same time axioms *A* can be *interpreted* as resource factors of organizational activity, and the relation *R* as the resource factors of technological activity for the formulas. Table 3 shows the results of this interpretation.

Based on the results of the interpretation of the content of the alphabet elements (see Table 3) the corresponding mathematical model has the following form for the factor representation of the process:

$$S_M =< \text{PFOA, PFTA, RFOA, RFTA} > \tag{9}$$

The architectures of the reviewed models (Figs. 5 and 7) have the same structure with the forms of relations established for the respective factors. Therefore, the conclusion could be done that the provisions of formal theory can be applied to the mathematical description of the architecture of these models.

For this purpose it is sufficient to interpret their elements (factors) as the alphabet B_a, and to consider the dialectical relations in the form of *"common"* $\lhd$ *"single"* and in the form of Cartesian multiplication ($\times$) as elements of the final set of relations R_a (see formula 7). The content of the alphabet B_a set is defined according to Table 3.

For the mathematical model based on formal theory it is introduced the concept of "signature" in [18]: the signature of a model is a set of names of relations in this model and the capacity of the corresponding relation must be specified.

The signature R_a in formula (7) is a double relation of $\lhd$ and $\times$.

The concept of "*isomorphism*" is very important for the model of formal theory. In [18] the definition of this concept is mentioned as it is possible only for models with the same signature. Since the model signatures R_a are the same so the isomorphism of these architectures is possible.

From the mentioned above follows that it is possible to establish an isomorphic architecture with clear definition of the relationships between factors in the form of dialectical and causal (functional) relations for four-factor logic models of knowledge of activity.

Based on that proposed to develop the knowledge management model for information security management system using ISO/IEC 15408, ISO/IEC 18045 standards.

4 The Model of Knowledge Management for Standards ISO/IEC 15408 and ISO/IEC 18045

4.1 Development of the Model

We propose the usage of four-factor knowledge structuring models to establish relationships between the elements of the standards of the ISO/IEC 15408 series and the ISO/IEC 18045 standard.

The Microsoft Office Excel spreadsheet was use as a simple tool for the demonstration of practical implementation of the proposed method of knowledge structuring. The advantage of this approach is that each cell in the Excel table may contain data in different forms: numeric, text, formulas for calculations, hyperlinks to other cells of the table, other pages of the corresponding book, hyperlinks to files of different formats, and most importantly to folders with sets of different documents and folders.

These spreadsheet properties allow you to place data, information, and knowledge in relevant cells that are relevant to certain element relationships.

Preliminary analysis showed that it is possible to establish special links between elements of these standards as shown in Fig. 8.

For the knowledge management model of the generated matrices, the special metric is proposed to be calculated in the next way. The total power of the Cartesian product for adjacent pairs of vectors of factors is determined:

$$|E^{(18045)} \times E^{(15408-1)}|; \quad |E^{(15804-1)} \times E^{(15408-2)}|; \tag{10}$$

$$|E^{(15804-2)} \times E^{(15408-3)}|; \quad |E^{(15804-3)} \times E^{(18045)}|. \tag{11}$$

After that, the Cartesian product capacities for the adjacent pairs of factor vectors with non-zero elements are calculated;

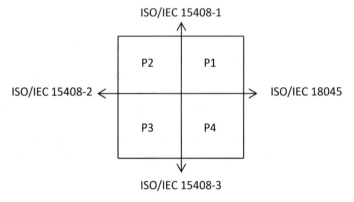

Fig. 8 Knowledge management model

$$|\Delta(E^{(18045)} \times E^{(15408-1)})|; \quad |\Delta(E^{(15804-1)} \times E^{(15408-2)})|; \tag{12}$$

$$|\Delta(E^{(15804-2)} \times E^{(15408-3)})|; \quad |\Delta(E^{(15804-3)} \times E^{(18045)})|. \tag{13}$$

Based on the results of these calculations, the percentage of Cartesian output with non-zero elements relative to the total Cartesian output is calculated.

$$M_E = \frac{|\bigcup_{i,j} \Delta E^{(i),(j)}|}{\bigcup_{i,j} (E^{(i)} \times E^{(j)})}. \tag{14}$$

Figure 9 shows a screenshot of the automation tool developed as a Microsoft Excel spreadsheet. Figure 10 shows a screenshot of the Microsoft Office Excel spreadsheet (quadrant P4 in Fig. 8).

The proposed knowledge management model *is an integrated one* because it includes different sources of knowledge about the subject area (content of standards). Establishing additional relationships between these sources provides for the formation *of new knowledge* about the chosen subject area and, in fact, it is *the result of knowledge management.*

The contents of all abbreviations given in Figs. 9 and 10 is defined in the relevant standards, which are indicated for the respective axes, namely:

- ISO/IEC 15408-1:2009 Information technology—Security techniques—Evaluation criteria for IT security—Part 1: Introduction and general model (the acronyms are indicated in the Fig. 6);
- ISO/IEC 15408-2:2008 Information technology—Security techniques—Evaluation criteria for IT security—Part 2. Security functional components (*Classes of functional requirements*: FAU—Security audit, FCO—Communication, FCS—Cryptographic support, FDP—User data protection, FIA—Identification and authentication, FMT—Security management, FPR—Privacy, FPT—Protection of

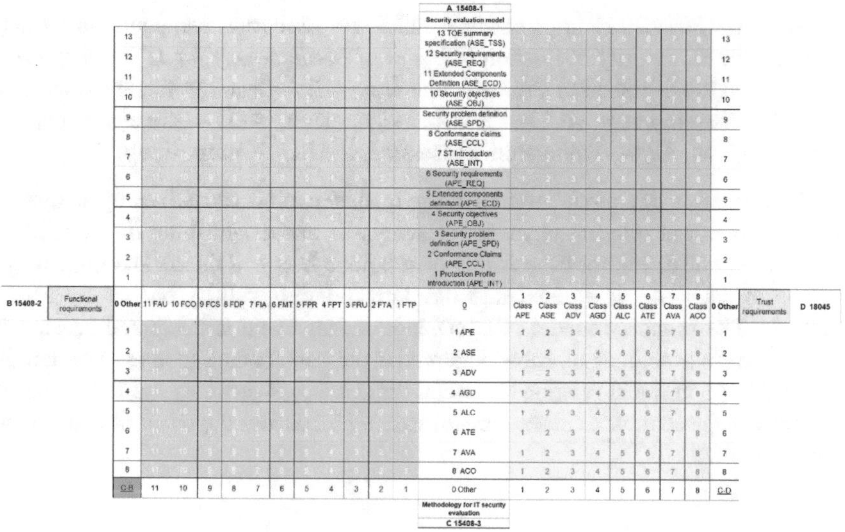

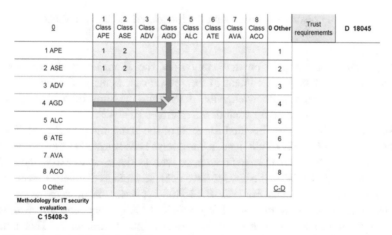

Fig. 9 Screenshot of the automation tool spreadsheet

Fig. 10 Screenshot of the fragment of automation tool spreadsheet (quadrant P4 in Fig. 5)

the TSF, FRU—Resource utilization, FTA—TOE (Target of Evaluation) access, FTP—Trusted path/channels);

- ISO/IEC 15408-3:2008 Information technology—Security techniques—Evaluation criteria for IT security—Part 3: Security assurance components (*Classes of trust requirements*: APE—Protection Profile evaluation, ASE—Security Target evaluation, ADV—Development, AGD—Guidance documents, ALC—Lifecycle support, ATE—Tests, AVA—Vulnerability assessment, ACO—Composition);

- ISO/IEC 18045:2008 Information technology—Security techniques—Methodology for IT security evaluation (*Classes of Methodology for IT security evaluation*: APE—Protection Profile evaluation, ASE—Security Target evaluation, ADV—Development, AGD—Guidance documents, ALC—Life-cycle support, ATE—Tests, AVA—Vulnerability assessment, ACO—Composition).

The formation of a model of knowledge management is carried out by analyzing the content of the relevant elements of standards and establishing the content of knowledge that determines the forms of relations between them. The knowledge formed in such way is entered into the corresponding cell by one of the above methods. The proposed model of knowledge management is integrated because it includes various sources of knowledge about the subject area (content standards). Establishing additional relations between these sources provides the formation of *new knowledge* about the chosen subject area, which is also the result of knowledge management.

4.2 Application of the Model

The sequence of constructing and using the integrated model is as follows.

The First Stage
The formation of composition and content of the elements of the relations is performed by analysing the content of the relevant elements of the standards and establishing the content of knowledge that determines the forms of relations between them with a non-zero value for each of the matrices P1–P4.

Structural elements are allocated for each of the standards (see Fig. 5). For the standard like an ISO/IEC 15408-1, such items have been selected as the composition and contents of the *security specification*, as well as the *security profile specification*. For ISO/IEC 15408-2, the classes have been selected as elements that describe *functional security* elements. For ISO/IEC 15408-3, classes that describe the *security assurance* are selected as *elements*. For ISO/IEC 18045, *subsystems for evaluation the security* of information technology have been selected as elements. This method of integration is based on the functional representation of the enterprise. It involves the integration of enterprise management systems and production process management systems, that means, the integration of the two control systems of parts of the enterprise into a whole one.

The correlation of the elements of the respective pairs is performed in the following sequence:

- ISO/IEC 15408-1–ISO/IEC 15408-2;
- ISO/IEC 15408-2–ISO/IEC 15408-3;
- ISO/IEC 15408-3–ISO/IEC 18045;
- ISO/IEC 15408-1–ISO/IEC 18045.

In case where the presence of relations is established for the corresponding pair of elements, an appropriate *content of knowledge* about these relations is formed. A document containing the content of this knowledge should be associated with the corresponding cell through a hyperlink. To access the contents of this document it is enough to go over the hyperlink of the corresponding cell. Thus, the formed matrices are reference. For each of the P1–P4 matrices, the corresponding M_{E1e}–M_{E4e} metrics, which are subsequently used as benchmarks, are calculated.

After the formation of knowledge for all relations is done, application of the formed model of knowledge management is possible.

The Second Stage

The information security system formation at the enterprise is carried out by forming a description of the object of assessment and forming a security profile for specific information technologies that are planned for use. This requires knowledge of the requirements that are formed in the elements of the planes P2 and P3 (see Fig. 5). For each of the matrices P2 and P3, the corresponding metrics M_{E2} and M_{E3} are calculated.

The Third Stage

The knowledge, which is formed in the documents, the messages to which are contained in the cells of planes P1 and P4, is used to carry out the assessment of *information technology security* according to ISO/IEC 18045.

At the same time, the composition and content of the actual measures that are being implemented are checked. The knowledge thus formed is entered into the corresponding cell by one of the above methods. The contents of the cells are then compared with non-zero values. The degree of compliance of the implemented measures with the requirements set out in the standard is established. Full compliance is rated "one". The discrepancy is rated zero. Intermediary estimates are possible. After that, real metrics M_{E1} and M_{E4} are calculated for the P1 and P4 matrices. The comparison with the benchmark of the relevant metrics establishes the degree of compliance of the actual security measures for the particular assessment entity with the ideal requirements.

For current information technology, the total value of M_E metrics is calculated, which is further used to assess the current level of information technology security.

The Knowledge Management Model, which is shown in Fig. 8 also corresponds to the four-factor logical model of the representation of knowledge of activity according to Fig. 5. This follows from the fact that dialectical relations can be established between the factors of the knowledge management model (see Fig. 8).

Compared with known methods of requirements profiling, in particular, requirements for functional security based on facet-hierarchical structures, their modifications using semantic codes, as well as the analysis of the evolution of quality models [22–24], the proposed approach is more strategic because it allows better determine the essential components of the requirements for profiling.

4.3   Factorial Representation of the Components of ISO/IEC 15408 and ISO/IEC 18045

For ISO/IEC 15408-1 standard, factors are the *composition and content* of the security target and the specification of the protection profile.

On the other hand, for ISO/IEC 15408-3, factors are selected that describe the classes of trust in security. These two forms of factors are proposed to be considered as resource factors. They are resource factors that provide for the implementation of activities (processes) for the formation of functional security elements (ISO/IEC 15408-2) and security assessment activities (ISO/IEC 15408-2).

It is obvious that security assurance classes are common to the security task specification as well as protection profile specifications that are specific to the security assurance classes. Accordingly, they can be identified as:

- security assurance classes are resource factors of organizational activity (RFOA)—general;
- specifications of the security target as well as the specification of the protection profile are resource factors of technological activity (RFTA)—specific.

For standard ISO/IEC 15408-2 factors are classes that describe the elements of functional security. For standard ISO/IEC 18045, factors are the subtypes of information technology security assessment activities. These two forms of factors are proposed to be considered as *process factors*. They are process factors that implement the activities for the formation of functional security elements (ISO/IEC 15408-2) and security assessment activities (ISO/IEC 15408-2).

It is clear that functional safety elements are general in relation to the subtypes of security assessment activities that are specific to functional safety elements.

Accordingly, they can be identified as:

- elements of functional safety are Process Factors of Organizational Activity (PFOA) (i.e. general);
- sub-types of security assessment activities are Process Factors of Technological Activity (PFTA) (i.e. specific).

From the above, the following dialectical relationships follow for the considered factors of the knowledge management model according to Fig. 8:

- ISO/IEC 15408-1 (RFOA) $\lhd$ ISO/IEC 15408-3 (RFTA);
- ISO/IEC 15408-2 (PFOA) $\lhd$ ISO/IEC 18045 (PFTA).

Thus, the investigated four factor models of structuring knowledge about activity have theoretical ground which is based on the establishment of forms of process and resource forms of factors that shape knowledge about activity, and forms of relations between these factors.

5 Conclusions

This paper represents extended version of previously published [25]. It substantiates the choice of the method of Knowledge Management Model formation for ISO/IEC 15408 and ISO/IEC 18045:2008 standards. It is proposed to choose four-factor model of knowledge management. It meets the requirement of open architecture, is accessible to the user, as well as universal in relation to the subject area.

The developed model can be applied both for the analysis of existing knowledge for the chosen subject domain and for the synthesis of new knowledge. In the latter case, it is sufficient to form the composition and content of the elements of coordinate meshes.

The architecture of *four-factor models* is proposed to apply also for the formation of knowledge management models in the organization of the information security management system in accordance with the standards of the series ISO/IEC 27000.

It is found that the main mechanisms of combining certain factors are the mechanism of dialectical unity of the concepts of "general" ⊲ "single", as well as the mechanism of cause and effect relations.

It is also found that the mechanism of dialectical unity is a primary one in relation to the mechanism of cause and effect relations. Thus, the architecture of the information model of the process (see Fig. 5) corresponds to the architecture of the information model of knowledge based on the central pattern of integrative brain activity (see Fig. 7).

It is hypothesized that since the activity of the natural intellectual system is realized using the above-defined forms of factors, then the human brain must form a *project of the future result of activity* for the appropriate moment of time on the basis of processing these factors [17].

The method of solving non-formalized problems used in this paper refers to a method that is based on the introduction of a hierarchy of spaces (precise, abstract, and meta-spaces) [21].

The simplest of these methods is based on the ability to factorize the space of solutions. Establishing the composition and content of four factors that determine knowledge about technological and semantic activity in models (see Figs. 5 and 7) provided a theoretical justification for the isomorphism of these models.

The developed conception and models have been adopted and implemented at the PC "RPC Radiy" in Kropyvnytskyi, Ukraine and PrJSC FED in Kharkiv, Ukraine.

The task of forming a reference model is planned for further research. The future steps can be dedicated to development application of the described model considering integrated safety and security management system [26], some parts of which were discussed in studies for critical application systems [27] and in international educational projects [28].

Acknowledgements This work was supported by the ECHO project which has received funding from the European Union's Horizon 2020 research and innovation programme under the grant agreement no 830943.

The authors very appreciated to scientific society of consortium and in particular the staff of Department of Computer Systems, Networks and Cybersecurity of National aerospace university "Kharkiv Aviation Institute" for invaluable inspiration, hardworking and creative analysis during the preparation of this paper.

References

1. An Introduction to the Business Model for information Security. https://www.isaca.org/Knowledge-Center/Research/Documents/Introduction-to-the-Business-Model-for-Information-Security_res_Eng_0109.pdf. Access date: Dec 2019.
2. Mejias, R.: An integrative model of information security awareness for assessing information systems security risk. In: Proceedings of the Annual Hawaii International Conference on System Sciences, pp. 3258–3267 (2012)
3. Princely, I.: Understanding information systems security policy compliance: an integration of the theory of planned behavior and the protection motivation theory. Comput. Secur. **31**(1), 83–95 (2012)
4. ISO/IEC 15408-1:2009: Informational technology—security techniques—evaluation criteria for IT security. Part 1: Introduction and General Model (2009)
5. ISO/IEC 15408-2:2008: Information technology—security techniques—evaluation criteria for IT security. Part 2: Security Functional Components (2008)
6. ISO/IEC 15408-3:2008: Informational technology—security techniques—evaluation criteria for IT security. Part 3: Security Assurance Requirement (2008)
7. ISO/IEC 18045:2008: Information technology—security techniques—methodology for IT security evaluation (2008)
8. ISO/IEC 27000:2018: Information technology—security techniques—information security management systems—overview and vocabulary (2018)
9. Alkaffaf, M., Muflih, M., Al-Dalahmeh, M.: An integrated model of knowledge management enablers and organizational creativity: the mediating role of knowledge management processes in social security corporation. Jordan J. Theor. Appl. Inf. Technol. **96**(3), 677–700 (2018)
10. Lawson, S.: Examining the relationship between organizational culture and knowledge management. Doctoral dissertation, Nova Southeastern University (2003). Retrieved from Nova Southeastern University dissertation database. UMI No. 3100959. Access date: Dec 2019
11. Lee, H.: Choi: knowledge management enablers, process, and organizational performance: an integrative view and empirical examination. J. Manage. Inf. Syst. **20**(1), 179–228 (2003)
12. Shannak, R.O.: Measuring knowledge management performance. Eur. J. Sci. Res. **35**(2), 242–253 (2009)
13. An Introduction to the Business Model for information Security. Printed in the United States of America. https://www.isaca.org/Knowledge-Center/Research/Documents/Introduction-to-the-Business-Model-for-Information-Security_res_Eng_0109.pdf. Access date: Dec 2019
14. Raudeliūnienė, J., Davidavičienė, V., Jakubavičius, A.: Knowledge management process model. Entrepreneurship Sustain. Issues **5**(3), 542–554 (2018). https://doi.org/10.9770/jesi.2018.5.3(10). Access date: Dec 2019
15. Steinberg, V.E.: Theory and Practice of Multi-dimensional Teaching Technology. National Education, Moscow (2015)
16. Krogerus, M., Tscheppeler, R.: 50 Erfolgsmodelle. Kleiner Handbuch für strategische Entschheidungen, Kein &Aber, AG Zürich, 200 p (2008)
17. Dotsenko, S.I.: Theoretical foundations for development of intelligent computer support systems for managing energy saving organizations. Dissertation for doctor of sciences degree 05.13.06, Kharkiv Petro Vasylenko National Technical University of Agriculture, Kharkiv, 369 p (2017)

18. Shreider, Y.A., Sharov, A.A.: Systems and models, radio and communication. In: Cybernetics, 152 p (1982)
19. Dotsenko, S.I.: On determining the content of categories of semantic thinking. Energy Comput. Integr. Technol. Agroind. Complex **1**(4), 23–27 (2016)
20. Dotsenko, S.I.: Modeling domain knowledge based on the central pattern of integrative brain activity. Technol. Audit Prod. Reserves **2/2**(28), 33–41 (2016)
21. Popov, E.V.: Expert systems: solving informal tasks in a dialogue with a computer, Moscow. Science 288 p (1987)
22. Kharchenko, V., Gordieiev, O., Fedoseeva, A.: Profiling of software requirements for the pharmaceutical enterprise manufacturing execution system 2016. In: Applications of Computational Intelligence in Biomedical Technology, pp. 67–92. Springer, Cham (2016)
23. Gordieiev, O., Kharchenko, V., Vereshchak, K.: Usable security versus secure usability: an assessment of attributes interaction. In: Proceedings of International Conference ICT in Education, Research, and Industrial Applications, ICTERI 2017, pp. 727–740 (2017)
24. Gordieiev, O., Kharchenko V., Fusani, M.: Evolution of software quality models: green and reliability issues. In: Proceedings of International Conference ICT in Education, Research, and Industrial Applications, ICTERI 2015, pp. 432–445 (2015)
25. Kharchenko, V., Dotsenko, S., Illiashenko O., Kamenskyi, S.: Integrated cyber safety and security management system: industry 4.0 issue. In: Proceedings of the 10th IEEE Dependable Systems, Services and Technologies Conference, DESSERT 2019, pp. 197–201 (2019)
26. Dotsenko, S., Illiashenko, O., Kamenskyi, S., Kharchenko, V.: Integrated model of knowledge management for security of information technologies: standards ISO/IEC 15408 and ISO/IEC 18045. Inf. Secur. Int. J. **43**(3), 305–317 (2019)
27. Kharchenko, V., Illiashenko, O., Brezhnev, E., Boyarchuk, A., Golovanevskiy, V.: Security informed safety assessment of industrial FPGA-based systems. In: Proceedings of the Probabilistic Safety Assessment and Management Conference, PSAM 2014 (2014)
28. Kharchenko, V., Illiashenko, O., Boyarchuk, A., Sklyar, V., Phillips, C.: Emerging curriculum for industry and human applications in internet of things. In: Proceedings of the 2017 IEEE 9th International Conference on Intelligent Data Acquisition and Advanced Computing Systems: Technology and Applications, IDAACS 2017, 8095220, pp. 918–922 (2017)

Embedding an Integrated Security Management System into Industry 4.0 Enterprise Management: Cybernetic Approach

Sergiy Dotsenko, Oleg Illiashenko, Sergii Kamenskyi, and Vyacheslav Kharchenko

Abstract The paper contains the results of the analysis of methodologies and standards obtaining the requirements to security management systems of enterprises including modern enterprises implementing Industry 4.0 principles. Key standards ISO/IEC 7498, 15408, 18045, 20000, 27000 have been analyzed to suggest an approach to the development of integrated security and safety management system structure considering threats of intrusion into physical, information and signal spaces. This system is part of the enterprise management system and based on cybernetic principle of control. These subsystems check and control according with individual and general objectives for physical, information and signal spaces and requirements-based models. The goal of the paper is to analyze the methodologies for developing an integrated security management system structure as a component of enterprise management systems in the context of Industry 4.0. The obtained results and recommendations for enhancing and implementation these systems are discussed.

Keywords Security · Safety · Enterprise management system · Control system · Standards · Integrated security management system · Industry 4.0

S. Dotsenko · S. Kamenskyi
Department of Specialized Computer Systems, Ukrainian State University of Railway Transport, Feijerbakha Square, 7, Kharkov 61050, Ukraine
e-mail: sirius_3k3@ukr.net

S. Kamenskyi
e-mail: mimonarch@gmail.com

O. Illiashenko (✉) · V. Kharchenko
Department of Computer Systems, Networks and Cybersecurity, National Aerospace University "KhAI", 17 Chkalova Street, Kharkov 61070, Ukraine
e-mail: o.illiashenko@csn.khai.edu

V. Kharchenko
e-mail: v.kharchenko@csn.khai.edu

© The Author(s), under exclusive license to Springer Nature Switzerland AG 2021   279
T. Tagarev et al. (eds.), *Digital Transformation, Cyber Security and Resilience of Modern Societies*, Studies in Big Data 84,
https://doi.org/10.1007/978-3-030-65722-2_17

1 Introduction

The task of developing and implementing the enterprise security management systems is becoming increasingly important at the point of view of the losses associated with a security breach resulting from management [1, 2]. Additional challenges for creating security management systems take place while developing the Industry 4.0 strategy [2–4]. The goal of implementing the enterprise security management systems is to have insurance that only authorized personnel can make changes to the process or influence to production in a permissible way [5].

According to ISO/IEC 7498-2:99 to achieve the goal the following tasks have to be completed [6]:

- granting the physical security by limiting access to the objects;
- realization of the control under the information flow that comes out of objects to protect intellectual property;
- realization of the control under the information flow that comes out of objects to control data transmissions;
- avoidance of influencing the production process by unauthorized remote access.

In this standard it is recommended to distribute the control object for the enterprise security management system into two components, specifically:

- technological process, that are realized in real-time mode;
- organizational processes that are implemented beyond the boundaries of technological processes (physical processes) and provide the organization of technological processes.

In the theory of automatic control, the development of an appropriate control system begins with the study of the model of the control object. Having investigated the model of the control object, decisions are made on the application of the control laws of the corresponding control object. The basic control law is the negative feedback control.

On the other hand, in the theory of management for the organization of the control system it is enough to define the elements of the management cycle. A classic example of such a system is the quality management system according to the standards of the ISO 9000 series. In this methodology a management cycle is used, known as the Deming-Shewhart cycle, specifically: "*Plan–Act–Check–Implement*". At the same time, the objects of management are products, processes and systems.

Another situation arises during enterprise security system management. In this case, all possible aspects of the organization and activities of the enterprise have to be potentially analyzed as a security system. At the same time, it is previously unknown which of the aspects of the organization and activity of the enterprise has the greatest importance.

The special attention should be paid to the following circumstances. Formation of the enterprise security system should begin to be engaged at the stage of formation of the enterprise. After all, security problems could be already identified at the stage of enterprise formation.

So, it is necessary to compare two methodologies for the formation of a security control system. The object of control for each of these methodologies has to be the enterprise security system. Therefore, the task is the formation of a control object in the form of a security system. On the basis of its analysis, it will become clear which control methodology should be implemented to provide the required quality for the management of security system. Additionally, it is important to determine the features of its construction for modern enterprises in the era of Industry 4.0 and integration in the overall structure of the management of such enterprises.

The goal of this research is to analyze the methodologies for developing an integrated security management system structure as a component of enterprise management systems in the context of Industry 4.0.

2 Analysis of Methods of Forming Enterprise Security Systems

Enterprise security systems are based on the approach introduced in a series of ISO/IEC 15408 standards [7–9] and is intended to protect information from unauthorized disclosure, modification or loss of its usability. Security categories related to these three types of security breaches are commonly called privacy, integrity, and availability.

This system of standards refers to the security system of information technology objects that the consumer intends to put in their own activities. This standard series introduces a different approach to presenting enterprise security. Of the five security services that are dealt with in the standard ISO/IEC 7498-2 examines three types of security breaches that are relevant to the specified services, specifically: confidentiality, integrity, availability.

According to ISO/IEC 15408-1 security is concerned with the protection of assets. Many assets are represented in the form of information that is stored, processed and transmitted by IT products to meet the requirements laid down by the owners of the information. Availability, distribution and modification of any such information have to be strictly controlled and the assets have to be protected from threats by countermeasures. In this security system, the main object of control is the risk. To provide a specified level of risk, an assessment of the Target of Evaluation (TOE) and development of appropriate countermeasures should be done. TOE according to ISO/IEC 15408-2 is defined as a set of software and firmware complexes accompanied by user and administrator guidance documentation.

The evaluation is carried out by special construction, specifically: security target (ST). According to ISO/IEC 15408-1 the ST begins with describing the assets and the threats to those assets. The ST then describes the countermeasures (in the form of Security Objectives) and demonstrates that these countermeasures are sufficient to counter these threats: if the countermeasures do what they claim to do, the threats are countered.

To unify the activities for the development of ST in the standard [7] the universal design in the form of a Protection Profile (PP) is proposed. Term "Security functional requirements" provided on the base of "Functional requirements paradigm" [8]. Functional safety components are implemented based on functional safety requirements.

According to ISO/IEC 15408-2 TOE is concerned primarily with ensuring that a defined set of security functional requirements (SFRs) is enforced over the TOE resources. The SFRs define the rules by which the TOE governs access to and use of its resources, and thus information and services controlled by the TOE.

From this thesis it follows that the TOE manages the use and access to its resources. It follows that the structure of the TOE should include appropriate methods of management. The security mechanism is the implementation of the TOE Security Functionality (TSF), which are defined by the SFR at the stage of formation and implemented through the mechanisms that follows the established rules. The implementation of these rules provides security capabilities.

From the abovementioned follows that the provision of information security on the stage of development and implementation of the TOE is an important part of the management of enterprise security system.

From the above analysis of the methods of forming enterprise security systems it follows that there are two independent methods for the formation of such systems, specifically:

- formation of the security system of information technologies, which are implemented and operated at the enterprise (ISO/IEC 7498);
- formation of the security system of information technologies, which are at the development stage (ISO/IEC 15408, ISO/IEC 18045).

In investigated security systems the information technologies are considered as assets. In ISO/IEC 15408 for information technologies the functional security requirements (ISO/IEC 15408-2) are described and the level of assurance is determined. Functional requirements relate to the relevant functions that need to be implemented in the activity.

It is clear that any security system requires proper management, so the task of analysing existing management methodologies, which is recommended for using, is appeared.

3   Analysis of Methods for Managing Security Systems

Information security management at the stage of development and implementation of information technologies is based on the requirements of ISO/IEC 27001 and ISO/IEC 27002 standards as specified in ISO/IEC 15408-1.

The methodological basis of this set of standards is the methodology of the formation of management systems. The most famous system of this class is the quality management system (QMS) according to the standards of the ISO 9000 series.

The management of the providing of information security services at the stage of development and implementation of information technologies is based on the requirements for management system for providing security services. This management system is based on the requirements of a set of ISO/IEC 20000 standards. This standard requires an integrated process approach at the time of planning, development, deployment, operation, monitoring, review, support and improvement of the service management system. According to the standard, a system of service provision is introduced that can be applied to provide services to the company to ensure its security. But the form of this security is not defined, that is, the model of the control object for which this system is formed is not defined.

From the mentioned above it follows that there are two approaches of ensuring enterprise security, specifically:

- security system for the reference model of interconnected open systems (ISO/IEC 7498-99);
- providing security of information technologies that are used in enterprises in the form of providing functional security and assurance to its evaluation.

At the same time, for the aforementioned approaches to ensure the enterprise security, respectively, different management methods are used, specifically:

- administrative management of system security, security services, security mechanisms, as well as the system of administrative management of security of interconnected open systems;
- management system in two alternative variants: according to ISO/IEC 20000 or to ISO/IEC 27000.

The question arises how these two security management methods are related. To answer this question, the existing methods of integration of management systems and enterprise management systems are considered.

4 Integration of Management Systems and Enterprise Security Management Systems

4.1 Integration of Management Systems

Integration of enterprise management systems is based on the publicly available specifications PAS 99: 2006. ISO Guide 72 for standards developers includes the basis for common requirements set in standards for management systems. The public technical specifications PAS 99: 2006 are applicable to the standards of the ISO/IEC

27001 series, ISO/IEC 20000-1, ISO/IEC 20000-2, that is, to the standards that are applied to ensure the information security of the enterprise.

While choosing for application the standards of series ISO/IEC 27000 or ISO/IEC 20000 the following uncertainties are arisen. The ISO/IEC 20000 series of standards is explicitly based on the ISO 9000 series methodology, for which the method of forming a management object model is not defined. The ISO/IEC 27000 series of standards is implicitly based on the ISO 9000 series methodology, but the management object model is explicitly defined in the form of "activities (as processes) for providing information security" for it.

4.2 Integrated Enterprise Security Management System

The integration of enterprise management systems is based on the series of standards IEC 62264-1-2014. In this case, integration means the following [10]: «Successfully addressing the issue of enterprise-control system integration requires identifying the boundary between the enterprise and the manufacturing operations and control domains (MO&C). The boundary is identified using relevant models that represent functions, physical equipment, information within the MO&C domain, and information flows between the domains».

This method of integration is based on the functional representation of the enterprise. It involves the integration of enterprise management systems and production process management systems, that is, the integration of the two control systems of parts of the enterprise into a whole one.

From the analysis of the mentioned integration methods it follows that at present time two methods of integration of enterprise management systems are proposed, specifically:

- integration of management systems based on a single element set of the management cycle (ISO Guide 72);
- integration of two forms of management systems, specifically: the enterprise management system and production process management system (IEC 62264-1-2014).

The basis of these two forms of integration is the functional representation of the enterprise.

However, based on the requirements of IEC 62264-1-2014, from a security point of view, the most significant should be the methodologies of enterprise modeling, in which the physical, informational and cybernetic (in the form of data transmission) representations should be presented in an explicit form.

On its basis, Fig. 1 presents an integrated enterprise security management system, which is proposed in [11]. The architecture of each channels of this management system is similar to the architecture of the operation management system [12].

For the composition of the system it is proposed to introduce three mutually connected areas of security on the levels of physical, information and signal spaces.

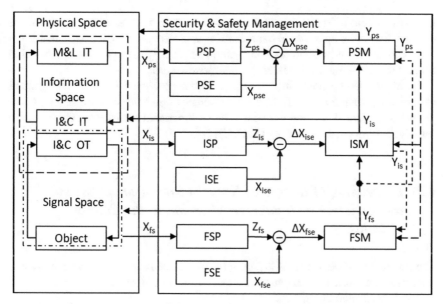

M&L – management & logistic; I&C – instrumentation & control;
OT – operation technologies; IT – information technologies;
PS – physical security; IS – information security; P – processing;
E – etalon; M – maker

Fig. 1 Integrated safety-security management system

Let's examine the work of the system on the example of the safety management channel "Physical Space". Signals about the enterprise security state as a physical object (Xps) are transmitted to the PSP block where the appropriate diagnosis (Zps) is formed. This diagnosis is transferred to the adder. In the adder it is compared with the reference value (Xpse), which is formed in the PSE block, and the formation of the control signal as the difference $(\Delta Xpse) = (Zps) - (Xpse)$ is provided. Under the action of the resulting signal in the PSM block, a control action is formed that guides to the processes in the "Physical Space". A similar algorithm is implemented in the "Information Space" and "Signal Space" channels.

The integration of control channels is carried out by transferring control signals from the "Physical Space" (PSM block) channel to the inputs of the ISM and FSM units. Due to this, "Information Space" and "Signal Space" channels are controlled taking into account the state of the "Physical Space" management channel.

Additionally, the control commands from the ISM and FSM units proceed to the PSM block. For this reason control action in the PSM block is formed taking into account the state of the channels "Information Space" and "Signal Space".

The enterprise security management system, which is shown in Fig. 1 corresponds to the principle of constructing the hierarchical control systems based on the integration of the appropriate channels of the control system. By the content of the control law, this system refers to cybernetic control systems with feedback, so it can be described with the appropriate mathematical apparatus. This will ensure its formation as an automated system of dialog enterprise security management, or a decision support system in the management of enterprise security.

It should also be noted that the usage of the developed security management system has certain features. It differs for enterprises that are software developers or project designers. They don't have a level of functional security (I&C OT) and a sub-level of information security (I&C IT).

4.3 Embedding of Integrated Security Management System into Enterprise Management Systems in the Context of Industry 4.0

The considered system (Fig. 1) is part of the overall enterprise management system. Integrated enterprise management system with enterprise security management system is presented in Fig. 2.

The integration of the enterprise security management system into the enterprise management system involves the interaction of the subsystems of production process management with the security management subsystems. Similar interaction exists between production subsystems and subsystems that describe the signalling, information and physical security levels.

It should be noted that the formation of security subsystems the indicated levels can be carried out using various methods of forming management systems, and management, which were described above.

It has been shown above that the developed and applied information technologies are based on a functional representation. At the same time, the standard of the IEC series 62264-1-2014 establishes that the most significant should be the methodology of modelling the enterprise in which the physical, informational and cybernetic (in the form of data transmission) views should be presented in an explicit form. This requirement is especially important for Industry 4.0. Global industry digitization raises the problem of cybernetic threats for any of the information processes implemented with the use of digital technologies for receiving, transmitting, storing and presenting data and information.

4.4 Mathematical Models of Enterprise Management and Security Systems

Figure 2 shows a general block diagram of an integrated enterprise management system with the enterprise security management system. A detailed analysis of the methodological bases of formation of enterprises information-control systems in

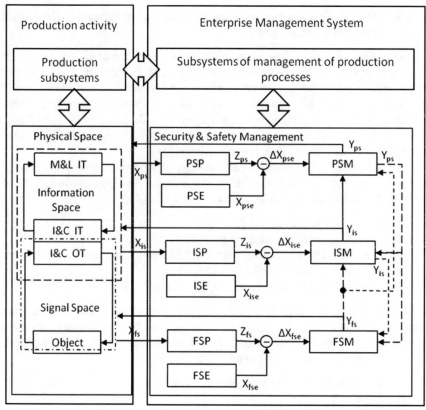

M&L – management & logistic; I&C – instrumentation & control;
OT – operation technical; IT – information technical; PS – Physical Security
IS – Information Security; P – processing; E – Etalon; M - Maker

Fig. 2 Integrated enterprise management system with the enterprise security management system

Industry 3.0 and an assessment of possibilities of their application is made in work
[13].

Methodological basis for the formation of programmatic and target management
is management of operations [14]. By doing this, the organization is considered as an
organizational system to which the principle of hierarchy is applied. This principle
applies both in the form of "tree" and in the form of more complex graphs.

As mentioned in [14]: underlying the method of constructing HCS <*hierarchical
control systems*>, that is under consideration, are two principles:

1. The principle of spatio-temporal aggregation of the original control problem with
   large dimension into a problem with smaller dimension for the higher level;
2. The principle of spatio-temporal decomposition of the original control problem
   into subproblems with smaller dimensions (the solutions for which are carried

out independently), which determines the method of disaggregation for optimal problems from the upper levels.

At the same time, the following position of the theory of hierarchical systems is important: in the hierarchical system, as already stated, the system itself is hierarchical, due to the fact that the overall purpose of the operation is achieved no different than by the execution of a hierarchical set of specific operations of different ranks. Therefore, the graph of goals and tasks of operations is identical to the graph of operations, where the vertices are matched to the operation and the arcs are the relation between the operations and their respective goals [14].

However, implementation of the methodology of formation of hierarchical organizational structures at the enterprise and organization level, as shown in [15], is connected with a number of serious problems, specifically:

- definition of the system hierarchy…;
- choice of management organization principles…;
- optimal distribution of functions performed between tasks, as well as between decision-makers (DM) and software and hardware;
- defining models and methods of solving system problems.

To solve these problems, it is proposed to apply integration within the organization on the basis of architectural approach [16]. On the basis of architectural approach to the structuring of the dialogical management production system, the following main ways of its decomposition are considered [15, 3]:

- control parameters and layers of task complexes;
- critical systems;
- phases of the management process.

The main advantage of this approach is that the architecture of the structural graph of the system does not depend on the organizational structure of the studied organization [15]. The methodological basis for the formation of control parameters for this approach to the formation of hierarchical systems is "The general theory of systems: mathematical foundations" by Mesarovic and Takahara [17]. It should be noted that this methodology is also the basis for the development of integrated enterprise management systems based on program-based management in the form of operations management.

For further description, it is proposed to apply a program-based methodology. The methodology of the architectural approach, according to [15], is advisable to apply at the integration of security systems in the production management systems for each of the layers of management. The proposed Integrated Safety-Security Management System has three control channels. For the structure, these channels are similar, but the content of the control processes in these channels is different.

Signal Space channel is characterized by the real-time mode. Typically, this is the level of technology management systems (SCADA systems). Therefore, it is recommended to design a security management system for this channel as an automatic control system.

On the contrary, Information Space and Physical Space channels are characterized by the mode of human participation. Therefore, in its formation, it is proposed to apply the principles of operations management in the form of operational management [14]. It should be also considered that formally automatic control circuits and operation control circuits are similar [14].

To move to the mathematical description of the integrated enterprise security management system presented in Fig. 1, we introduce the common notations for control signals according to Table 1. This will provide a transition to the generalized scheme for one security management channel presented in Fig. 3 according to [14]. This scheme is quite similar to the control channel diagrams in Fig. 1, but is more detailed.

It should be noted that the connection between the channels is made by submitting the control to the system input except for control as an *error function*, and control actions of adjacent channels $u(t)_{сум}$ (see Fig. 3).

For automatic control systems with negative feedback equation input-state-output for the object and the measurement system in the integral form have the form [14]:

$$x(t) = F_x(x(t_0), t, u_{[t_0,t]}, v_{[t_0,t]}) \tag{1}$$

$$y(t) = g_x(x(t)) \text{ or } y(t) = C_x x(t), \tag{2}$$

$$\xi(t) = F_\xi(\xi(t), t, y_{[t_0,t]}, v_{[t_0,t]}), \tag{3}$$

$$\eta(t) = g_\xi(\xi(t)) \text{ or } \xi(t) = C_\xi \xi(t), \tag{4}$$

where

$x(t)$ vector-function for the parameters of system status;
$y(t)$ vector-function for the control parameter;
t time parameter;
$\xi(t)$ vector-function for the measuring system state;
$\eta(t)$ vector-function for the measuring system output;
v vector-function of random perturbations.

In this case, the feedback control assumes that the output $\eta(t)$ or $\eta[t_0, t]$ is compared with the desired value $y^e(t)$ or $y^e[t_0, t]$ and the value of the error is fixed [14, p. 77]:

$$\varepsilon(t) = y^e(t) - \eta(t). \tag{5}$$

Subsequently, such control u as the error function $\varepsilon(t)$ and the adjacent channels control actions $u(t)_{сум}$, which minimizes this error and the control action of the adjacent channels, are selected.

Table 1 Common notations for control signals

Object	Input control signal	Output signal	Output of the measuring channel	Goal settings	Control error	External perturbation
Physical security (PS block)	Y_{ps}-management of the personnel of the physical protection service	X_{ps}	Z_{ps}-parameters of data flows in security surveillance and access systems	X_{pse}-prevention of the unauthorized entry and external malicious activity	ΔX_{pse}	n/a
Informational security (M&L IT)	Y_{is}-management of counteraction means to external cyber attacks on information-management systems of the enterprise		Z_{is}-data flow parameters in the information and telecommunication network	X_{ise}-prevention of external cyber attacks through information and telecommunication network on information-management systems of the enterprise	ΔX_{ise}	n/a
Functional security (I&C OT)	Y_{fs}-management of counteraction means to external cyber attacks on enterprise SCADA systems	X_{fs}	Z_{fs}-physical parameters of technological processes	X_{fse}-prevention of external cyber attacks through information and telecommunication network on enterprise SCADA systems	ΔX_{fse}	n/a
Control system model	$u(t)$	$x(t)$	$\eta(t)$	$y^e(t)$	$\varepsilon(t)$	$v_1, v_2, \ldots v_m$

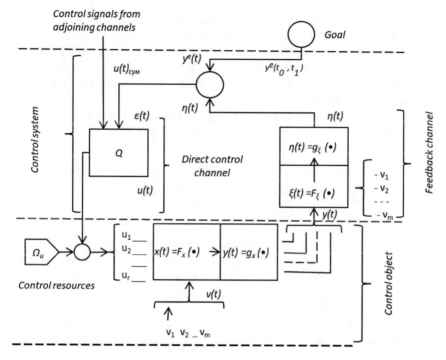

Fig. 3 Generalized scheme for one security management channel

It should be noted that the content of control actions in the integral form, the dependence of control on error is written in the form [14]:

$$u(t) = Q(u(t_0), t, \varepsilon_{[t_0,t]}, u(t)_{cym}), \qquad (6)$$

where

Q called control operator (algorithm);
$u(t)_{cym}$ control actions of adjacent channels.

In automatic control, the program $y^e[t_0, t_1]$ have to be entered in the regulator. This type of control is characteristic of the signal layer (the level of functional security) that must be implemented for automated process control systems. The control problem can be presented in the following form [14, p. 77]

$$z = \min_{Q \in R[Q]} \left[\begin{array}{l} \Phi(\varepsilon_{[t_0,t_1]}, Q_{[t_0,t_1]}, v_{[t_0,t_1]}) | \varepsilon(t) = y^e(t) - y(t); \, y(t) = g(x(t)); \\ x(t) = F_x(x(t_0), t, u_{[t_0,t]}, v_{[t_0,t]}); \, u(t) = Q(u(t_0), t, \varepsilon_{[t_0,t]},); \, y^0; \, y^1; \, t \in [t_0, t_1]] \end{array} \right],$$
$$(7)$$

where

Φ criterion (evaluation) of quality of management.

In automatic control, Eq. (7) displays the model of the control process at the design and creation stage of the control system with the operator Q [14].

In case of automated human–machine control, information about programs $y^e[t_0, t_1]$ and $y[t_0, t_1]$ is pre-processed by computing devices and presented to the person in the optimal form, and the transformation operations for $\varepsilon(t)$ and are $u(t)$ formalized and performed by automatons [18].

For *Information Space* and *Physical Space* channels, the task of automated security management can be presented as follows [14, p. 116]

$$\Psi = \min_{Q \in R[Q]} \left[\begin{array}{l} \overline{\Phi}(x^e{}_{[t_0,t_1]}, y_{[t_0,t_1]}, Q_{[t_0,t_1]}, v_{[t_0,t_1]}) \mid \varepsilon(t) = y^e(t) - y(t); y(t) = L(x_{[t_0,t]}, n^1{}_{[t_0,t]}); \\ x(t) = F_x(x^0, t, u_{[t_0,t]}, v_{[t_0,t]}); u(t) = Q(x^e{}_{[t_0,t]}, y_{[t_0,t]}, v_{[t_0,t]} u(t)_{cym}, t); \\ v(t) = N(v_{[t_0,t]}, n^2{}_{[t_0,t]}); x^0; x^{e1} \in R[x] t \in [t_0, t_1] \end{array} \right] \tag{8}$$

Function F characterizes the costs of operational control, which is minimized by the choice of control algorithm $Q[t_0, t_1]$. From the above it follows that the fundamental task of forming each of the channels of the security management system is the task of forming the control algorithm $Q[t_0, t_1]$ itself. It is considered as a decision problem [14].

Since the automated control system belongs to the class of purposeful systems (decision-making systems), M. D. Mesarovic gives the following definition to such systems [17, pp. 297–299]:

In this case, the system is described not directly, but by some decision-making task. In fact, such a system $S = X \times Y$ is determined by requiring that the pair (x, y) belongs to S if and only if y is a solution to the decision problem given by the element x ....

In [17] two types of problems are considered, specifically:

- the general task of optimization;
- the general task of satisfactory.

The General task of optimization is formulated in the next way [17]:

Let g: X → V be a function that maps an arbitrary set X to a set V, which is assumed to be linear or partially ordered by ≤ .

Then the general problem of optimization is as follows.
For a given subset $X^f \subseteq X$, find $\hat{x} \in X^f$ such that for all $x \in X^f$

$$g(\hat{x}) = g(x) \tag{9}$$

In this case the following naming could be used:

- The set X—the solutions set;
- The set X^f—the set of admissible solutions;
- The function g—objective function;
- The set of estimates—V.

In these terms, the joint optimization problem is given by a pair (g, X^f). The element $\hat{x} \in X^f$, which satisfies condition for all X from X^f is called *the solution of the optimization problem given by the pair* (g, X^f).

Often function g is defined by two functions:

$$P : X \rightarrow Y; \quad G : X \times Y \rightarrow V, \tag{10}$$

$$g(x) = G(x, P(x)) \tag{11}$$

In this case, the function P is called *the output function* or *model* (control object), the function G is *the quality function* or *the evaluation function*. Thus *the optimization problem* can then be defined by a triple (P, G, X^f) or a pair (P, G) if $X^f = X$.

It is interesting to know about M. D. Mesarovich's opinion on the model of the object of control [17]:

> What is called as the "Management Object Model", or P means the following. The three-factor optimization problem (P, G, Xf) is actually determined by the system to be controlled and described by the function P. However, in the general case, we are not obliged to assume that the model P and the real control object are related with any specific relationships. Moreover, the very hypothesis of the existence of a control object as a whole thing can be questioned and used only to set the task of optimization, the determining system through the decision-making process.

However, for the practical implementation of the integrated system under study, it is extremely advisable to develop management object models for certain control channels. The presence of such a model allows us to consider the system under study as Model-driven DSS. Therefore, it is urgent to develop appropriate management security models (as management objects) for each of the Integrated Safety-Security Management System channels.

The proposed model of integrated enterprise management system with enterprise security management system has both theoretical and practical significance.

From a theoretical point of view, this model:

- reveals the content of the model of functional architecture of the enterprise security management system, which should be considered as part of the overall integrated enterprise management system;
- allows one to determine the composition and content of the functions that must be implemented when creating a security management system.

From a practical point of view, based on this knowledge next actions are possible:

- development of a technical specification for the development of a security management system;
- development of a feasibility study of creating a security management system;
- analysis of compliance of the developed security management system with the requirements of international standards for functional security (IEC 61508) and information security (ISO/IEC 15408 and 18045, as well as ISO/IEC 27000).

From a practical point of view, based on this knowledge it is possible to the following:

- development of a technical specification for the development of the security management system;
- development of a feasibility study of for the security management system;
- an analysis of the compliance of the developed security management system with the requirements of international standards for functional safety (IEC 61508) and information security (ISO/IEC 15408 and 18045, as well as ISO/IEC 27000).

5 Conclusions and Future Work

There are two independent methods of developing security systems for enterprise systems such as:

- formation of a security system for information technologies, that is implemented and operated by an enterprise (ISO/IEC 7498-2: 99);
- formation of a security system for information technologies, which are at the stage of development (ISO/IEC 15408, ISO/IEC 18045: 2008).

At the point, for the aforementioned approaches to ensure the security of enterprises different management methods are used, respectively:

- *administrative* management of system security, security services, mechanisms of security, as well as the system of administrative control of interconnected open systems;
- *management system* in two alternatives: according to ISO/IEC 20000-1 or ISO/IEC 27000.

The enterprise security management system, which is shown in Fig. 1 corresponds to the principle of constructing the hierarchical control systems based on the integration of the corresponding channels of the control system. By the content of the control law, this system refers to cybernetic control systems with feedback, so it can be described with the appropriate mathematical apparatus. This will ensure its formation as an automated system of enterprise security dialogue management, or a decision support system in the management of enterprise security.

The subsystems of the integrated enterprise management system with the enterprise security management system should be formed taking into account the forms of production subsystems of the enterprise and security subsystems, specifically: signalling, information and physical views. For the effective implementation of the Industry 4.0 concept, it is expedient to integrate the integrated enterprise security system into the enterprise management system.

The developed safety and security management system conception and models have been adopted and implemented at the PC "RPC Radiy", Kropyvnytskyi, Ukraine and PrJSC FED Kharkiv Ukraine.

Acknowledgements This work was supported by the ECHO project, which has received funding from the European Union's Horizon 2020 research and innovation programme under the grant agreement no 830943.

The authors very appreciated to scientific society of consortium and in particular the staff of Department of Computer Systems, Networks and Cybersecurity of National aerospace university "KhAI" for invaluable inspiration, hardworking and creative analysis during the preparation of this paper.

References

1. Roberto, M.: An integrative model of information security awareness for assessing information systems security risk. In: Proceedings of the Annual Hawaii International Conference on System Sciences, pp. 3258–3267 (2012)
2. Systems Integration for Industry. https://www.automation.com/automation-news/article/systems-integration-for-industry-40. Access date: Dec 2010
3. Smarter Security for Manufacturing in The Industry 4.0. Era Industry 4.0 Cyber Resilience for the Manufacturing of the Future. White HITE paper https://www.symantec.com/content/dam/symantec/docs/solution-briefs/industry-4.0-en.pdf. Access date: Dec 2019
4. Kondiloglu, A., Bayer, H., Celik, E., Atalay, M.: Information security breaches and precautions on INDUSTRY 4.0. Technol. Audit Prod. Reserves**6/4**(38), 58–63 (2017)
5. ISO/IEC 7498-1:1999: Information technology. Open systems interconnection. Basic reference model. Part 1: The Basic Model (1999)
6. ISO/IEC 7498-2:1999: Information technology. Open systems interconnection. Basic reference model. Part 2: Security Architecture (1999)
7. ISO/IEC 15408-1:2009: Informational technology—security techniques—evaluation criteria for IT security. Part 1: Introduction and General Model (2009)
8. ISO/IEC 15408-3:2008: Informational technology—security techniques—evaluation criteria for IT security. Part 3: Security Assurance Requirement (2008)
9. ISO/IEC 18045:2008: Informational technology—security techniques—methodology for IT security evaluation (2008)
10. IEC 62264-1:2014: Enterprise-control system integration. Part 1: Models and Terminology (2014)
11. Kharchenko, V., Dotsenko, S., Illiashenko, O., Kamenskyi, S.: Integrated cyber safety and security management system: industry 4.0 issue. In: Proceedings of the 10th IEEE Dependable Systems, Services and Technologies Conference, DESSERT 2019, pp. 197–201 (2019)
12. Ackoff, R., Sasieni, M.: fundamentals of operations research hardcover. J. Nano Electron. Phys. https://doi.org/10.21272/jnep.11(2).02013. Access date: Dec 2019
13. Dotsenko, S., Illiashenko, O., Kamensky, S., Kupreishvili, D., Kharchenko, V.: Analysis of methodological foundations of enterprises' information-managing systems formation in industry 3.0: movement towards industry 4.0. Radioelectron. Comput. Syst. **2**(90), 9–44 (2019)
14. Pospelov, G., Irikov, V.: Program-Targeted Planning and Management (Introduction), p. 440. Sov. Radio, Moscow (1976)
15. Meltzer, M.: Interactive Production Management (Models and Algorithms), p. 240. Finance and Statistics, Moscow (1983)

16. Epshtein, V., Senichkin, V.: Language Tools of the ACS Architect, p. 136. Energy, Moscow (1979)
17. Mesarovic, M.D., Takahara, Y: General Systems Theory: Mathematical Foundations. Systems Research Center Cleveland, Ohio Case Western Reserve University Academic Press, New York San Francisco London (1975)
18. Ifinedo, P.: Understanding information systems security policy compliance: an integration of the theory of planned behavior and the protection motivation theory. Comput. Secur. **31**(1), 83–95 (2012)

Emerging Cybersecurity Technologies and Solutions

Cybersecurity in Next Generation Energy Grids: Challenges and Opportunities for Blockchain and AI Technologies

Notis Mengidis, Theodora Tsikrika, Stefanos Vrochidis, and Ioannis Kompatsiaris

Abstract Renewable energy sources and the increasing interest in green energy has been the driving force behind many innovations in the energy sector, such as how utility companies interact with their customers and vice versa. The introduction of smart grids is one of these innovations in what is basically a fusion between the traditional energy grid with the IT sector. Even though this new combination brings a plethora of advantages, it also comes with an increase of the attack surface of the energy grid, which becomes susceptible to cyberattacks. In this work, we analyse the emerging cybersecurity challenges and how these could be alleviated by the advancements in AI and blockchain technologies.

Keywords Cybersecurity · Blockchain · AI · Energy grid · Smart grid · Smart contracts · Consensus algorithms

1 Introduction

In past decades, the development of power grids has not been keeping pace with industrial and societal advancements that have created an increased demand of power supply. According to Ratner and Glover [39], during the period from 1950 to 2014, just in the US, energy production and consumption increased more than two and three times respectively. With this increased demand of electricity, issues like voltage spike and sags, blackouts, and overloads have increased as well, resulting in availability

N. Mengidis (✉) · T. Tsikrika · S. Vrochidis · I. Kompatsiaris
Centre for Research and Technology-Hellas (CERTH), Thessaloniki, Greece
e-mail: nmengidis@iti.gr

T. Tsikrika
e-mail: theodora.tsikrika@iti.gr

S. Vrochidis
e-mail: stefanos@iti.gr

I. Kompatsiaris
e-mail: ikom@iti.gr

© The Author(s), under exclusive license to Springer Nature Switzerland AG 2021
T. Tagarev et al. (eds.), *Digital Transformation, Cyber Security and Resilience of Modern Societies*, Studies in Big Data 84,
https://doi.org/10.1007/978-3-030-65722-2_18

issues which consequently lead to revenue losses for the energy industry. As an example, a study conducted by Knapp and Samani [25] indicated that the American economy loses annually approximately $150 billion due to power interruptions. Furthermore, the power industry alone produces up to 40% of United States' carbon dioxide emissions [28], a percentage slightly lower within the European Union [41].

To cope with the aforementioned shortcomings of the energy industry, the need to efficiently manage a variety of energy sources became evident. It also became clear that legacy power systems can no longer meet the requirements of modern society in terms of reliability, scalability, manageability, and cost-effectiveness. These needs gave birth to *smart grid*, a dynamic and interactive infrastructure with new energy management capabilities, which however inevitably created a system with potential vulnerabilities in terms of cybersecurity. In this paper, we present some of the most emerging cybersecurity challenges related to smart grid and discuss mitigation techniques based on blockchain and AI.

2   Background

Section 2 provides a detailed overview on what consists a smart energy grid, its main components and a high-level description on the communication protocols used by its elements. We also present how blockchain technology works and the different types of existing system architectures and consensus algorithms.

2.1   Overview of Smart Grids

The smart grid can be considered as the next evolution step in today's power grid technology and smart meters specifically are the corner stone of this evolution. In case an energy provider decides to shift towards a smart grid implementation, the first step is to install a smart meter in every customer and premises. Smart meters are devices that offer the capability both to the provider and to the customer real-time (or near real-time) monitoring of electricity consumption or production, in the case of e.g. photovoltaic cells. They also offer the possibility to read the measurements locally and remotely, and additionally allow the provider to limit or terminate the supply of electricity where appropriate.

The National Institute of Standard and Technology (NIST) defines the smart grid as a composition of seven domains: bulk generation, transmission, distribution, customers, markets, service providers, and operations [19]. The first three domains are responsible for the power flow, whereas the last four correspond to the part of the energy grid responsible for data collection and power management. In order to interconnect the aforementioned domains, a backbone network is required which can be broken down to smaller local-area networks. Figure 1 illustrates how this interconnection takes place in a logical as well as in a network level.

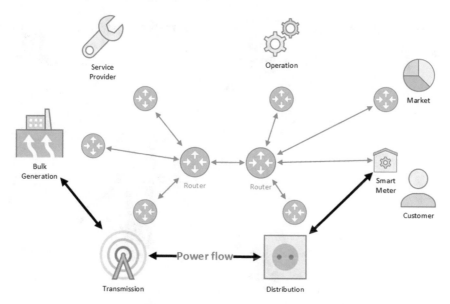

Fig. 1 Network architecture of the smart grid

On a higher level, a smart grid consists of four main components; the Advanced Metering Infrastructure (AMI), the Supervisory Control and Data Acquisition (SCADA), the plug-in hybrid vehicle (PHEV), and various communication protocols [21]. AMI's role is measuring and analyzing energy usage and allows a two-way communication between the consumer and the utility company.

Smart meters communicate with the AMI headend, which aggregates the information from a large number of meters, and relay the aggregated data to the Meter Data Management System (MDMS). Communication between the smart meters and the AMI headend is usually achieved through wireless links such as Wireless Sensor Networks (WSN) [26], cellular systems [32] or even cognitive networks [17].

As a result of the highly-distributed nature of the AMI network and the openness of the wireless communication medium, we are motivated to examine the cybersecurity challenges that arise due to the increased attack surface and investigate the opportunities that this early stage of smart meters' adoption has to offer.

2.2 Overview of Blockchain Technologies and Consensus Algorithms

The idea of cryptocurrencies was first perceived by David Chaum in his proposal for untraceable payments [9] where he described a system where third-parties are unable to determine payees and time or amount of payments made by an individual. He took his idea one step further in 1990 by creating the first cryptographic anonymous

Fig. 2 A single trusted
authority holding a copy of
the ledger

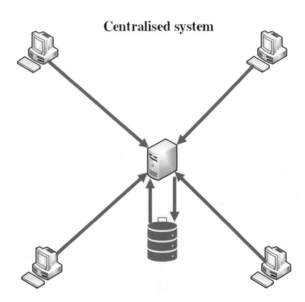

Centralised system

electronic cash system, known as ecash [10]. Later in 90s, a lot of startups emerged
trying to implement electronic cash protocols, attempts that ultimately failed.

Cryptocurrencies, as we know them today, are peer-to-peer decentralized digital
assets based on the principles of cryptography. Most cryptocurrencies use a
distributed database as the pillar of their system, known as blockchain, which allows
them to use it as a distributed public ledger without having to rely to any form of
centralized control similar to banking systems (Fig. 2).

The blockchain is the equivalent of a book maintained by a bank which contains
all the accounts and each transaction made. Of course, this is an oversimplification
and in reality, there are many differences, possibly the most noticeable being the
fact the bank's records are private whereas the blockchain is publicly available and
easily accessible by everyone. One of the most interesting aspects of blockchains
is that they contain the records of every transaction made since the beginning, also
known as genesis block, by using a peer-to-peer distributed timestamp server which
generates computational proof of the chronological order of the transactions [35].

Blockchains in general require a network to run and transmit their data. In the
context of cryptocurrencies, this transmission is equivalent of copying coins from
one electronic wallet to another, and here is where the biggest challenge lies; how to
ensure that every coin is spent only once. Whereas the traditional approach for such
a problem would be to rely on a centralised authority that would validate the status of
each transacting party, blockchain solves this problem by allowing everyone in the
network to have their own copy of the historic log of all transactions. This however
creates another issue, which is how we can make all transacting parties to agree upon
the validity of the state of the ledger. Depending on the type of blockchain used, there
are many proposed validation techniques, however in principal all these techniques
are called distributed consensus algorithms (Fig. 3).

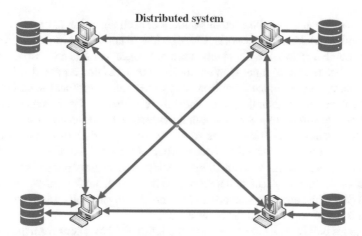

Fig. 3 A distributed platform where each node has a copy of the ledger

As shown by Baliga [5], a resilient consensus algorithm in terms of reliability, node failure and malicious activities can be a quite challenging process. There are many proposed consensus algorithms, each with its own strengths and weaknesses. In general, these algorithms can be classified into two broad categories, namely lottery-based and voting-based [1].

The first category of algorithms consists of proof-of-work (PoW) algorithms, which are the most commonly used ones by cryptocurrencies such as Bitcoin and Ethereum, and proof-of-stake (PoS) algorithms. In PoW algorithms, the consensus is achieved by solving cryptographic puzzles which depending on the computational power that each node offered, rewards them with their fair amount of votes in the network. On the other hand, in PoS algorithms, the weight of the vote is determined by the size of the stake that each node has in the network, e.g. the amount of cryptocurrency deposited or mined in a wallet.

The second generic category of consensus algorithms are the voting-based systems where the validation is achieved through a multi-round process where all nodes vote for the next block candidate to be included in the blockchain. As soon as the voting ends, the validating nodes have to agree on whether or not the voted block will be accepted in the network. Since the votes are transmitted in a potentially untrusty network and the trustworthiness of each node cannot be ensured, the design of such a system has to be carefully considered. Starting with Bitcoin's PoW, we are going to present the most popular consensus algorithms.

2.2.1 Proof of Work (PoW)

Hashcash [4] was the first time a PoW mechanism was used, even though it was developed for a different purpose than cryptocurrencies; the mitigation of denial of service attacks towards internet resources. Proof of work algorithms became more widely known when Bitcoin used one variant of such algorithm in order to validate newly added blocks in its network. The algorithmic approach of Bitcoin involves a random number (nonce) that it can be used only once and it is the hashed value of the block header. Each miner then competes in order to find a hashed value that it lower than the nonce. Since there is no way to determine whether the hashed value will be actually lower than the nonce, the only feasible action is through continuous trial and error, similar to a computationally expensive brute force attack.

When a transaction is transmitted to the network, it is then subjected to validity checks and it is not verified until it becomes part of the blockchain. New transactions constantly flow in the network and they get added to a memory pool of unconfirmed transactions handled by each node. Since the size of each block is finite, transactions have to deal with competition in order to be added in the new block and the selection criteria is based on who paid the highest fee.

As nodes build a new block, they add unconfirmed transactions from the memory pool to a new block and attempt to solve a computationally intensive problem to prove that the block is valid. This is the proof-of-work concept of Bitcoin and the process of solving it is called mining. Mining ensures that transactions are only confirmed if enough computational effort was spent on the blocks that contain them. More blocks mean more effort which subsequently means more trust [2].

To incentivize mining, each mining node includes a special transaction in its block containing a transaction that pays its own address a reward (currently 12.5 BTC per block) of newly created Bitcoins. If the node finds the solution before the other nodes in the network, then the block becomes valid, and it wins the reward since the block is added to the blockchain, thus the reward transaction becomes spendable. This reward transaction is the only exception to the rule that a transaction's outputs has to be smaller or equal to its inputs.

The block is then propagated throughout the network and contains a list of transactions that the node which created the block committed since the previous block [12]. To prevent denial-of-service attacks and spam, every node that receives this newly created block validates it before forwarding it further. If it determines that it is a valid block then it propagates it to its adjacent nodes, discards its previous mining efforts, applies the transactions from the current block and immediately starts working on building the next block.

At this point, the network has agreed on the validity of the transactions contained in the newly mined block and the transactions are confirmed and do not have to be reap-plied. The transactions that were not included will have to be validated again and reap-plied on top of the new block state.

One of the main disadvantages of PoW is that it requires large amounts of electricity. According to [37], Bitcoin could one day consume up to 60% of global electricity production, 13,000 TW h, equal to powering 1.5 billion homes. Another report from [13] claims that the increase in electricity consumption of the bitcoin network may lead to a draw of over 14 GW of electricity by 2020, equivalent to the total power generation capacity of a small country, like Denmark.

Another drawback in PoW design, is that if a mining entity (either a mining pool or an individual miner) managed to contribute more than half of the network's hash rate, then that entity would have total control of the network and would be able to manipulate the blockchain at will. This is often called the 51% attack even though it has been proved that actually less hash power suffices to perform this kind of attack [15].

To have a better understanding of this attack, let's assume two blockchains, which both have a common ancestor, with lengths n and m ($n > m$). If n is the honest chain and m the chain of the attacker (who has more than half of the network's hash rate), then both counterparts can create a chain with length $k > n$ with probability $p\hat{}(k - l)$, where l is the current chain length and p the percentage of the attacker's hash rate. Evidently, if the attacker picks a k large enough, he will have a bigger probability of finding a longer chain than the honest one.

2.2.2 Proof of Stake (PoS)

The inherent weaknesses and the subsequent criticism of PoW lead to the development of a new distributed consensus algorithm. In PoS the creator of the new block is selected through a combination of randomness and wealth or age which called stake. The chance of a node getting selected is usually proportional to the amount of wealth that the specific node has invented into the network, however a certain amount of randomness is also introduced in order to avoid the case where the wealthiest member of the network has an advantage and gets selected all the time.

This approach offers greatly reduced power consumption and are less susceptible to 51% attacks, however they arise the issue of *nothing-at-stake,* where the nodes that have nothing to lose, vote for multiple blockchain candidates hence preventing the chain to reach a consensus state. Since the cost for such attempts is little to none, some blockchains adopting PoS algorithms are prone to fake states attacks [23].

2.2.3 Delegated Proof of Stake (DPoS)

In Delegated Proof of Stake (DPoS) the nodes in the network, instead of voting for the validity of the blocks, vote to elect a number of so-called *witnesses,* that will generate blocks on their behalf. After a predetermined interval has elapsed, the witnesses are shuffled and new witnesses are allowed to produce blocks per n number of seconds (n depends on the implementation). After each round, witnesses are rewarded according to the amount of blocks they produced, however failure to do so means an increase in the probability of being voted out in the next round [3]. Some of the advantages of DPoS include greater scalability compared to PoW, faster verification times, and energy efficiency. However, since the number of witnesses in the network are somehow limited, the danger of centralization is apparent.

3 Cybersecurity Challenges

Cybersecurity poses one of the largest and multifaceted challenges that the smart energy grid and the IoT ecosystem in general will have to address in the years to come. Given the number of interconnected sensors, devices and networks that constitute a smart grid, it becomes evident that it is susceptible to online probing, espionage, and constant exploitation attacks by malicious actors aiming at disrupting the stable and reliable energy grid operation, obtaining sensitive customer information, as well as threatening the CIA triad (confidentiality, integrity and availability) of the network [42]. In order to have a clearer picture of the dangers posed by the integration of smart energy meters in the traditional energy grid, we will examine the security requirements of a smart grid and analyse the most high-profiled challenges from a cybersecurity perspective.

3.1 Cybersecurity Requirements and Objectives in the Smart Grid

According to NIST, the main criteria required to ensure the security of information in any given information system, thus smart grid as well, are *confidentiality, integrity* and *availability, also known as the CIA triad* [42]. It is also widely accepted that *accountability* is another important aspect of security, therefore it will also be included as an additional criterion below [27].

3.1.1 Confidentiality

Generally, confidentiality is the preservation of authorised restrictions on information access and disclosure, including means for protecting personal privacy and proprietary information. Once an unauthorised entity, individual, or process gains access to proprietary information, we consider that the confidentiality of the specific system is lost. In the context of the smart energy grid, information such as the past and present measurement values of a meter, consumption usage, and billing information are considered confidential and hence must be protected. Most utility providers nowadays offer electronic bills and some of them even web portals with real-time statistics of energy usage for each customer individually. With this increased accessibility of consumer data on the internet, confidentiality is starting to become increasingly significant [45].

3.1.2 Availability

Availability is defined as the provision of timely and reliable access to and use of information and services. In the case of the smart grid, availability can arguably be considered as the first priority since an availability loss in the grid can potentially have a serious adverse effect on organisational operations, organisational assets and individuals. An availability attack takes place in the form of traffic flooding, where the attacker aims to delay or disrupt message transmission [29], or buffer flooding where the malicious actor aims to overwhelm the AMI's buffer with false events [22]. Both attacks fall under the umbrella of Denial of Service (DoS) and the main objective of the attacker is to exhaust the computational resources of the smart grid and degrade the network communication performance of the grid.

3.1.3 Integrity

Integrity in smart grid is ensuring that there will be no kind of violation of data, including destruction, modification or loss of information while maintaining consistency and accuracy [43]. In smart grids, malicious alteration and tampering of critical data in sensors, meters, and command centers can be divided into three major categories. First, there is the integrity of the information in the network, which includes price information and power consumption. In addition, there is the integrity of the software running on the devices, and finally there is the integrity of the hardware which is somewhat of a more cyber-physical challenge. For instance, a set of compromised smart meters whose readings have been altered by the attacker can be considered as an integrity attack [18].

3.1.4 Accountability

Accountability is ensuring that every action in any given system can be traced back to the person or entity that performed it. This way, all the information can be used as evidence without anyone being able to dispute the chain of custody of the information or question the non-repudiation of the system. An example of an accountability attack concerns the monthly electricity bill of the consumers. Typically, a smart meter is able to determine and report the customer's power consumption on a daily basis. However, if a meter is under attack and its readings are altered, then the customer will end up with two separate readings, one from the meter and one from the utility company.

3.2 Cybersecurity Threats and Weaknesses

In this section, we will identify four of the most prevalent cybersecurity challenges that stem from the integration of IT with traditional energy grid systems. Also, we will see how most of the challenges emanate from our need to defend the CIA triad which we analysed in Sect. 3.1

3.2.1 Cyber-Attacks

Cyber-attacks on smart grids are a very commonly discussed topic due to the vulnerabilities existing in the grids' communication, networking, and physical entry points. Attacks in the smart grid environment can be categorised into two broad categories [8]:

- **Passive attacks**: these are attacks that do not intend to affect system resources and their sole purpose is to extract system information [11]. In these kinds of attacks, the attacker's objective is to learn or use information that it is transmitted, or to retrieve information stored in the system. Generally, passive attacks are relatively hard to detect, since no alteration of data takes place, thus the best defense against them is prevention through solid security mechanisms.
- **Active attacks**: these attacks are aimed towards a system's resources and attempt to either modify or disrupt them. The most common actors in these kinds of attacks are malicious users, spyware, worms, Trojans, and logic bombs [20]. According to Gai and Li [16], the most ordinary types of these attacks are device attacks, data attacks, network availability attacks, and privacy attacks, whereas [44] classify the attacks as those targeting availability, those targeting integrity, and finally those targeting confidentiality.

3.2.2 Trust

Varying requirements exist for operations performed in smart grids. The system consists of the power grid itself, the communication network, and the devices controlling the process [30]. Honesty and trustworthiness are essential behaviours in the relationship between the consumer and the utility company, thus the validity of the energy bill of the consumed energy is of vital importance from the consumer point of view, whereas the energy provider needs a trustworthy and fully auditable reporting tool for each operating device in the grid. These demands create new challenges that need to be addressed in an environment that all entities cannot be considered as trusted. Therefore, a trusted intermediary entity needs to decide upon the status validity of the devices and manage the access policies for the network, in a way that can authentically report the current state of the network to third parties.

3.2.3 Single Point of Failure

From a reliability perspective, it is well documented that a single point of failure is one of the biggest concerns in a master-slave architecture. In smart grids, a DDoS attack could disrupt, delay, or prevent the flow of data and eventually even collapse the AMI network. This denial of data exchange means a loss of control messages and may affect the power distribution to the customers in the smart grid.

In the UK, there were concerns regarding the way a proposed national Data Communication Company (DCC) was going to be set up, something that created significant delays to the rollout of the SMETS2 smart metering standard [31]. This centralized way of gathering smart meter data is a good paradigm of why a single data authority such as DCC, hence a single point of failure, should be avoided.

From a scalability perspective, the number of the clients is limited by the capacity of the AMI network in terms of bandwidth and routing capabilities, and the latency is determined by the round-trip time (RTT) between the AMI head-end and the devices in the network. In addition, as related research shows [40], there is an exponential growth of IoT devices, a trend that will likely be followed by smart energy meters as well. Therefore, scalability is emerging as one of the key factors for energy grid development and exploitation, considering the technical challenges connected with the geographical distribution over broad areas and the connectivity and resource availability in general [7].

3.2.4 Identity and Access Management

One issue with smart meters in smart grids is the management of the cryptographic keys that are required by every meter for cryptographic computations, such as the encryption of the transmitted data. Before the deployment of the AMI, the confidentiality of customer privacy and customer behaviour, as well as message authentication for meter reading, and control messages must be ensured. This can be solved by encryption and authentication protocols which depend on the security provided by cryptographic keys. The current industry standard is the use of a X.509 certificate for identification and for establishing a secure connection during data transmission. However, these cryptographic keys remain static for the whole life-cycle of the meter, and a key management mechanism that would allow manufacturers to periodically update or revoke them does not seem to be currently implemented. Furthermore, since such keys are also considered a form of strong device recognition, an attacker could possibly abuse the private key of the device [6] and enable access to the device by unauthorised parties, or even potentially impersonate the device in the network.

Based on the requirements set by NIST regarding cryptographic keys, e.g., a fixed cryptoperiod (i.e., expiration date) or the existence of a key recovery function [36], we consider that such a generic approach cannot be applied in an intelligent environment such as a smart grid, since the keys remain static and vulnerable and even though some functional requirements can be met, stricter security requirements cannot be fulfilled. A zero trust design philosophy is required in order to inspire confidence in the validity of the secure keys and certificates.

4 Opportunities

The emergence of technologies such as Blockchain and Artificial Intelligence (AI) has created a new field for research and innovation, while at the same time offering opportunities in the field of smart energy grids. In the following section, we will attempt to identify some of these opportunities and envision how to apply these technologies in order to countermeasure the aforementioned cybersecurity challenges.

4.1 Blockchain Application for Cyber Resiliency

Blockchain is defined as a distributed data base or *digital ledger* that records trans- actions of value using a cryptographic signature that is inherently resistant to modi- fication [38]. In a move towards a cyber-resilient energy grid, Blockchain could commoditise trust and also potentially support auditable multi-party transactions between energy providers and customers.

The blockchain is the equivalent of a book maintained by a bank, which contains all the accounts and each transaction made. One of the most interesting aspects of blockchains is that they contain the records of every transaction made since the beginning, also known as *genesis block*, by using a peer-to-peer distributed times- tamp server which generates computational proof of the chronological order of the transactions [34].

The use of blockchain presents numerous potential cybersecurity benefits to the electricity infrastructure:

- **Identity of Things**: As mentioned in Sect. 3.2.4, identity and access manage- ment of the devices in the grid is an issue that needs to be addressed efficiently. The ownership of a device can change during its lifetime or even be revoked in case a consumer is not consistent with his financial obligations towards the energy provider. Apart from ownership, there are also attributes that each device has, such as manufacturer, type, deployment GPS coordinates etc. Blockchain is able to address these challenges since it can register and provide identity to connected devices along with a set of attributes that can be stored on the blockchain distributed ledger in a fully auditable manner [24].
- **Data integrity**: As per blockchain's design, every transmitted block in the network, thus all data transmitted by the devices in the grid, are cryptograph- ically signed and proofed by the sender. Each node has its own unique public and private key and thereby it is ensured that the data are encrypted and cannot

be tampered. Finally, all blocks are recorded and timestamped on the chain and cannot be changed in a later time, therefore ensuring the accountability and the integrity as described in Sects. 3.1.4 and 3.1.3 respectively.

- **Securing communications**: The most commonly used network communication protocols, such as HTTP, MQTT and XMPP, are not secure by design and thus have to be wrapped within TLS at the application layer. However, protocols such TLS or IPSec rely on complicated and centralised certification authorities for the management of the keys, mainly through a public key infrastructure (PKI). With blockchain, there is no longer the need to rely on a centralised authority, since each node in the network receives a Universally Unique Identifier (UUID), as soon as it joins the network, and also creates an asymmetric key pair. This allows to simplify the handshake procedure and use light-weight protocols, such as TinyTLS, without handling and exchanging PKI certificates during the initial phase of the connection [24]. This way we are able to tackle the challenge described in Sect. 3.2.4 in an efficient manner without the added overhead of complex PKIs.

4.2 AI and Smart Contracts

Despite the fact that blockchain solutions add a layer of cryptography in communications and digital transactions, in complex IoT environments such smart energy grids, many complex cybersecurity challenges remain. An example is the patch management of the smart meters or their improper configuration. Especially in the first case, the timing between the discovery of a new vulnerability and the deployment of the patch to the affected devices is crucial. In such a scenario, a public repository could be queried periodically in order to check whether a new patch is available. The process could be performed with a blockchain-based *smart contract*, which would validate the transportation of the correct patch and provide an incentive for updating. Such a smart contract could operate on the basis of device-specific information, mainly model and firmware version of the device. According to this data, the contract would decide on whether an update is necessary and instruct the device to perform the update. In case the device is compromised and refuses to update, its trust score could start to decline and the energy provider would be notified regarding the misbehaving device.

Also, smart contracts could allow customers to directly trade with energy suppliers through autonomous trading agents without having to rely on middle-men. The agreement between the two transacting parties can be recorded in a smart contract which could also handle the automatic payment of the provided service. This way payments can occur automatically through the distributed ledger without the risk of financial data theft from data stored in energy retail supplier's databases [14]. Apart from that, smart contracts and automated transaction execution allows for real time settlement and accurate billing of payments overcome issues experienced in developing countries with delayed payments, debt and large numbers of unbanked population [1].

Whereas the distributed public ledger of blockchain may assist in increasing the trustworthiness, AI-enabled smart contracts could add unique value in the timely response to emerging cyber threats like an emergency response to a naturally occurring weather event or a cyber-physical hybrid attack [33]. That way, some functions of the power grid would become self-healing and resilient.

Additionally, through the combination of AI and blockchain, we could achieve an almost real-time security response to unauthorised attempts to change configurations or network and sensor settings. Anomaly-based intrusion detection systems assisted by Machine Learning (ML), could be an effective method to detect intrusions and attacks, which have not been previously detected. Such a system, combined with the immutability of blockchain, could reduce the overhead of the forensics investigation in case of a security incident, by providing a well-established timeline of events for evidence-analysis.

5 Conclusions

Smart grid is a system composed of various distributed components with the primary goal to intelligently deliver electricity, while at the same time allows the easy integration of new features and metrics in the traditional grid. Cybersecurity in the smart grid is a relatively new area of research and in this paper we presented an initial survey of security requirements and challenges. This was followed by a discussion on opportunities and mitigation techniques based on disruptive technologies such as blockchain and AI. Even though the proposed solutions still remain an uncharted territory in smart grid applications, the advancements in blockchain and AI make them the more attractive technologies thus far in the pursuit of building a secure and resilient smart grid.

Acknowledgements This work was supported by the ECHO project which has received funding from the European Union's Horizon 2020 research and innovation programme under the grant agreement no 830943.

References

1. Andoni, M., Robu, V., Flynn, D., Abram, S., Geach, D., Jenkins, D., McCallum, P., Peacock, A.: Blockchain technology in the energy sector: a systematic review of challenges and opportunities. Renew. Sustain. Energy Rev. **100**, 143–174 (2019)
2. Antonopoulos, A.M.: Mastering Bitcoin: unlocking digital cryptocurrencies.O'Reilly Media, Inc.(2014)
3. Bach, L., Mihaljevic, B., Zagar, M.: Comparative analysis of blockchain consensus algorithms. In: 2018 41st International Convention on Information and Communication Technology, Electronics and Microelectronics (MIPRO), pp. 1545–1550. IEEE (2018)
4. Back, A.: Hashcash-a denial of service counter-measure (2002)

5. Baliga, A.: Understanding blockchain consensus models. In: Persistent (2017)
6. Baumeister, T.: Adapting PKI for the smart grid. In: 2011 IEEE International Conference on Smart Grid Communications (SmartGridComm), pp. 249–254. IEEE (2011). https://doi.org/10.1109/SmartGridComm.2011.6102327
7. Bellavista, P., Zanni, A.: Towards better scalability for IoT-cloud interactions via combined exploitation of MQTT and CoAP. In: 2016 IEEE 2nd International Forum on Research and Technologies for Society and Industry Leveraging a better tomorrow (RTSI), pp. 1–6. IEEE (2016). https://doi.org/10.1109/RTSI.2016.7740614
8. Bou-Harb, E., Fachkha, C., Pourzandi, M., Debbabi, M., Assi, C.: Communication security for smart grid distribution networks. IEEE Commun. Mag. **51**(1), 42–49 (2013). https://doi.org/10.1109/MCOM.2013.6400437
9. Chaum, D.: Blind signatures for untraceable payments. In: Advances in Cryptology, pp. 199–203. Springer, USA (1983)
10. Chaum, D., Fiat, A., Naor, M.: Untraceable electronic cash. Adv. Cryptol. CRYPTO **88**, 319–327 (1990)
11. Cui, S., Han, Z., Kar, S., Kim, T.T., Poor, H.V., Tajer, A.: Coordinated data-injection attack and detection in the smart grid: a detailed look at enriching detection solutions. IEEE Signal Process. Mag. **29**(5), 106–115 (2012). https://doi.org/10.1109/MSP.2012.2185911
12. Decker, C., Wattenhofer, R.: Information propagation in the bitcoin network. Paper presented at the IEEE P2P 2013 Proceedings (2013)
13. Deetman, S.: Bitcoin could consume as much electricity as Denmark by 2020. Vice (2017). https://www.vice.com/en_us/article/aek3za/bitcoin-could-consume-as-much-ele ctricity-as-denmark-by-2020 (2019)
14. Deloitte: Blockchain Enigma. Paradox. Opportunity (2017)
15. Eyal, I., Sirer, E.G.: Majority is not enough: Bitcoin mining is vulnerable. Commun. ACM **61**(7), 95–102 (2018)
16. Gai, K., Li, S.: Towards cloud computing: a literature review on cloud computing and its development trends. In: 2012 Fourth International Conference on Multimedia Information Networking and Security (MINES), pp. 142–146. IEEE (2012)
17. Ghassemi, A., Bavarian, S., Lampe, L.: Cognitive radio for smart grid communications. In: 2010 First IEEE International Conference on Smart Grid Communications, pp. 297–302. IEEE (2010). https://doi.org/10.1109/SMARTGRID.2010.5622097
18. Giani, A., Bitar, E., Garcia, M., McQueen, M., Khargonekar, P., Poolla, K.: Smart grid data integrity attacks: characterizations and countermeasures π. In: 2011 IEEE International Conference on Smart Grid Communications (SmartGridComm), pp. 232–237. IEEE (2011)
19. Greer, C., Wollman, D.A., Prochaska, D.E., Boynton, P.A., Mazer, J.A., Nguyen, C.T., Fitz-Patrick, G.J., Nelson, T.L., Koepke, G.H., Hefner Jr, A.R.: NIST framework and roadmap for smart grid interoperability standards, release 3.0 (2014)
20. Gunduz, M.Z., Das, R.: Analysis of cyber-attacks on smart grid applications. In: 2018 International Conference on Artificial Intelligence and Data Processing (IDAP), pp. 1–5. IEEE (2018). https://doi.org/10.1109/IDAP.2018.8620728
21. Halim, F., Yussof, S., Rusli, M.E.: Cyber security issues in smart meter and their solutions. Int. J. Comput. Sci. Netw. Secur. **18**(3), 99–109 (2018)
22. Jin, D., Nicol, D.M., Yan, G.: An event buffer flooding attack in DNP3 controlled SCADA systems. In: Proceedings of the 2011 Winter Simulation Conference (WSC), pp. 2614–2626. IEEE (2011)
23. Kanjalkar, S., Kuo, J., Li, Y., Miller, A.: Short paper: I can't believe it's not stake! resource exhaustion attacks on PoS. In: International Conference on Financial Cryptography and Data Security. Springer, Berlin (2019)
24. Khan, M.A., Salah, K.: IoT security: review, blockchain solutions, and open challenges. Future Gener. Comput. Syst. **82**, 395–411 (2018). https://doi.org/10.1016/j.future.2017.11.022
25. Knapp, E.D., Samani, R.: Applied cyber security and the smart grid: implementing security controls into the modern power infrastructure. Newnes (2013)

26. Len, R.A., Vittal, V., Manimaran, G.: Application of sensor network for secure electric energy infrastructure. IEEE Trans. Power Deliv. **22**(2), 1021–1028 (2007). https://doi.org/10.1109/TPWRD.2006.886797
27. Liu, J., Xiao, Y., Gao, J.: Achieving accountability in smart grid. IEEE Syst. J. **8**(2), 493–508 (2014). https://doi.org/10.1109/JSYST.2013.2260697
28. Liu, J., Xiao, Y., Li, S., Liang, W., Chen, C.P.: Cyber security and privacy issues in smart grids. IEEE Commun. Surv. Tutorials **14**(4), 981–997 (2012)
29. Lu, Z., Lu, X., Wang, W., Wang, C.: Review and evaluation of security threats on the communication networks in the smart grid. In: 2010-MILCOM 2010 Military Communications Conference, pp. 1830–1835. IEEE (2010)
30. McDaniel, P., McLaughlin, S.: Security and privacy challenges in the smart grid. IEEE Secur. Priv. **7**(3), 75–77 (2009). https://doi.org/10.1109/MSP.2009.76
31. Meadows, S.: Only 80 second generation smart meters have been installed—as rollout stalls again the Telegraph (2018). https://www.telegraph.co.uk/bills-and-utilities/gas-electric/80-second-generation-smart-meters-have-installed-rollout-stalls/
32. Mohagheghi, S., Stoupis, J., Wang, Z.: Communication protocols and networks for power systems-current status and future trends. In: 2009 IEEE/PES Power Systems Conference and Exposition, pp. 1–9. IEEE (2009). https://doi.org/10.1109/PSCE.2009.4840174
33. Mylrea, M.: AI enabled: blockchain smart contracts: cyber resilient energy infrastructure and IoT. In: 2018 AAAI Spring Symposium Series (2018)
34. Nakamoto, S.: Bitcoin: a peer-to-peer electronic cash system (2008)
35. Nakamoto, S.: Bitcoin: a peer-to-peer electronic cash system (2009)
36. NIST: Recommendation for key management (2016). https://doi.org/10.6028/NIST.SP.800-57pt1r4
37. Pilkington, M.: 11 Blockchain technology: principles and applications. In: Research Handbook on Digital Transformations, p. 225 (2016)
38. Radziwill, N.: Blockchain revolution: how the technology behind Bitcoin is changing money, business, and the world. Qual. Manage. J. **25**(1), 64–65 (2018). https://doi.org/10.1080/10686967.2018.1404372
39. Ratner, M., Glover, C.: U.S. Energy: Overview and Key Statistics (2014)
40. Rodrigues, L., Guerreiro, J., Correia, N.: RELOAD/CoAP architecture with resource aggregation/disaggregation service. In: 2016 IEEE 27th Annual International Symposium on Personal, Indoor, and Mobile Radio Communications (PIMRC), pp. 1–6. IEEE (2016). https://doi.org/10.1109/PIMRC.2016.7794607
41. Rootzén, J.: Reducing carbon dioxide emissions from the EU power and Industry sectors—an assessment of key technologies and measures (2012)
42. SGC Committee: Smart grid cybersecurity strategy architecture and high-level requirements. The smart grid interoperability panel. Technical Report (2014)
43. Siozios, K., Anagnostos, D., Soudris, D., Kosmatopoulos, E.: IoT for Smart Grids. Springer, Berlin (2019)
44. Wang, W., Lu, Z.: Cyber security in the smart grid: survey and challenges. Comput. Netw. **57**(5), 1344–1371 (2013). https://doi.org/10.1016/j.comnet.2012.12.017
45. Yang, Y., Littler, T., Sezer, S., McLaughlin, K., Wang, H.: Impact of cyber-security issues on smart grid. In: 2011 2nd IEEE PES International Conference and Exhibition on Innovative Smart Grid Technologies, pp. 1–7. IEEE (2011). https://doi.org/10.1109/ISGTEurope.2011.6162722

The Use of Genetic Algorithms for Cryptographic Keys Generation

Michal Turčaník and Martin Javurek

Abstract On the base of Vernam's cipher security conditions the encryption key must be greater than or equal to the open text we want to encrypt. On the other hand, this key must not be repeated in another encryption. Also, each change of the encryption key adds security to the encryption process. If a cipher is changed several times while encrypting a single open text, it becomes very difficult to decrypt the message. Therefore, our goal was to design a method to generate an encryption key for a Tree Parity Machine by Genetic Algorithm. This method is able to create the same encryption keys on both sides that enter the encryption process. These keys could be changed during encryption. One of the first tasks is to create an input population for the genetic algorithm from the synchronized Tree parity machine. In this paper, the modified genetic algorithm for cryptography system was implemented and demonstrated. Genetic Algorithm is used to produce new population of cryptographic keys. The paper starts explaining the basic theory of the Tree Parity Machine and the proposed method of cryptographic key generation by Genetic Algorithm is presented and then the results will be analyzed.

Keywords Cryptographic keys generation · Genetic algorithms · Tree parity machine

1 Introduction

Security of the internet is still more and more important along with spreading the internet connections over the world [8]. Availability of commercial communication systems cause the potential risk of their misusing from adversary. There are several scenarios how communication can be misused. One of them is to coordinate of adversary attacks against friendly units during operations [5, 9].

The motivation of the paper is to create a method to generate cryptographic keys for Tree Parity Machine. Tree Parity Machine (TPM) is a special type of multilayer

M. Turčaník (✉) · M. Javurek
Armed Forces Academy of General M.R. STEFANIK, Liptovsky Mikulas, Slovakia
e-mail: michal.turcanik@aos.sk

© The Author(s), under exclusive license to Springer Nature Switzerland AG 2021
T. Tagarev et al. (eds.), *Digital Transformation, Cyber Security and Resilience of Modern Societies*, Studies in Big Data 84,
https://doi.org/10.1007/978-3-030-65722-2_19

315

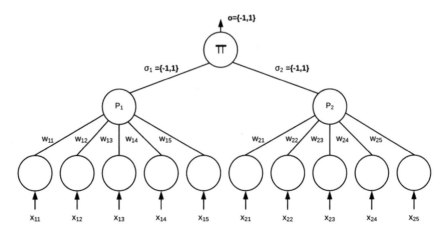

Fig. 1 Tree parity machine

forward neural network that is used in cryptography. The neural network is composed of neurons and synapses. The behavior of TPM can be changed by changing the values of synapses. For secure communication, TPM must be set on both sides of the chain the same way. 4 We do not intend to send all parameters (cryptographic keys) by insecure channel but we would like to create them by genetic algorithms independently on both sides of the communication chain. Results of the genetic algorithm generation on both sides of the chain must be identical and the process must be deterministic. The algorithm and the results of the new method of cryptographic key generation by the genetic algorithm are presented in the paper (Fig. 1).

2 Tree Parity Machine

Tree Parity Machine (TPM) is a special type of multilayer forward neural network that is used in cryptography. It is an artificial neural network topology for neural cryptography. The basis of TPM topology cryptography is the use of two identical artificial neural networks that are able to synchronize after mutual learning. TPM consists of K hidden neurons, N inputs to each hidden neuron and one output o (Fig. 1). Each entry into a hidden neuron has a generated random weight that can take values from $-L$ to $+L$. 1 These weights are changed by the learning rules so that after synchronization, the synapses values, between the respective inputs to each hidden neuron in both synchronizing TPMs, are the same. This means that the weights values are the same in both TPMs after synchronization. Furthermore, these weights may be used directly as an encryption key or may be used in any of the encryption key creation algorithms. The number of TPM weights depends on the number of hidden neurons and from inputs into each hidden neuron [1, 2].

The basic assumption for using TPM in cryptography is that synchronizing TPMs are initially identical. This means that the number of hidden neurons, inputs into each hidden neuron, as well as the possible values that can gain weights are identical and are kept secret from attackers. The weights values are different due to their random generation but must be kept secret [3].

The TPM synchronization process consists of generating random input x to be used as input to both TPMs to calculate the output in both TPMs. Only if the output value of both TPMs is the same, some of the learning rules is applied. This learning rule must be the same for both TPMs. On the basis of the learning rule, there will be updates, that is, a change in weights. The learning rule is applied until the both TPMs have the same values of weights [6].

3 Genetic Algorithms

Genetic algorithms (GAs) are adaptive heuristic algorithms mainly applied in the tasks of the optimization and the searching. They are based on principles of natural selection and natural genetics. They belong to the class of evolutionary algorithms and they are using biological evolutionary tools (operator of the mutation, crossover and selection). GA's heavily rely on inheritance to find solutions for optimization problems [4, 10].

The main idea involved in the GAs is to replicate the randomness of nature where the population of individuals adapts to its surroundings through the natural selection process and behaviour of the natural system. This means that the survival and reproduction of a specimen are promoted by the elimination of weak features. GAs creates a population in such a way that the feature which is dominant that has higher fitness value is replicated more likewise rest of the population. Evolution makes GAs a good candidate for the process of generating keys [7, 11].

The main task of GAs is to create set of keys, which will be used for securing communication, on the side of source and the destination of communication simultaneously. From set of keys the final user will have a possibility to choose the same key on both sides which will be used for setting of parameters of TPM. When transfer of data is done new key can be potentially applied for new communication.

4 Algorithm of Cryptographic Key Generation

The algorithm of key generation for TPM is divided to 3 basic steps (Fig. 2). Last two steps are repeated until final condition is met. The 2nd step is repeated until the expected size of population is reached (e.g. 80 or 100 genotypes). The 3rd step uses all three basic genetic operations (selection, crossover and mutation) to produce final population, which is used to define parameters of TPM. The final condition for 3rd step is represented by realisation of defined numbers of iterations.

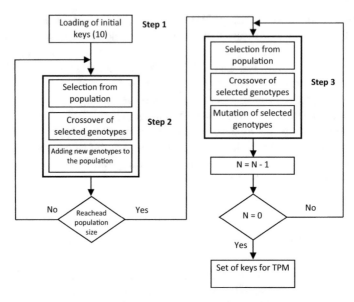

Fig. 2 Algorithm of cryptographic key generation by genetic algorithm

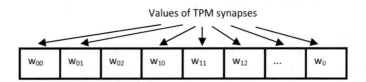

Fig. 3 Structure of the genotype for TPM

4.1 Loading of Initial Keys Set (Step 1)

The initial key set is input to the process of key generation. The initial key set consists of fixed number of keys (e.g. 10) and each key is composed of values of TPM synapses (Fig. 3). The range for every synapse is from -7 to 7.

4.2 Population Size Boosting (Step 2)

Traditional genetic algorithms store genetic information in the set of genotypes. Every genotype is represented by a bit array. On the base of the loaded key set the initial population of genotypes is created. Each genotype represents one key.

The crossover operator is really the main tool of the genetic algorithms. The crossover operator combines the characteristics of two solutions (parents) in order to generate a new solution (offspring). This operator is inspired in the way in which

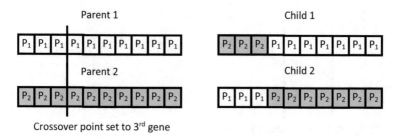

Fig. 4 Single point crossover operation

the genetic code of one individual is inherited to its descendants in nature. After that selected genotypes are used to create new population with the same size by single point crossover operator, which creates two new offspring genotypes on the base two of parent's genotype (Fig. 4). The only one crossover point was set after 3rd gene of the parent's genotypes. The 1st child is created on the base of first three genes of 1st parent and 4th until 10th genes of 2nd parent. The 2nd child is created from remaining parts of parents genotypes. During the execution of the algorithm, the position of the crossover point is changed in the range between 1st and 9th genes.

Input to the step 2 is population created in step 1. At first they are selected two genotypes for crossover operation. After that single point crossover operator is realized over two parent genotypes and creates two new offspring genotypes. The only one crossover point was set after randomly generated position in the parent´s genotypes. This operation is repeated for whole population. The new created genotypes are added to the actual population. The step 3 is repeated until the final size of population is reached (e.g. 80 genotypes) and it can be set by user. The optimal size of population is 80 genotypes which represents 80 possible setting for TPM.

4.3 Population Diversification (Step 3)

Input to the step 3 is population created in step 2. At first they are selected two genotypes for crossover operation. After that two point crossover operator is realized over two parent genotypes and creates two new offspring genotypes. In two-point crossover, the positions of two crossover points are picked randomly from the parent genotypes. The bits in between the two points are swapped between the parent organisms (Fig. 5). This operation is repeated for whole population. The new created genotypes are added to the actual population.

The next realised genetic algorithm operation is mutation. Mutation may be defined as a small random change in the genotype, to get a new solution. It is used to maintain and introduce diversity in the genetic population and is usually applied with a low probability—p_m. If the probability is very high, the GA gets reduced to a random search (Fig. 6).

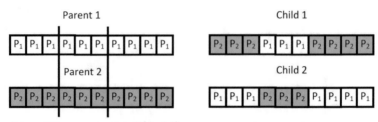

Two point crossover set to 3rd and 6th genes

Fig. 5 Two point crossover set operation

Fig. 6 Mutation operation

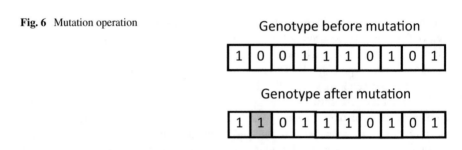

Mutation is the essential part of the genetic algorithm which is related to the "exploration" of the search space. It has been observed that mutation is essential to the convergence of the GA while crossover is not.

The mutation rate is set to value 0.002. It can be used several types of mutation: flipping of bits, boundary mutation, non-uniform mutation, uniform mutation and Gaussian mutation.

The size of population after realisation of 3rd step is the same (no new genotypes are added to the population).

5 Results

For verification of proposed algorithm of cryptographic key generation by genetic algorithm initial population of crypto key was created. Each key represents one configuration of TPM (values of synapses). In Fig. 7a there is shown the initial population consist of 10 genotypes. Each row represents one genotype. Every row (genotype) has assigned one colour. In the next step over this population the operation of single point crossover is realized (after the second gene). The result of given operation is depicted in Fig. 7b. Each genotype (row) is now composed from genes of two parent's genotypes which are depicted by two different colours in the each row.

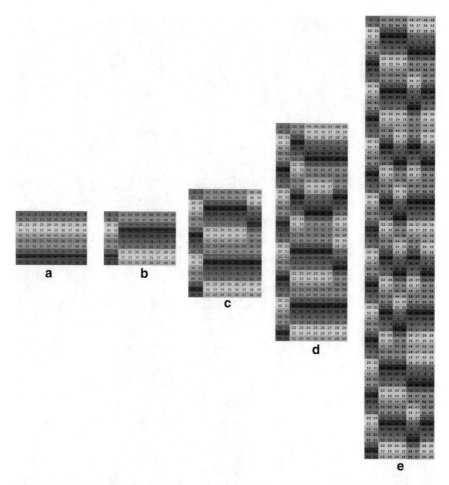

Fig. 7 Population size boosting **a** initial population, **b** initial population after single point crossover, **c** doubling the size of initial population, **d** double size increase of previous population (from step **c**), **e** increase of the previous population (from step **d**)

The population size is doubled in the next step and the size is 20 genotypes. After that another crossover operation over population is realized (after the 8th gene). The results are shown in Fig. 7c.

To reach the final size of the population we need to increase size of the population two times. At first it has to be changed from 20 to 40 genotypes (see Fig. 7d) and after that from 40 to 80 genotypes (Fig. 7e). Each change of population is followed by operation of crossover over all genotypes from population. In the boosting from 20 to 40 genotypes single point crossover was set to 4th gene and for other change from 40 to 80 genotypes was set to 6th gene.

The change of population of cryptographic keys due to diversification is shown in the Fig. 8. The input population from step 2 (Population size boosting) of the

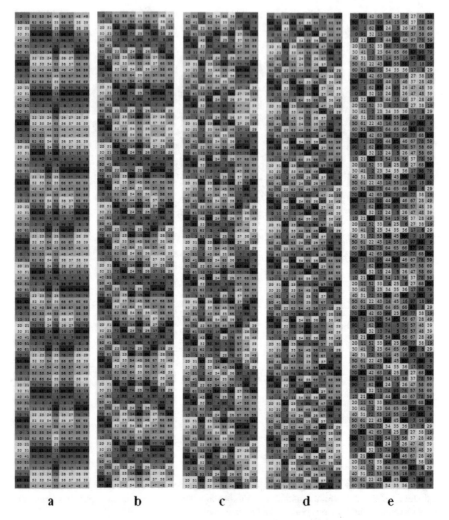

| a | b | c | d | e |

Fig. 8 The population diversification **a** population from the second step, population of cryptographic keys after **b** 5 generations, **c** 10 generations, **d** 15 generations, **e** 20 generations of GA

proposed method is shown in the Fig. 8a. We can track how parts of genotypes from the initial population distributed through all population and create required diversity.

The population of cryptographic keys for TPM after realisations of 5 generations of genetic algorithm is shown in the Fig. 8b, after realisations of 10 generations of genetic algorithm is shown in the Fig. 8c, after realisations of 15 generations of genetic algorithm is shown in the Fig. 8d and after realisations of 20 generations of genetic algorithm is shown in the Fig. 8e.

Because of using different colours to differentiate between given genotypes we can see appropriate level of diversity of cryptographic keys for TPM after realisation of 20 generations. If we increase the number of generations to 50 or 100 the level of diversity will not rise proportionally. Appropriate results and optimal time for finding reasonable solution we can achieve after realisation 20 or 25 generation for this method.

6 Frequency Test of Generated Cryptographic Keys

Frequency test tells us whether the output of the genetic algorithms fulfils the requirements expected from the new cryptographic keys. Moreover, the quality of new cryptographic keys can be commented considering test results. Any new cryptographic key is composed of some parts of the initial population which was received during initialization of the communication chain. The major requirement is continuous uniform distribution of values of final population. Each member of the analyzed population is equally probable for the continuous uniform distribution.

The result of the frequency test of the proposed method after 10 iterations of the GAs is shown in Fig. 8. In this figure ten different colours are used to depict the origin of the cryptographic key values. Every colour is equal to the one of the initial keys which was received in the beginning. Whole string of synapses (cryptographic key) is represented by the one column in Fig. 8. Every value from the final population has frequency of appearance 1%, which is an expected result.

7 Conclusions

In today communication technology, data encryption algorithms are used to provide secure communication between users. 4 So in this study, the possibility of using of genetic algorithms for cryptographic keys generation was analyzed.

For a mechanism of generation of the cryptographic key using Tree Parity Machine and genetic algorithm that will be identical on both sides of the communication chain for the encryption algorithm, we have designed new method to create a sufficiently large population without using too many synchronizing TPMs. By doing so, we can ensure a great variety of weights that can be used to create the encryption key. Thus, the encryption key can be changed quickly while encrypting a single message, making it difficult for the attacker to decrypt the encrypted message. Our next effort will be aimed to create a model of a cryptographic system that will use the resulting population.

References

1. Javurek, M., Turčaník, M.: Synchronization of two tree parity machines. In: 2016 New Trends in Signal Processing (NTSP), Demanovska Dolina, pp. 1–4 (2016)
2. Kanter, I., Kinzel, W.: The theory of neural networks and cryptography. Quantum Comput. Comput. Ing. **5**(1), 130–140 (2005)
3. Kinzel, W., Kante, I.: Neural cryptography. In: 9th International Conference on Neural Information Processing, vol. 3, pp. 1351–1354, Singapore (2002)
4. Michalewicz, Z.:Genetic Algorithms + Data Structures = Evolution Programs, 3rd edn. Springer, Berlin(1996)
5. Platenka, V., Mazalek, A., Vranova, Z.:The transfer of hidden information in data in the AMR-WB codec. In: 2019 Communication and Information Technologies (KIT), pp. 1–5, Vysoke Tatry, Slovakia (2019)
6. Ruttor, A., Kinzel, W., Kanter, I., Dynamics of neural cryptography. Phys. Rev. E Stat. Nonlinear Soft Matter Phys. **75**, 056104 (2007)
7. Ruttor, A., Kinzel, W., Nach, R., Kanter, I.: Genetic attack on neural cryptography. Phys. Rev. E **73**, 036121 (2006)
8. Santhanalakshmi, S., Sangeeta, K., Patra, G.K.: Design of stream cipher for text encryption using soft computing based techniques. IJCSNS Int. J. Comput. Sci. Netw. Secur. **12**(12), 149–152 (2012)
9. Stoianov, N.: One approach of using key-dependent S-BOXes in AES. In: Dziech, A., Czyżewski, A. (eds.) Multimedia Communications, Services and Security. MCSS 2011. Communications in Computer and Information Science, vol. 149. Springer, Berlin(2011)
10. Turcanik, M.: The optimization of the artificial neural network and production systems by genetic algorithms. In: MATLAB 2002: Proceedings of the Conference, pp. 562–568, Prague. ISBN 80-7080-500-5
11. Turcanik, M., Javurek, M.: Hash function generation by neural network. In: 2016 New Trends in Signal Processing (NTSP), pp. 1–5, Demanovska Dolina (2016)

An Alternative Method for Evaluating the Risk of Cyberattack Over the Management of a Smart House with Intuitionistic Fuzzy Estimation

Tihomir Videv, Boris Bozveliev, and Sotir Sotirov

Abstract The development of digital technologies has led to their introduction into all scopes of life, including people's homes. The need for effective management and control of a Smart House has also increased. With the remote-control capabilities that give the user extreme freedom, there is also the danger of hacking and taking control of the Smart House by people with malicious intents. This chapter is a follow-up to the Modelling of Smart Home Cyber System with Intuitionistic Fuzzy Estimation article, which focuses on cyber-threats in the so-called Smart House, which use fuzzy estimations for analysing possible ways and places to penetrate the security system and intervene in the functioning of the Smart House control systems.

Keywords Generalized nets · Cyber systems · Intuitionistic fuzzy estimation · Smart house

1 Introduction

This paper examines the Generalized Network (GN) [1–3], a Smart House model that is remotely controlled based on various identifications, channels and auto-mated internal and external systems, as well as cloud-based applications. We are also looking into the possibility of intruders entering the system to take control of the Smart House and the corresponding malicious activities in it [4]. We will also include here several types of grades using intuitionistic fuzzy estimations. Generalized nets and index matrices [5] are tools with which we can describe the processes running through the system very detailed [6]. Here you can model processes that

T. Videv · B. Bozveliev · S. Sotirov (✉)
"Prof. Dr. Assen Zlatarov" University, "Prof. Yakimov" Blvd, 8010 Burgas, Bulgaria
e-mail: ssotirov@btu.bg

T. Videv
e-mail: tvidev@abv.bg

B. Bozveliev
e-mail: bozveli@gmail.com

T. Tagarev et al. (eds.), *Digital Transformation, Cyber Security and Resilience of Modern Societies*, Studies in Big Data 84,
https://doi.org/10.1007/978-3-030-65722-2_20

work in parallel. The Smart House system we are looking at is implemented through remote control modules via telephone, tablet, laptop, PC and other devices via a cloud platform [7].

The applications on the client's end connect to the cloud platform, verify themselves and submit applications to its systems. Automated communication channels are also connected in the same way. On its side, the cloud structure communicates with the Smart House controller and through it controls the devices and systems. The possibilities of penetration into the system [8, 9] are through the applications on the client's end, the cloud structure, through the Smart House controller and through the communication channels between the cloud structure and the Smart House, as well as between the cloud structure and the end devices. When a user logs into the Smart House System Management application, they are verified using a password or other authentication methods. After successful verification, the application connects to the cloud platform. The software in the system also, through a similar method of verification, communicates with the controller that operates the systems in the Smart House.

As noted above, the capabilities of intrusion into the system, we add processes implemented through intuitionistic fuzzy estimations [10, 11], which in turn will monitor verification and analyse successful and unsuccessful verification attempts. Failures will be divided into deliberate and unintentional. Here we will add four more grades of the possibilities for penetration into the Smart House system, namely: Highly optimistic, optimistic, pessimistic and highly pessimistic.

In this article we look at a common Smart House model [12]. It is a follow-up to the article on Modelling of Smart Home Cyber System with Intuitionistic Fuzzy Estimations. The GN model will help us easily and clearly understand the primary mode of communication of systems in Smart House and the possibility of penetration so that we can improve security, troubleshoot and to better analyse the whole process. In the above context, we will evaluate the opportunities for communication penetration into the Smart House system through the grades mentioned above. Here we will use fuzzy sets (IFS). Intuitionist fuzzy sets (IFS), see [10, 11] represent an extension of the concept of fuzzy sets, showing the function $\mu_A (X)$ defining the presence of an element x to the set A, graded in the interval [0; 1]. The difference between fuzzy sets and intuitionist fuzzy sets (IFS) is in the presence of a second function $\mu_A (x)$, which determines the absence of the element x in the set A, where $\mu_A (x) \in [0; 1]$, $v_A (x) \in [0; 1]$ under the condition $\mu_A (x) + v_A (x) \in [0; 1]$.

The IFS itself is formally denoted by

$$A = \{\langle x, \mu_A(x), v_A(x)\rangle | \quad x \in E\}.$$

Strongly optimistic formula:

$$\langle \mu_1, v_1 \rangle + \langle \mu_2, v_2 \rangle + \cdots + \langle \mu_n, v_n \rangle = \sum_{i=1}^{n} \langle \mu_i, v_i \rangle = \langle 1 - \prod_{i=1}^{n}(1 - \mu_i), \prod_{i=1}^{n} v_i \rangle;$$

Optimistic formula:

$$\max(\langle \mu_1, \nu_1 \rangle, \ldots, \langle \mu_n, \nu_n \rangle) = \langle \max(\mu_1, \ldots, \mu_n), \min(\nu_1, \ldots, \nu_k) \rangle;$$

Pessimistic formula:

$$\min(\langle \mu_1, \nu_1 \rangle, \ldots, \langle \mu_n, \nu_n \rangle) = \langle \min(\mu_1, \ldots \mu_n), \max(\nu_1, \ldots, \nu_k) \rangle;$$

Strongly pessimistic formula:

$$\langle \mu_1, \nu_1 \rangle . \langle \mu_2, \nu_2 \rangle \ldots \langle \mu_n, \nu_n \rangle = \prod_{i=1}^{n} \langle \mu_i, \nu_i \rangle = \langle \prod_{i=1}^{n} \mu_i, 1 - \prod_{i=1}^{n} (1 - \nu_i) \rangle.$$

2 GN—Model

The system of Smart house facilitates the users. Initially the following tokens enter in the generalized net: For facilitation we separate the following types of tokens:

α Intruder $i \in [1 \div n]$;
β Database;
Υ Verification;
μ Person;
Ω Channel of the protocols and the management;
ω IFE.

The GN model of Smart House system (Fig. 1) is introduced by the set of transitions:

$$A = \{Z_1, Z_2, Z_3, Z_4, Z_5, Z_6\},$$

where the transitions describe the following process:

Z_1 Actions of the intruder;
Z_2 Cloud platform management;
Z_3 Actions of the users;
Z_4 Systems management;
Z_5 IFE Evaluation;
Z_6 Management of protocols.

$$Z_1 = \langle \{L_1, L_5\}, \{L_3, L_4, L_5\}, R_1, \wedge(L_1, L_5) \rangle$$

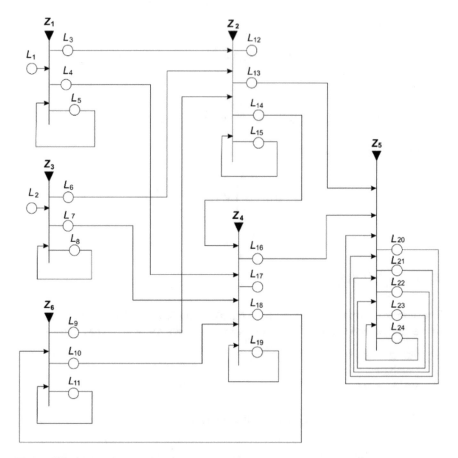

Fig. 1 GN model of the smart house system

$$R_1 = \begin{array}{c|ccc} & L_3 & L_4 & L_5 \\ \hline L_1 & False & False & True \\ L_5 & W_{5,3} & W_{5,4} & False \end{array}$$

where

$W_{5,3}$ The Intruder manages to break in the verification of cloud platform.

$W_{5,4}$ The Intruder manages to break in the verification of systems management.

The α token that enters place L_1 obtains the characteristic "Intruder".

The Υ_1 token that enters place L_3 obtains the characteristic "The Intruder is breaks verification of the cloud platform".

The Υ_2 token that enters place L_4 obtains the characteristic "The Intruder is breaks verification of the Systems management".

The β_1 token that enters place L_5 obtains the characteristic "Data base of the Intruder".

$$Z_2 = \langle \{L_3, L_6, L_9, L_{15}\}, \{L_{12}, L_{13}, L_{14}, L_{15}\}, R_2, \vee(L_3, L_6, L_9, L_{15}) \rangle$$

$$R_2 = \begin{array}{c|cccc} & L_{12} & L_{13} & L_{14} & L_{15} \\ \hline L_3 & False & False & False & True \\ L_6 & False & False & False & True \\ L_9 & False & False & False & True \\ L_{15} & W_{15,12} & True & W_{15,14} & False \end{array}$$

where

$W_{15,14}$ The verification is successful.
$W_{15,12}$ $W_{15,14}$.

The μ_2 token that enters place L_{12} obtains the characteristic "Exit".

The ω_1 token that enters place L_{13} obtains the characteristic "IFE identification".

The Ω_3 token that enters place L_{14} obtains the characteristic "Management of the controller".

The β_4 token that enters place L_{14} obtains the characteristic "Data base of the cloud platform".

$$Z_3 = \langle \{L_2, L_8\}, \{L_6, L_7, L_8\}, R_3, \vee(L_2, L_8) \rangle,$$

$$R_3 = \begin{array}{c|ccc} & L_6 & L_7 & L_8 \\ \hline L_2 & False & False & True \\ L_8 & W_{8,6} & W_{8,7} & False \end{array}$$

where

$W_{8,6}$ The verification is successful.
$W_{8,7}$ The intruder has destroyed the access database.

The μ token that enters place L_2 obtains the characteristic "Person".

The Υ_3 token that enters place L_6 obtains the characteristic "Verification to enter in the cloud platform".

The Υ_4 token that enters place L_7 obtains the characteristic "Verification in smart house controller".

The β_2 token that enters place L_8 obtains the characteristic "The person's data base".

$$Z_4 = \langle \{L_4, L_7, L_{10}, L_{14}, L_{19}\}, \{L_{16}, L_{17}, L_{18}, L_{19}\},$$
$$R_4, \vee(L_4, L_7, L_{10}, L_{14}, L_{19}) \rangle,$$

$$R_4 = \begin{array}{c|cccc} & L_{16} & L_{17} & L_{18} & L_{19} \\ \hline L_4 & False & False & False & True \\ L_7 & False & False & False & True \\ L_{10} & False & False & False & True \\ L_{14} & False & False & False & True \\ L_{19} & W_{19,16} & W_{19,17} & True & False \end{array}$$

where

$W_{19,16}$ The identification is successful;
$W_{19,17}$ There is a problem in the system.

The ω_2 token that enters place L_{16} obtains the characteristic "IFE identification".
The μ_3 token that enters place L_{17} obtains the characteristic "Exit".
The Ω_4 token that enters place L_{18} obtains the characteristic "The protocols canal".
The Ω_5 token that enters place L_{19} obtains the characteristic "Canals of the real system".

$Z_5 = \langle\{L_{13}, L_{16}, L_{20}, L_{21}, L_{22}, L_{23}, L_{24}\}, \{L_{20}, L_{21}, L_{22}, L_{23}, L_{24}\}, R_5, \vee(L_{13}, L_{16}, L_{20}, L_{21}, L_{22}, L_{23}, L_{24})\rangle,$

$$
R_5 = \begin{array}{c|ccccc}
 & L_{20} & L_{21} & L_{22} & L_{23} & L_{24} \\
\hline
L_{13} & True & True & True & True & True \\
L_{16} & True & True & True & True & True \\
L_{20} & True & False & False & False & False \\
L_{21} & False & True & False & False & False \\
L_{22} & False & False & True & False & False \\
L_{23} & False & False & False & True & False \\
L_{24} & False & False & False & False & True \\
\end{array}
$$

The ω_3 token that enters place L_{20} obtains the characteristic "IFE identification".
The ω_4 token that enters place L_{21} obtains the characteristic highly optimistic "IFE identification".
The ω_5 token that enters place L_{22} obtains the characteristic optimistic "IFE identification".
The ω_6 token that enters place L_{23} obtains the characteristic pessimistic "IFE identification".
The ω_7 token that enters place L_{24} obtains the characteristic highly pessimistic "IFE identification".

$$Z_6 = \langle\{L_{11}, L_{18}\}, \{L_9, L_{10}, L_{11}\}, R_6, \vee(L_{11}, L_{18})\rangle$$

$$
R_6 = \begin{array}{c|ccc}
 & L_9 & L_{10} & L_{11} \\
\hline
L_{11} & W_{11,9} & W_{11,10} & False \\
L_{18} & False & False & True \\
\end{array}
$$

where

$W_{11,9}$ The verification is successful.
$W_{11,10}$ $W_{11,9}$

The Ω_1 token that enters place L_9 obtains the characteristic "Control Smart device".
The Ω_2 token that enters place L_{10} obtains the characteristic "Connection with home controller".
The β_3 token that enters place L_{11} obtains the characteristic "Data base of the protocols".

3 Conclusion

The "Smart House" system is used to assist the user. The model is presented with its generalized network and shows the processes that are going on in the system as well as possible errors that can be caused. The possibility of malicious intrusion into the system is also being considered, using fuzzy grades through IFE. The GN model helps us analyse possible problems or simulate other problems so that we can optimize the system's behaviour.

Acknowledgements This work was supported by the Bulgarian Ministry of Education and Science under the National Research Programme "Information and Communication Technologies for a Digital Single Market in Science, Education and Security," approved by DCM # 577/17.08.2018.

References

1. Atanassov, K.: Generalized Nets. World Scientific, Singapore (1991)
2. Atanassov, K.: On Generalized Nets Theory. "Prof. M. Drinov" Academic Publishing House, Sofia (2007)
3. Atanassov, K., Sotirova, E.: Generalized Nets Academic Publishing House. "Prof. M. Drinov", Sofia (2017)
4. Perišić, A., et al.: A smart house environment-the system of systems approach to model driven simulation of building (house) attributes. In: 2015 IEEE 1st International Workshop on Consumer Electronics (CE WS), pp. 56–59. IEEE (2015)
5. Atanassov, K.: Index Matrices: Towards an Augmented Matrix Calculus. Springer, Cham (2014)
6. Bureva, V., Sotirova, E., Bozov, H.: Generalized net model of biometric identification process. In: Proceedings of the 20th International Symposium on Electrical Apparatus and Technologies (SIELA), pp. 1–4 (2018)
7. Zhou, Y., et al.: A design of greenhouse monitoring & control system based on ZigBee wireless sensor network. In: 2007 International Conference on Wireless Communications, Networking and Mobile Computing, pp. 2563–2567. IEEE (2007)
8. Karnouskos, S., et al.: Monitoring and control for energy efficiency in the smart house. In: International Conference on Energy-Efficient Computing and Networking. Springer, Berlin (2010)
9. Tseng, S.-P., et al.: An application of internet of things with motion sensing on smart house. In: 2014 International Conference on Orange Technologies, pp. 65–68. IEEE (2014)
10. Atanassov, K.: Intuitionistic Fuzzy Sets. Springer, Heidelberg (1999)
11. Atanassov, K.: Intuitionistic Fuzzy Logics. Springer, Cham (2017)
12. Videv, T., Bozveliev, B., Sotirov, S.: Modeling of smart home cyber system with intuitionistic fuzzy estimation. Digit. Transformation Cyber Secur. Resilience **43**(1), 45–53 (2019). ISSN: 0861-5160 (print), ISSN:1314-2119 (online)

A New Approach to Assess the Risk of Cyber Intrusion Attacks Over Drones Using Intuitionistic Fuzzy Estimations

Boris Bozveliev, Sotir Sotirov, Tihomir Videv, and Stanislav Simeonov

Abstract Drones or UAVs are unmanned aircraft systems, which include a UAV, a ground-based controller, and a system of communications between the two. The flight of UAVs can be autonomous, controlled by onboard computers, or controlled remotely by a human operator. In both cases the system may include GPS sensors. Drones can also be controlled by computerized satellites, mainly used by the military forces and, in some cases, for some civil missions. For military purposes, the communication control may be heavily encrypted by specially designed protocols. This chapter presents a method utilizing intuitionistic fuzzy estimations to assess the possible risk of cyber-attack over an UAV.

Keywords Drone · Generalized nets · Intuitionistic fuzzy sets · Open TX · Receiver · Telemetry · Transmitter · UAV

1 Introduction

In this article, with the help of Generalized Net (GN) [1, 2] model, we are going to use a new different way to assess the risk of cyber-attack intrusion [3, 4] over the communication control of a UAV [5, 6]. This article is preceded by a previous one, estimating the possibility of a cyber-theft over a drone [7].

Drones or UAVs are unmanned aircraft system, which include a UAV, a ground-based controller, and a system of communications between the two. The flight of UAVs can be autonomous, it can be controlled either by a human operator or autonomously by onboard computers [8]. These may include GPS sensors. Drones

B. Bozveliev (✉) · S. Sotirov · T. Videv · S. Simeonov
"Prof. Dr. Assen Zlatarov" University, "Prof. Yakimov" Blvd, 8010 Burgas, Bulgaria
e-mail: bozveli@gmail.com

S. Sotirov
e-mail: ssotirov@btu.bg

T. Videv
e-mail: tvidev@abv.bg

© The Author(s), under exclusive license to Springer Nature Switzerland AG 2021
T. Tagarev et al. (eds.), *Digital Transformation, Cyber Security and Resilience of Modern Societies*, Studies in Big Data 84,
https://doi.org/10.1007/978-3-030-65722-2_21

can also be controlled by computerized satellites, mainly in military applications and, in some cases, for some civil missions [9, 10].

For military purposes, the communication control may be heavily encrypted by specially designed protocols [3, 4, 11].In order to operate the UAV, we need some equipment, mainly a Radio Receiver that is located in the Drone itself and a Transmitter (Tx), usually located in the remote control, in some cases this control can be cloud or satellite based (military drones).

The transmitter is an electronic device that uses radio signals to transmit commands wirelessly through a pre-set radio frequency to a radio receiver that is located on the UAV. The receiver receives the commands and sends them to the drone's processor, which translates our drone commands into movement.

The UAV radio transmitter commonly uses the some of the following frequencies: 1.3 GHz, 2.4 GHz and 33 MHz. The 900 MHz and 1.3 GHz are typically long-range frequencies used in other systems.

The 2.4 GHz is most popular frequency. In this paper, we are going to use an open source firmware "Open TX". We are using it because it is widespread and used by many controllers.

Its main advantages are:

1. It is highly configurable and comes with a bundle of handy features, some of these features are:
2. It can store a large number of models on the radio, supports direct flashing, inflight audio/speech feedback etc.
3. Some logical switches and special functions can be easily programmed (low battery voltage, consumption) and much more.

Telemetry is the relevant information about our drone's flight. Information such as altitude, power consumption, our flight path, temperature, "RSSI" (Radio Signal Strength Indication) and even more. Either in real-time, or recorded, telemetry measures and keeps track of many layers of information about what our drones are doing. Real time telemetry data can be displayed on our radio transmitter screen. Some telemetry data can be recorded and downloaded, to a specific software on our computer in order to give us a millisecond view of our flight over time. This telemetry data can be displayed on the telemetry screen (in Open TX), and can be customized as audio warnings.

The radio receiver is installed on the UAV and it is capable of receiving signals from the radio transmitter, interpreting the signals via the flight controller where those commands are converted into specific actions controlling the aircraft.

TX Protocols are specific to some brands FrSky, Spektrum etc. Whereas RX protocols are usually universal PCM, PWM, PPM, SBUS. They are also used by some specific brands like TBS, Graupner, FrSky etc.

This means that a receiver must be compatible with a transmitter in order to establish communication. Frequencies should also be the same on both TX and RX receiver. For example, a 1.2 GHz receiver works only with a 1.2 GHz transmitter. In order to establish communication, a transmitter and a receiver must be paired together. A drone radio transmitter transmits commands via channels. Each channel is an individual action command that is being sent to the aircraft. Yaw, Throttle, Roll and Pitch are the four main inputs necessary for a drone control. Each of them uses one channel, so there is minimum of four channels required.

Every switch, slider or knob on the transmitter uses one channel to send the information through to the receiver. So at least 6 channels are used for the control [6] of cheaper models and more on the high-end ones. For our model, we are going to use the most common 2.4 GHz controller with an Open TX firmware, with channels selection module and a protocol selection DB. We will show how the TX protocol communicate with the receiver and also show how the signals are transmitted to the RX receiver, and a telemetry channel for RX for backwards communication with monitor exit on the transmitter controller that receives signals from the drone receiver. We will take a look at the UAV's processor and how is working and translating commands to the system. In this model, we are going to emulate an intruder who may scan our working radio signal communication ant try to take over any of the communication channels and secure protocols [4].

Our GN model is going to accumulate information from all input and output tokens and will send it to our intuitionistic fuzzy estimations algorithm which is going to assess the risk of cyber intrusion attacks, and will make an estimate whether the communication control [6] may be taken over by a cyber-attack.

The possible cyber intrusion attack may come from the interference of the radio communication between the controller and the receiver of the UAV [12].

It this paper, we are going to use fuzzy sets (IFS). The Intuitionistic Fuzzy Sets (IFSs, see [13–16]) represent an extension of the concept of fuzzy sets, as defined by Zadeh [17], exhibiting function $\mu_A(x)$ defining the membership of an element x to the set A, evaluated in the [0; 1][0; 1]-interval. The difference between fuzzy sets and intuitionistic fuzzy sets (IFSs) is in the presence of a second function $\nu_A(x)$ defining the non-membership of the element x to the set A, where $\mu_A(x) \in [0; 1]$, $\nu_A(x) \in [0; 1]$, under the condition

$$\mu_A(x) + \nu_A(x) \in [0; 1].$$

The IFS itself is formally denoted by:

$$A = \{\langle x, \mu_A(x), \nu_A(x)\rangle | x \in E\}.$$

We need (IFS), in order to evaluate the possible intrusion of the communication. The estimations are presented by ordered pairs $\langle \mu, \nu \rangle$ of real numbers from set [0, 1], where:

$$S = S_1 + \cdots + S_7$$
$$\mu = \frac{S_1 + S_2}{S},$$

where

S All the possible TX-communication attempts.

S_1 All the tokens from $\{L_3, L_6, L_9\}$ that enter in place L_{15}.

S_2 The number of error attempts when the token in place L_{12} enters in place L_{16}.

$$v = \frac{S_3 + S_4}{S},$$

where

S_3 All the tokens from place $\{L_3, L_6, L_9\}$ that enter in place L_{16}.

S_4 The number of error attempts when the token in place L_{12} enters in place L_{15}.

S_5 number of intruder attempts to take over communication.

S_6 All the successful attempts made to take over communication.

S_7 All the errors.

$$\pi = \frac{S_5 + S_6 + S_7}{S},$$

The degree of uncertainty $\pi = 1 - \mu - v$ is all the packets of information in the communication that go to their destination and all the possible manipulated entries by an external source.

In addition to our preceding paper [7] here we are going to use four additional ways to assess the risk of cyber-intrusion attack [12, 18–21] over the communication protocols of a UAV they are:

Strongly optimistic formula:

$$\langle \mu_1, v_1 \rangle + \langle \mu_2, v_2 \rangle + \cdots + \langle \mu_n, v_n \rangle = \sum_{i=1}^{n} \langle \mu_i, v_i \rangle = \langle 1 - \prod_{i=1}^{n} (1 - \mu_i), \prod_{i=1}^{n} v_i \rangle$$

Optimistic formula:

$$\max(\langle \mu_1, v_1 \rangle, \ldots, \langle \mu_n, v_n \rangle) = \langle \max(\mu_1, \ldots, \mu_n), \min(v_1, \ldots, v_k) \rangle;$$

Pessimistic formula:

$$\min(\langle \mu_1, v_1 \rangle, \ldots, \langle \mu_n, v_n \rangle) = \langle \min(\mu_1, \ldots, \mu_n), \max(v_1, \ldots, v_k) \rangle;$$

Strongly pessimistic formula:

$$\langle \mu_1, v_1 \rangle . \langle \mu_2, v_2 \rangle ..... \langle \mu_n, v_n \rangle = \prod_{i=1}^{n} \langle \mu_i, v_i \rangle = \langle \prod_{i=1}^{n} \mu_i, 1 - \prod_{i=1}^{n} (1 - v_i) \rangle$$

2 Radio Communication Protocol Between Transmitter and Receiver

The transmitter of the radio control provides the communication between UAV's onboard receiver, and then the drone's CPU processes the signal commands translating these in to movements [11, 19].

Initially the following tokens enter in the generalized net:

β User;
Λ Transmitter, Channel processing, Receiver processing and Protocol processing databases;
δ Drone CPU.
α Intruder.
χ IFS estimation algorithms.

3 GN—Model

GN model of common ъ process between a UAV radio transmitter Tx and a Rx radio receiver (Fig. 1) is introduced by the set of transitions:

$$A = \{Z_1, Z_2, Z_3, Z_4, Z_5, Z_6, Z_7, Z_8\},$$

where the transitions describe the following processes:

Z_1 Transmission signal processing;
Z_2 Communication protocols processing;
Z_3 Channel selection processing;
Z_4 Radio communication cyber-attack.
Z_5 Receiver commands processing;
Z_6 UAV controller processing commands;
Z_7 Flying drone;
Z_8 Intuitionistic fuzzy estimations algorithm assessing the risk of cyber-intrusion.

GN model of a radio communication control between a receiver and a transmitter and backwards communication with the receiver showing a possible cyber-attack over the UAV.

$$Z_1 = \langle \{L_1, L_2, L_7, L_{10}, L_{18}\}\{L_2, L_3, L_4, L_5\}, R_1, \vee(L_1, L_2, L_7, L_{10}, L_{18})\rangle$$

$$R_1 = \begin{array}{c|cccc} & L_2 & L_3 & L_4 & L_5 \\ \hline L_1 & true & false & false & false \\ L_2 & true & true & W_{2,4} & false \\ L_7 & true & false & false & false \\ L_{10} & true & false & false & false \\ L_{18} & false & false & W_{18,4} & true \end{array}$$

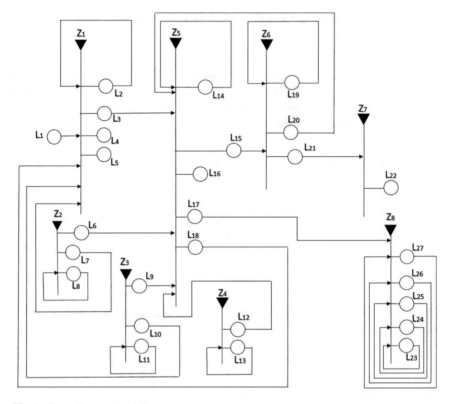

Fig. 1 Generalizen net model

where

$W_{2,4}$ There is an error with communications commands;
$W_{18,4}$ There is a telemetry Rx signal error;

The token that enters place L_2 obtains the characteristic "Database command transmitter";

The token that enters place L_3 obtains "signal commands sent to receiver";

The token that enters place L_4 obtains the "Error exit";

The token that enters place L_5 obtains the characteristic "Audio/video screen signal transmitter";

$$Z_2 = \langle\{L_8\}\{L_6, L_7, L_8\}, R_2, \wedge(L_8)\rangle$$

$$R_2 = \frac{\begin{array}{c|ccc} & L_6 & L_7 & L_8 \\ \hline L_8 & true & true & true \end{array}}{},$$

The token that enters place L_6 obtains the characteristic "Signal command sent to the transmitter";

The token that enters place L_7 obtains the characteristic "Signal command sent to the receiver";

The token that enters place L_8 obtains the characteristic "Signal commands processing database";

$$Z_3 = \langle \{L_{11}\}\{L_9, L_{10}, L_{11}\}, R_3, \wedge(L_{11})\rangle$$

$$R_3 = \frac{\begin{array}{c|ccc} & L_9 & L_{10} & L_{11} \\ \hline L_{11} & true & true & true \end{array}}{},$$

The token that enters place L_9 obtains the characteristic "Channel selection signal sent to receiver";

The token that enters place L_{10} obtains the characteristic "Channel selection signal sent to transmitter";

The token that enters place L_{11} obtains the characteristic "Channel processing database";

$$Z_4 = \langle \{L_{13}\}\{L_{12}, L_{13}\}, R_4, \vee(L_{13})\rangle$$

$$R_4 = \frac{\begin{array}{c|cc} & L_{12} & L_{13} \\ \hline L_{13} & true & true \end{array}}{},$$

The token that enters place L_{12} obtains the characteristic "Random cyber-attack over communication link";

The token that enters place L_{13} obtains the characteristic "Drone communication channel/link scanner";

$$Z_5 = \langle \{L_3, L_6, L_9, L_{12}, L_{14}, L_{20}\}\{L_{14}, L_{15}, L_{16}, L_{17}, L_{18}\},$$
$$R_5, \wedge(L_3, L_6, L_9, L_{12}, L_{14}, L_{20},)\rangle$$

$$R_5 = \begin{array}{c|ccccc} & L_{14} & L_{15} & L_{16} & L_{17} & L_{18} \\ \hline L_3 & true & false & false & false & false \\ L_6 & true & false & false & false & false \\ L_9 & true & false & false & false & false \\ L_{12} & true & false & W_{12,16} & true & false \\ L_{14} & true & W_{14,15} & W_{14,16} & true & false \\ L_{20} & false & false & W_{20,16} & false & true \end{array}$$

where

$W_{14,15}$ There is a TX signal sent OK;
$W_{12,16}$ The cyber-attack is unsuccessful;
$W_{14,16}$ $W_{14,15}$;
$W_{20,16}$ There is a telemetry signal error;

The token that enters place L_{14} obtains the characteristic "Receiver processing database";

The token that enters place L_{15} obtains the characteristic "Successful translation of signal to processor";

The token that enters place L_{16} obtains the characteristic "Error/exit";

The token that enters place L_{17} obtains the characteristic "accumulated information form tokens L_{15}, L_{16}—IFS estimate unknown";

The token that enters place L_{18} obtains the characteristic "Rx—backwards communication telemetry monitor";

$$Z_6 = \langle \{L_{15}, L_{19}\}\{L_{19}, L_{20}, L_{21}\}, R_6, \wedge(L_{15}, L_{19}) \rangle$$

$$R_6 = \begin{array}{c|ccc} & L_{19} & L_{20} & L_{21} \\ \hline L_{15} & true & false & false \\ L_{19} & true & W_{19,20} & W_{19,21} \end{array},$$

where

$W_{19,20}$ There is a telemetry back communication signal sent to the transmitter;
$W_{19,21}$ There is a translated command to the drone;

The token that enters place L_{19} obtains the characteristic "Drone CPU".

The token that enters place L_{20} obtains the characteristic "Receiver telemetry".

The token that enters place L_{21} obtains the characteristic "Translated signal into a drone movement".

$$Z_7 = \langle \{L_{21}\}\{L_{22}\}, R_4, \vee(L_{21}) \rangle$$

$$R_7 = \begin{array}{c|c} & L_{22} \\ \hline L_{21} & true \end{array}$$

The token that enters place L_{22} obtains the characteristic "Drone is in operational state".

$$Z_8 = \langle \{L_{17}, L_{23}, L_{24}, L_{25}, L_{26}, L_{27}\}\{L_{23}, L_{24}, L_{25}, L_{26}, L_{27}\},$$
$$R_8, \vee(L_{17}, L_{23}, L_{24}, L_{25}, L_{26}, L_{27}) \rangle$$

$$R_8 = \begin{array}{c|ccccc} & L_{23} & L_{24} & L_{25} & L_{26} & L_{27} \\ \hline L_{17} & True & True & True & True & True \\ L_{23} & True & False & False & False & False \\ L_{24} & False & True & False & False & False \\ L_{25} & False & False & True & False & False \\ L_{26} & False & False & False & True & False \\ L_{27} & False & False & False & False & True \end{array}$$

The token that enters place L_{17} obtains the characteristic "accumulated information form tokens L_{15}, L_{16}—IFS estimate unknown". The token that enters place L_{23} obtains the characteristic "IFS estimations $\langle \mu_k, \nu_k \rangle$".

Initially when no information has been derived from places L_4, L_{13}, L_{16} L_{17}, all estimates take initial values of $\langle 0, 0 \rangle$.

When ≥ 0, the current $(k + 1)$-st estimation is calculated on the basis of the previous estimations according to the recursive formula (as before):

$$\langle \mu_{k+1}, \nu_{k+1} \rangle = \langle \frac{\mu_k k + \mu}{k + 1}, \frac{\nu_k k + \nu}{k + 1} \rangle,$$

where $\langle \mu_k, \nu_k \rangle$ is the previous estimation, and $\langle \mu, \nu \rangle$ is the latest estimation of the possible communication intrusion, for $\mu, \nu \in [0, 1]$ and $\mu + \nu \leq 1$. This way the token in place L_{21} forms the final estimation of the accumulated information from all the input and output tokens on the basis of previous and the latest events.

The token that enters place L_{23} obtains the characteristic "IFE identification".

The token that enters place L_{24} obtains the characteristic highly optimistic "IFE identification".

The token that enters place L_{25} obtains the characteristic optimistic "IFE identification".

The token that enters place L_{26} obtains the characteristic pessimistic "IFE identification".

The token that enters place L_{27} obtains the characteristic highly pessimistic "IFE identification".

4 Conclusion

Assessing the risk of a successful cyber-attack over a drone is a complex task, consisting of a detailed analysis of the possible threats over the communications management.

In this article, we tried to make an estimate using the IFE method, using a model that included 5 different estimate grades. Accordingly, the more estimate models we add the more accurate the results will come out. With this analysis, we showed all the processes of the communication signals and commands of the transmitter and the receiver in an UAV system. In addition, IFC analysis gave us a comprehensive view and a very good risk estimation.

Acknowledgements This work was supported by the Bulgarian Ministry of Education and Science under the National Research Programme "Information and Communication Technologies for a Digital Single Market in Science, Education and Security" approved by DCM # 577/17.08.2018.

References

1. Atanassov, K.: Generalized Nets. World Scientific, Singapore, New Jersey, London (1991)
2. Atanassov, K.: On Generalized Nets Theory. Prof. M. Drinov Academic Publ. House, Sofia (2007)
3. Tsvetanov, T., Simeonov, S.: Applying pattern detection network security against denial-of-service attacks. In: WMSCI 2006—The 10th World Multi-Conference on Systemics, Cybernetics and Informatics, Jointly with the 12th International Conference on Information Systems Analysis and Synthesis, ISAS 2006—Proceedings, vol. 7, pp. 314–319 (2006). ISBN:9806560728
4. Tsvetanov, T., Simeonov, S.: Securing a campus network. In: MIPRO 2006—29th International Convention Proceedings: Digital Economy—3rd ALADIN, Information Systems Security and Business Intelligence Systems, vol. 5 (2006). ISBN: 9532330186
5. Lin, Y., Saripalli, S.: Collision avoidance for UAVs using reachable sets. In: 2015 International Conference on Unmanned Aircraft Systems (ICUAS), pp. 226–235 (2015)
6. Simeonov, S., Kostadinov, T., Belovski, I.: Implementation of collision sense and orientation system. In: 2019 16th Conference on Electrical Machines, Drives and Power Systems, ELMA 2019—Proceedings, 1 June 2019. ISBN: 978-1-7281-1413-2
7. Bozveliev, B., Sotirov, S., Videv, T.:Generalized net model of possible drone's communication control cyber theft with intuitionistic fuzzy estimations43(1), 35–44 (2019)
8. Kardasz, P., Doskocz, J.: Drones and possibilities of their using. J. Civ. Environ. Eng. 6 (2016). https://doi.org/10.4172/2165-784X.1000233
9. Cassara, P., Colucci, M., Gotta, A.: Command and Control of UAV Swarms via Satellite (2018). https://doi.org/10.1007/978-3-319-76571-6_9
10. Singhal, G., Bansod, B., Mathew, L: Unmanned Aerial Vehicle Classification, Applications and Challenges: A Review (2018). https://doi.org/10.20944/preprints201811.0601.v1
11. Simeonov, S., Georgieva, P., Germanov, V., Dimitrov, A., Karastoyanov, D.: Computer system for navigating a mobile robot. In: ISCI 2011—2011 IEEE Symposium on Computers and Informatics 2011, pp. 183–187
12. Vardeva, I., Valchev, D.: generalized net model for building a standard ad-hoc on demand distance vector routing in a wireless network, development in fuzzy sets, intuitionistic fuzzy sets. In: Generalized Nets and Related Topics. Applications, vol. II, pp. 303–310. System Research Institute, Polish Academy of Science. Warsaw (2010). ISBN: 13 9788389475305
13. Atanassov, K.: Intuitionistic fuzzy sets. In: Proceedings of VII ITKR's Session, Sofia, June 1983 (in Bulgarian)
14. Atanassov, K.: Intuitionistic fuzzy sets. Fuzzy Sets Syst. 20(1), 87–96 (1986)
15. Atanassov, K., Intuitionistic fuzzy sets: theory and applications. Physica, Heidelberg (1999)
16. Atanassov, K.: On Intuitionistic Fuzzy Sets Theory. Springer, Berlin (2012)
17. Zadeh, L.A.: Fuzzy sets. Inf. Control 8, 333–353 (1965)
18. Vardeva, I.: Modeling net attacks for listening and break-down encrypted messages by using generalized nets model. In: Ninth International Workshop on Generalized Nets, Sofia, vol. 1, pp. 20–27 (2008). ISSN: 1310-4926
19. Vardeva, I., Staneva, L.: Intuitionistic fuzzy estimations of establishing connections with file transfer protocol for virtual hosts. In: 1st International Workshop on IFSs, Mersin, 14 Nov 2014. Notes on Intuitionistic Fuzzy Sets, vol. 20, no. 5, pp. 69–74 (2014). ISSN: 1310–4926
20. Sandip, P., Zaveri, J. (2010). A risk-assessment model for cyber attacks on information systems. J. Comput. 5. https://doi.org/10.4304/jcp.5.3.352-359
21. Chedantseva, Y., Burnap, P., Blyth, A., Eden, J., Soulsby, S.: A review of cyber security risk assessment methods for SCADA systems. Comput. Secur. 56, 1–27 (2016). https://doi.org/10.1016/j.cose.2015.09.009.

Footwear Impressions Retrieval Through Textures and Local Features

Joseph G. Vella and Neil Farrugia

Abstract The proposed artefact applies pre-processing filters, extracts key features, and retrieves the relevant matches from a shoeprint impression repository. Two functions were utilized for matching impressions. One function is texture based and creates an MPEG-1 movie out of two input images and employs the size of the output movie as a similarity measure. The other function is local feature based and uses SURF feature extraction and MSAC for matching. For pre-processing of the prints, a set of well-known techniques were employed. Also, we implemented a technique to facilitate better matching through splitting the input prints into smaller prints and then matching on these. FID 300 is a publicly available dataset of footwear impressions in greyscale. It comes with 1175 reference prints (e.g. sole images from tip to heel), and 300 prints lifted from real crime scenes, the latter being incomplete and with low image quality. The evaluation was done over various options and always against all reference prints in the FID 300. Clearly the evaluation results are affected by the quality of the lifted images. Evaluations were done in three batches (each having different pre-processing): first, all crime scene prints with the texture function got an average accuracy of 61%; second, a sample of 43 lifted prints with the texture function got 65% average accuracy; third, all crime scene prints and the local feature function applied got 50% average accuracy.

Keywords Digital forensics · Footwear impressions · Texture based similarity · Feature based similarity

J. G. Vella (✉) · N. Farrugia
Department of Computer Information Systems, University of Malta, Msida, Malta
e-mail: joseph.g.vella@um.edu.mt

N. Farrugia
e-mail: neilf95@gmail.com

© The Author(s), under exclusive license to Springer Nature Switzerland AG 2021
T. Tagarev et al. (eds.), *Digital Transformation, Cyber Security and Resilience of Modern Societies*, Studies in Big Data 84,
https://doi.org/10.1007/978-3-030-65722-2_22

343

1 Introduction

Footwear impressions are left by the outside sole of a shoe on a surface whilst its wearer is walking. In most Scene of the Crime (SoC) environs, a number of footwear impressions are lifted, albeit partial when compared to a shoe sole's design [5]. In fact, it is estimated that over 30% of burglaries produce viable footwear prints that can be efficiently obtained from a SoC [2]. In the majority of cases, footwear impressions, and occasionally foot prints [11], are more commonly found at SoC than other biometric data such as fingerprints. This could be so because fingerprints are usually prevented from being left behind by using gloves, but covering is not prevalent with footwear.

From the lifted footwear impressions investigators can gain insights regarding the SoC and its dynamics. The variety of prints can determine the number of people involved, the sharpness of the print can tell if a person was walking, running, sneaking, and even show the path a perpetrator traversed. The lifted prints can also link multiple crimes together.

New shoes of the same model will leave behind the same or very similar prints. However, as time passes, natural erosion known as "wear" occurs on the shoe, which damages it in a distinctive way. Where and to what extent the wear occurs depends on the shoe's material, which surfaces it has been in contact with, and owner's use of the shoe. A famous judicial case where footprints were involved during the investigation was the OJ Simpson Trial of 1995 [27]. In this case the investigators found many blood-soaked shoeprints at the SoC and these matched to a pair of designer Bruno Magli shoes owned by Simpson [12].

2 Aims and Objectives

There is a real need for automating footwear impression; i.e. a system that can sift prints accurately and efficiently with minimal subjective intervention by investigators. The aim of this study is to create a system that can automatically compare footwear impressions and return relevant matches similar to the footwear impression of interest. Since prints from real crime scenes are not clear and contain high amounts of noise, the system must be robust too. It needs to cope with varying amounts of noise including a shoe's wear, occultation and print overlaps.

An important objective is to have a good and a realistic dataset. The lifted prints should be real and the reference print dataset is in number and variety.

The matching of prints is to be based on a similarity measure and it is expected that two techniques are made available. One technique is based on texture, and the second based on local features. Whereas the first is novel, the second allows to compare to other and similar projects to gauge its effectiveness.

Techniques that help better image quality and reduce noise are to be made available.

3   Background and Related Work

3.1   Footwear Impressions

A footwear impression, known variously as a shoeprint, imprint and footmark, is an outline or indentation left by a foot on a surface [5]. These patterns are mostly particular for every shoe which makes them very useful hints for evidence collaboration by the SoC investigators. It is also known that some shoe manufacturers share sole patterns. The difference in the manufactured patterns can depend on many factors such as the material of the sole, the brand and also the shoe's size.

Impressions obtained from new shoes in a laboratory, e.g. sampled form a manufacturer, are usually of good quality and are then encoded in a raster format. These prints are called reference prints in contrast to lifted prints extracted from a SoC. Also these reference impressions are devoid of a walker's gait and walking effects. The number of reference prints per shoe sole are usually limited; only a few cases from a variety left/right soles and sizes are available.

Each and every shoe, except newly purchased ones, is said to be distinctive in some way. Even if two shoes are of the same brand and size, when worn by two different people each will have their own unique imprint given a substantial amount of use. This is because as a shoe is worn, it is naturally damaged and eroded as a result of particular use and aging [6]. The area of most wear on the sole of a shoe will be particular for every person due to the wearer's gait, the environment they walk in, and the material of the shoe. Also acquired marks on a shoe can be very discriminating.

Footprints lifting from crime scenes use two generic techniques: 3D data profile left by a shoe onto a soft surface (e.g. soil, mud, snow); and 2D data profile left by a shoe on harder surface by leaving accumulated material (e.g. dust, blood, liquids like oil). Furthermore SoC lifted prints are rarely complete; these are usually partial, overlapping, and occulated. This is a limiting factor when compared to the better conditions in a laboratory to obtain a reference print (e.g. sessile, complete, and unworn shoes). SoC prints also need to be converted into a raster format and pre-processing is needed to clean and orient the prints.

Due to deterioration of a shoe's patterns, investigators are interested in shoe prints as collaborating evidence. A study by Adair et al. [1] asserts that the characteristics of each footwear impression (even that of the same shoe brand and model) are distinctive to a specific print at that specific point in time. As time passes a sole's pattern is further degraded and thus changing its print.

Quality shoe impressions which are lifted during a forensic investigation and accurately matched to a reference print can be as crucial to a case as other strong biometric data such as fingerprints [3].

3.2   Existing Systems

In academia there are prototypes of automated shoe print analysis, however it is difficult to assess whether these can develop into actual systems. Presently, the systems that are commercially available are not fully automated, meaning they still rely on human interaction to classify shoeprint instances. Advances of semi-automatic methods that had been made [3, 24] involve manual interaction of the domain expert to create a description of a print using a "codebook" of basic shapes such as circles, lines, and wavy patterns. The expert must perform the same procedure for each lifted image (called query image in some cases). SoleMate FPX [26] is an offering from Foster Freeman that has such an approach. https://www.fosterfreeman.com/qdelist/trace-list/511-fpx-shoeprint-identification.htmlThe lifted print image is adorned with codebook entries by an expert which may lead to it being inconsistency compared to reference images as an expert's encoding is subjective. The main drawback here is that the process is tedious and very time-consuming. These problems, and other requirements, push the need for automatic footprint analysis systems to be further researched, developed, and deployed.

Many different studies have been conducted regarding automatic shoeprint recognition systems [13, 14, 16, 20, 22, 28, 30] include cross domain image matching paper. Determining which implementation is fit for a purpose is very perplexing because all the models and their evaluations were conducted on different datasets, using different types of matching methods, and adopting different testing environments [23]. Also many report their findings on unavailable datasets and furthermore their SoC prints are based on edits of their reference datasets (e.g. some of the reference prints are impaired with noise, masking, and disorientated).

A paper by Luostarinen and Lehmussola [16], compare and contrast seven different automated shoe impression recognition prototypes in order to determining which is the most reliable; this paper also contains author's own contributions too. The evaluated approaches were set-up closely following the respective primary literature source. The approaches are Power Spectral Density (PSD), Hu's Moment Invariants, Gabor Transform, Fourier-Mellin Transform, Mahalonobis Distance Map, Local Interest Points with RANSAC, and Spectral Correspondence of Local Interest Points. The authors tested these seven methods under the same environment in order to get a fair estimate of how they perform and thus have a comparable baseline. The authors collected 499 pairs of reference shoe prints and used these to generate and partition them into three datasets marked as 'good', 'bad' and 'ugly' over which the seven methods were applied. Each dataset differed in quality from excellent quality in the good set, to noisy and damaged images (similar to actual crime scene lifted prints) in the ugly set. The results showed significant differences in the system's accuracy when applied on the good and bad datasets however they all performed poorly when making use of actual prints taken from a SoC (i.e. using lifted prints). Results for the bad dataset range from 10 to 85% probability of good match when comparing 5% of the dataset. Another contribution of this paper is the manual cropping of a lifted print into five sub-regions, e.g. one is the heel, that correspond to regions actually found in a SoC print and use a cropped region for comparison (rather than the whole incomplete lifted print).

Another system proposal is by Kortylewski et al. [14] who describe an unsupervised system to extract periodic patterns from the footprints and then attempt to match the prints based on these patterns. The authors make use of pre-processing techniques such as local and rotation normalization. The system is said to convert the periodic patterns which are extracted using a Fourier Transform and compare these transformations in order to check if two footwear impressions are similar. Results range from 10 to 55% probability of good match when comparing 5% of the dataset for crime scene impressions.

In a recent paper by Wang et al. [30] the authors use a multi-layer feature extraction and matching method (mainly the wavelet-Fourier-Mellin transform and similarity). Of interest here is the authors modelling of a print into partitions that are related to a foot's anatomy and integrating this into their retrieval system after user intervention. Another interesting facet of their reported work is their lifting process also measures pressure values due to gait. The authors build their own datasets and assert that: their reference prints are complete and concise (i.e. with little to no noise) and are related to pressure measures exerted by the shoe wearer; lifted images from a SoC also include contours that they are able to measure by using specialized equipment. Stated results are very good, when compared to parallel methods they run, albeit their method comes with a heavier computational load.

Rida et al. [23] compiles a survey on a good number of systems (and a subset of which are also mentioned in [16]). The authors consider forty works from the literature and divides these in their modes of recognition on the whole impression (called holistic) or based on a local feature. This surveys reports, citing the respective primary source, an accuracy ranging from 27 to 100% for 1% request retrieval. The authors indicate that the primary sources vary in methodology, datasets, and reporting of accuracy.

The area of automatic recognition of crime-scene shoeprints is still a developing area and which requires more attention and thorough research. Some attention must be devoted to datasets and indicative requirements of the functionalities expected.

3.3 Datasets

The dataset selected to support this study is maintained by Kortylewski et al. [14]. The dataset is called the Footwear Impression Database (FID 300). This public dataset was created by the German State Police which confirms the validity of the lifted prints being from real crime scenes and therefore are comparable with what investigators actually lift from a SoC.

The dataset consists of three different classes of footwear impressions in greyscale. These classes being Real Crime Scene Prints labelled as 'tracks_original'. Processed Crime Scene Prints labelled as 'tracks_cropped', and the Reference Prints labelled as 'references'. The Crime Scene Prints dataset consists of 300 prints lifted from

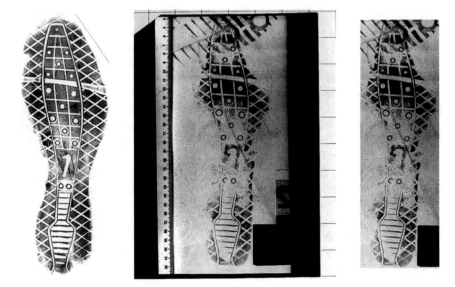

Fig. 1 Shoe print specimen from the FID 300 [14]; reference, SoC, and cropped SoC print

various SoCs. These prints were extracted by forensic experts using gelatine filters or by photographing the impression. The Processed Crime Scene Prints are the same exact prints of the Real Crime Scene Prints but with adjustments made to them; e.g. the prints have all been rotated into an upright position, and cropped to relatively the same size and are in a JPEG file format. The Reference Prints set consists of 1175 specimens of very similar image dimensions and are in a PNG file format (see Fig. 1). The reference prints have been collected from known shoe manufacturers including Nike, Reebok and Adidas.

The dataset comes with a class labeling of each lifted print to a reference print and is the basis for our evaluation later in this paper. It is being assumed that the labelling is correct (Kortylewski et al. [14]). The reference and lifted sets (i.e. cropped and original) vary in image size.

It needs to be emphasized that the lifted prints are real; specifically these are not a sub-set of reference prints that have been rotated, polluted with noise, or partially occulted. Using Luostarinen and Lehmussola [16] classification of prints, the lifted prints in this dataset would be labelled as 'ugly'.

4 Requirements and Design

The management and investigation of shoe impressions in a digital forensics context is best served if a Content Based Image Retrieval (CBIR) model is adopted. The design and development of the system must be influenced by operational praxis of its practitioners and based on robust techniques that have a predictable accuracy of retrieval and a manageable computational load. All the systems reviewed [3, 13–16, 20, 24, 28, 30] produced promising results and are therefore good options for entrenching them into the retrieval and inference modules of a CBIR. However, these results are based on various datasets, and consequently evaluation techniques are strongly tied to it.

The proposed system for this study aims to provide a CBIR with automatic and robust impressions retrieval with the goal of making the work of a forensic analyst faster, less laborious, less subjective, and more efficient. Specifically, any input of a lifted print is processed and a set of possible candidate reference prints is provided in a descending similarity measure. We expect the lifted print is extracted from a SoC and that a reasonable set of reference prints are available for matching. It is expected that pre-processing filters and transformations are applied to prints and these steps should be applied with a minimal of subjective intervention and are able to be composable (i.e. applied in sequence).

When undertaking image matching there should be two distinct methods to choice from and, in either case, proper reporting on effort to retrieve a match is given to an investigator.

As this project is mainly a research oriented one, an added feature is required to bulk test an array of lifted prints and their overall matching performance is given across the batch job with any possible set-up options recorded. Consequently a non-functional requirement is introduced here in that most operations and evaluations are data driven and results recorded in detail (and in a database). This repository is available, i.e. to read and write, from any programming environment adopted.

5   Implementation

Two main tools were required in order to undertake this study. PostgreSQL managed the shoe prints crime scene database. PostgreSQL was primarily chosen because of its ability of loading images and its effective use of stored procedures for managing batch jobs on the server. Also its query constructs enabled reporting and statistical computation in a straightforward manner.

MATLAB R2019a [17] was used to implement the majority of the system's coding and it was primarily chosen due its Image Processing Toolbox; this is an extensive module based on well-known and respected algorithms. It was also selected because of its ease of use and its active community.

5.1   Implementation of Image Matching

Two distinct measuring techniques have been adopted here. The first is image texture based, an image composed of visual patterns related in a spatial world, and works on comparing two images together and it is called CK1 [8]. The second is based on local feature extraction, e.g. blobs, of each image and then matching these blobs on their computed features across comparing images; the modalities used are based on spectral Speeded-up Robust Features Method (SURF) [4] and M-estimator SAmple Consensus (MSAC) algorithm [29] (a variant of RANSAC). Both image matching techniques are used on the same FID 300 [14] lifted and reference prints datasets.

The CK1 Distance Measure

This distance function is provided by Campana and Keogh [8]. The function, named the CK1 Distance Function, was selected because of three reasons. Firstly because of the wide variety of datasets that the function was applied to. These varied from clustering sets of insects such as moths [18] to classification of wood by their grain [25]. The second is that it had not been used with footwear impressions offering an opportunity to analyze and test the robustness of this function for this area. The third is that the function is parameter-light; actually the implementation adopted here had no parameter setting requirements. The benefit of having no parameters is that it makes the function open to more general use rather than specific to one scenario. Setting a parameter improperly can also seriously attenuate accuracy.

This CK1 Distance function is used to calculate the similarity of two footwear impressions. It takes two images as input, i.e. i_1 and i_2, and returns a value for their distance, i.e. $distance_{CK1}$. Another feature of this function is that it is a symmetric; but it is not a metric distance measure. The CK1 Distance Function works by creating a movie out of the two images passed in, which is rendered using an MPEG-1 encoder in MATLAB [17]. Once a movie is generated the following calculation of distance is performed.

$$distance_{CK1} = \frac{(\text{mpegSize}(i_1, i_2) + \text{mpegSize}(i_2, i_1))}{(\text{mpegSize}(i_1, i_1) + \text{mpegSize}(i_2, i_2))} - 1$$

Campana and Keogh [8] relate the CK1 measure, obtained through video compressors, to an approximation of the Kolmogorov complexity. The concept behind using video size is that, if two images are similar then the movie created by these two frames would be small in size. Therefore, it is indicated that the smaller the distance returned by the function the more similar are the images. The authors asserts that the function is robust and it can detect subtle differences in textures too.

The SURF Identification and MSAC Based Matching as a Distance Measure

The technique adopted here, which is motivated by Nibouche et al. [20], is to extract local features based on their spectral descriptors found in an impression print. In our case the SURF [4] method uses the determinant maxima points of a Hessian matrix to build a vector based descriptor of a local point. The SURF neighborhood size can be specified for each run and a larger filter size is indicated for detecting large blob features. The distance function adopted is Euclidean. The MATLAB [17] image processing library functions are used to implement this section.

Once a SURF invocation returns a set of possible points, the top points are retained; i.e. referred to as the 'strongest' based on their computed point features. This number is set proportional to the size of the impression's area and therefore there is a maximum of detected points.

Once each print has its strongest SURF points selected one can match two impression's detected points for correspondence. The preferred matching method is based on exhaustive pairwise distance comparison between all point's feature vectors. Also in each comparison the sum of squared differences is used. The procedure is restricted to consider only unique matchings; i.e. no feature can be matched with more than one other feature. This matching does allocate a number of outlier correspondences, and their number is substantial if either image has degraded quality.

To attenuate the outliers, a process is spawned to sift them out from the matched sets. The process is based on MSAC [29] employing an affine transformation. MSAC not only focuses on inlier selection but also gives a transformation matrix (i.e. with x and y offsets). The praxis is to use the number of inliers returned and the size of x and y offsets to compare the similarity of two pairs of matched images. The matching with the greater the number of inliers is the better one, and the smaller the affine offsets are the better the matching is (see Fig. 2). Nonetheless it is very important to note that MSAC is not a deterministic measure; i.e. each run can give different number of inliers and offsets.

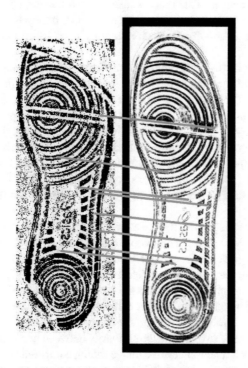

Fig. 2 MSAC [29] matching local features found in a SURF [4] run (left is a FID 300 SoC print)

SURF and MSAC have been chosen for a number of reasons. Firstly, SURF is computationally efficient (relative to SIFT). Secondly, SURF favors blob like features. Thirdly, and this is important in our case, local features extraction remains effective even when parts of an image are occulated, evident clutter is present, and differing lighting conditions in each image is noted. Lastly SURF, and the related SIFT, are used by a number of projects in Footwear recognition and therefore we have a way to relate results through a common method.

5.2   Implementation of Pre-processing and Feature Extraction

Pre-Processing and Feature Extraction are two extremely important processes that must be utilized in any system that involves digital image processing especially where querying by images is incorporated. In particular each of the two matching techniques adopted here require specific pre-processing techniques.

Preprocessing for and Computing the CK1 Distance Measure

The two required processes where grouped into one stage for convenience and more efficient coding practices. This process takes in two images, applies techniques of pre-processing and feature extraction with the aim of improving the input images' quality. The processing consists of three main phases (third phase calculates the similarity):

The first phase involves the application of the pre-processing techniques. The first technique that is applied to the images is Local Normalization. Local Normalization was applied following the implementation Patil et al. [22] that showed its usefulness when processing of footwear impressions as it helps in improving the quality of the footprint processing. The other pre-processing technique adopted at this point is filtering. Applying filters is effective for removing noise from the images which results in improved accuracy of the classification system. The filter chosen for the purpose of the footwear impressions was the Wiener Filter [19] as it known to have promising results on grayscale images; our chosen dataset impressions are in greyscale.

The second phase deals with feature extraction and is performed on the processed images of the previous phase. The feature extraction chosen to be implemented is edge detection. Edge detectors extracts patterns from footwear impressions as they enhance the lines of the print making detection less complicated and more accurate. The detector that was implemented was the Roberts Cross Edge Detector [21]. This detector is utilized also because of its excellent results on grayscale images.

The final phase involves the image similarity measurement. Here the distance function, CK1, is called to calculate the similarity between two input images. Before applying the distance function, the two images being compared are divided into three equal sections. This is called tessellation and was a concept created for the purpose of this study. The separation of each image into three parts allows for the distance function to analyze the image in more detail since it will be focusing on a smaller area of the image rather than the whole and hopefully texture variety is addressed. The distance function is run on each segment individually and an average is calculated.

Given that novelty of CK1 measure a number of test have been done to understand the behavior of the function when calculating distances. For example, the CK1 function is quite robust to minor changes in one of the source images; but this quickly breaks once major edits to an image are done. Varies noise additions and major rotations of an image do affect the CK1 value too. During tests it transpired that the larger the image the better the matches get. Varies test with image tessellations indicate that a good similarity between opposite tiles can effectively increase the similarity when compared to the whole image similarity measure.

Pre-processing for and Computing the SURF and MSAC Distance Measure
To ensure better local feature extraction from the impression images a few pre-processing steps have been undertaken. Given the quality of the impressions, especially the low quality of the lifted prints, and the nature of the feature extraction adopted (i.e. SURF) the binarization technique for greyscale images based on thresholding was adopted. Furthermore the thresholding was computed on local mean intensity of a pixel rather than whole image; this technique is attributed to Bradley and Roth [7]. Since the shoe imprint is expected to be dark then the pre-processing has to be adjusted to have the darker pixels as foreground. Finally, to attenuate dark background pixels being left as part of the output image, the sensitivity threshold was lowered at the expense of darker blobs being slightly smaller. These filters have been applied to the 'tracks_cropped' part of the dataset.

The similarity measure between two images based on our local feature extraction method is rather straightforward. It is computed from the offset distances of the affine transformation, i.e. square root of sum of squares, and the number of inlier matches between images.

Both the pre-processing and the similarity measures required testing to ensure an adequate set-up for our dataset and similarity evaluation. Furthermore these filters and measures and are known in the literature and many of these references give insights on what values to set these parameters.

5.3 Results Computation

To expedite the analysis of the results, a batch method was adopted to automatically run and calculate the accuracy from the results for each of the techniques used; i.e. texture and local feature based.

The approach taken to calculate the accuracy of the image matching on the footwear impression, is to use a table of results provided in the dataset as a benchmark to compare our results to. Every individual Crime Scene Print is compared to all the Reference Print instances and results read into the database. The results are available in ascending order by distance, from lowest distance being the most similar, to the highest distance being the least similar.

Thus, to calculate the accuracy of the current matching retrieval the following formula is computed (this is scaled to 100%). The letter r_i denotes the rank of the correct match, p_n is the number of reference prints (in this case $p_n = 1175$).

$$100 \times \left[\frac{p_n - (r_i - 1)}{p_n} \right]$$

The average of the whole run is given by the following formula (this is scaled to 100%). The letter c_n denotes the number of lifted prints being checked (in this case $c_n = 300$).

$$\frac{1}{c_n} \sum_{i=1}^{i=c_n} 100 \times \left[\frac{p_n - (r_i - 1)}{p_n} \right]$$

6   Results and Evaluations

The following section explains the methods and their merits with regards to the performance matching done. Several tests have been run, which have placed the system under different working conditions in order to analyze how these affect its behavior. The results of these tests are evaluated and discussed.

6.1   Evaluation of the System with CK1 and Without Pre-processing

During this run, every single Crime Scene Print (i.e. one of the 300) was compared to all the Reference Print (i.e. 1175) without the application of pre-processing and feature extraction but with tessellating the impressions into three parts. The accuracy was then calculated for each individual Crime Scene Print with the CK1 function. As stated earlier we have two sets of lifted prints: the first as is (i.e. 'tracks_original'), and the second has been cleaned through editing by the dataset owners (i.e. 'tracks_cropped').

As it can be seen in Fig. 3, the results of this run on 'tracks_original' where quite promising considering no pre-processing was being applied. Several prints, (i.e. 17) are in the upper 1% bracket accuracy (i.e. within the range of 100–99% accuracy); and 31 prints in the upper 5% bracket accuracy (i.e. within the range of 100–95% accuracy). There were also low accuracies that were reported, however, the overall accuracy achieved is recorded at 54%.

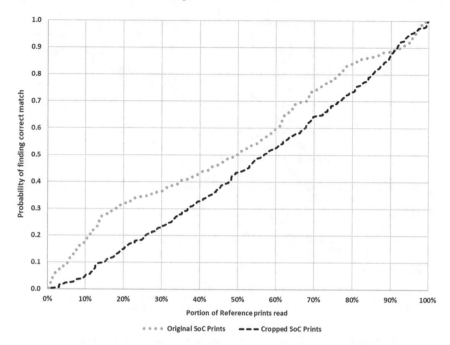

Fig. 3 Cumulative match probabilities for CKI applied on FID 300 without pre-processing

The run on 'tracks_cropped' yielded an average accuracy 46% but the ranking numbers variance compared to the above was smaller.

6.2 Evaluation of the System with CK1 and with Pre-processing

This test builds upon the previous one. The same run (i.e. as in Sect. 6.1) is conducted but pre-processing is applied before and the target dataset 'tracks_cropped' is processed. Results are shown in Fig. 4. The average accuracy was computed at 61%. Three lifted prints were in the in the upper 1% bracket and 19 prints in the upper 5%.

The run on 'tracks_original' with pre-processing gave an average of 45%. Both runs had very similar ranking numbers variance.

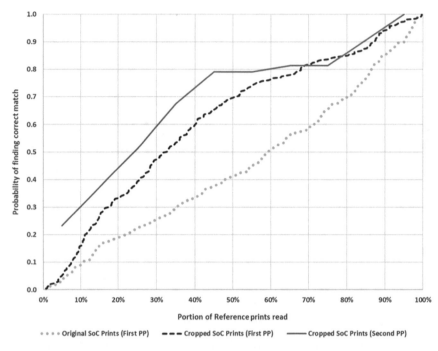

Fig. 4 Cumulative match probabilities for CKI applied on FID 300 with pre-processing. Second pre-processing option applied on 43 specimens from FID 300

6.3   Evaluation of Other Pre-processing CK1 Based Method

Another pre-processing method was created with the goal of improving the overall accuracy of the system. This method makes use of different pre-processing and feature extraction techniques, which are the median filter [15] for pre-processing and the Canny edge detector [9] for feature extraction.

Unlike the initial tests, this was run on a subset of the original dataset; specifically a sample of 43 prints. The highest achieved accuracy using this method was of 99%. The overall average accuracy was at 65% which is an increase of 4% when compared to the other pre-processing method and an increase of 11% when compared to using no pre-processing. Results are also shown in Fig. 3.

6.4   Comparison of CK1 Based Results

When investigating these results two observations are apparent. When pre-processing is applied to prints that achieved high accuracy without it, the distance is increased, and when it is applied to prints which have achieved a high accuracy the distance is decreased. At this point one can posit that pre-processing should be applied to prints that achieved a lower accuracy during testing without pre-processing; i.e. apply pre-processing when results are not encouraging. From the indication obtained through the application of different pre-processing it is clear that the CBIR must offer a number of these and consequently applied in relation to the image quality and characteristics of the lifted prints.

6.5 Evaluation of the System with Local Features

The SURF and MSAC technique, when run with its indicated pre-processing on the 'tracks_cropped' dataset gave an average accuracy of 50% when maximum SURF points, before sifting for the strongest, was set to 2500. The upper 1% had 7 instances and the upper 5% had 25 instances; refer to Fig. 5 for results. This techniques results, we have to repeat, are non-deterministic as the similarity measure does change with every run due to the inlier and outlier separation being based on a simulation. Nonetheless the local feature matching fared well and it computational load is marginal. A positive aspect of this technique is its explanation potential due to possibility of visualizing the SURF point's correspondence between the lifted to the reference image.

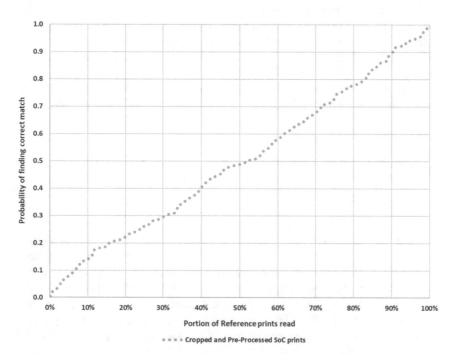

Fig. 5 Cumulative match probabilities for SURF/MSAC applied on FID 300 with pre-processing

6.6   Summary of Results

From the various runs that have been conducted here on the FID 300 [8] dataset it can be concluded that the CK1 Distance function is sensitive to noise, size, rotation and alterations present within the image. Likewise the efficacy of local feature matching, here based on SURF and MSAC, suffered from image quality issues too.

The results obtained here, and when these are qualitatively compared to other studies, indicate the acute difficulty of dealing with real life shoe impressions collected from a SoC. Indeed the introduction of local feature analysis allowed us to better relate our results with those of other studies.

Considering these data quality issues, our system based on CK1 has performed well with an accuracy of 54% (300 lifted prints against 1175 reference prints) even when no pre-processing was applied to the prints. Through testing, it has been found that pre-processing is useful in certain scenarios but not in all; this is further supported in the run were different pre-processing was applied. If the prints have achieved a good accuracy score, then there is no need for pre-processing as it might interfere with the print identification. However, when the prints did not achieve acceptable accuracy, pre-processing techniques can improve the accuracy of the print identification. From the two different pre-processing techniques implemented for CK1, the second proved to be more accurate overall with an accuracy of 65% (on a dataset of 43 randomly selected prints).

The local feature method, i.e. with SURF and MSAC, performed well at an average accuracy of 50%. This result is comparable to other papers that use local features for matching but use different datasets. If we compare with Luostarinen and Lehmussola [16] we find RANSAC average accuracy at 35% when evaluating over an ugly set (i.e. typical quality from a SoC lifted print).

When we compared our rankings for one of the CK1 runs and the SURF and MSAC run, over the FID 300 dataset, we found there was no rank correlation.

7   Conclusions and Future Work

7.1   Future Works

As repeatedly indicated in the literature, e.g. by Rida et al. [23], a current problem impeding the fruition of research on automated shoe-print impressions identification and classification is the lack of publicly available data and well-defined evaluation techniques. With that being said, the fact that various systems have been proposed is a step in the right direction. On top of a good dataset, what is also needed is precise indications of what functionalities are expected for a system that uses this dataset in terms of both features and performance. Efforts such as those by the DigForASP Cost network [10] helps with defining such functionality through an establishment of a benchmark.

The use of texture, as a basis for a distance measure, should be applied to other Digital Forensic thematic areas such as motor vehicle wheel tracks. Also other alternatives for distance measure between textures should be pursued and evaluated.

An important improvement is optimizing the use of the computational resources required and attaining favorable time to compute pre-processing and matching tasks.

7.2 Conclusions

The main scope of this study was to develop a CBIRs in the area of Digital Forensics for Footwear Impressions for matching lifted prints from a SoC. It has been widely asserted in the literature that the evaluation of such a system must be realistic. Hence our adoption of FID 300 dataset meets that requirement.

The resulting artefact that has been developed here manages to match real SoC footprints at the best average accuracy of 54% (with no pre-processing) and 65% (with particular pre-processing methods). The accuracy of the matching is strongly dependent on the completeness of the image and the amount of noise present.

The Campana Keogh distance function (CK1), based on a footprint's texture, was appropriate for the purpose of this study. It has been demonstrated to be have a tenable basis and this is reflected in the results that were achieved. The average accuracy of our SURF and MSAC mechanism, at 50%, also confirms the viability of this local feature approach; furthermore the results obtained are comparable to those found in the literature for images with similar quality.

MATLAB proved to be a very effective tool. It provides a consistent and robust development environment and its rich Image Processing Toolbox was also extensively used. The third-party libraries which are also made available for MATLAB users helped in making the implementation more efficient. PostgresSQL was an effective data handling tool; e.g. querying results was aggressively facilitated our analysis.

References

1. Adair, T.W., Lemay, J., Mcdonald, A., Shaw, R., Tewes, R.: The Mount Bierstadt study: an experiment in unique damage formation in footwear. J. Forensic Ident. **57**(2), 199–205 (2007)
2. Alexandre, G.: Computerized classification of the shoeprints of Burglar's soles. Forensic Sci. Int. **82**, 59–65 (1996)
3. Ashley, W.: What shoe was that? The use of computerised image database to assist in identification. Forensic Sci. Int. **82**, 7–20 (1996)
4. Bay, H., Ess, A., Tuytelaars, T., Van Gool, L.: SURF: speeded up robust features. Comput. Vis. Image Underst. **110**(3), 346–359 (2008)
5. Bodziak, W.J.: Footwear impression evidence: detection, recovery and examination, 2nd edn. CRC Press, Boca Raton (1999)
6. Bodziak, W.J., Hammer, L., Johnson, G.M., Schenck, R.: Determining the significance of outsole wear characteristics during the forensic examination of footwear impression evidence. J. Forensic Ident. **3**, 254–262 (2012)

7. Bradley, D., Roth, G.: Adapting thresholding using the integral image. J. Graph. Tools **12**(2), 13–21 (2007)
8. Campana, B.J.L., Keogh, E.J.: A compression based distance measure for texture. J. Stat. Anal. Data Min. **3**, 381–398 (2010)
9. Canny, J.: A computational approach to edge detection. IEEE Trans. Pattern Anal. Mach. Intell. **8**(6), 679–698 (1986)
10. Digital forensics: evidence analysis via intelligent systems and practices (DigForASP)— CA17124.https://digforasp.uca.es/. Last accessed 29th Feb 2020
11. Footprints can play an important role in solving crimes. Available: https://www.crimemuseum. org/crime-library/forensic-investigation/footprints/. Last accessed on 29th Feb 2020
12. Footwear News Portal. https://footwearnews.com/2016/business/media/oj-simpson-murder-trial-bruno-magli-shoes-bloomingdales-188994/. Last accessed 29th Feb 2020
13. Kong, B., Supančič, J., Ramanan, D., Fowlkes, C.C.: Cross-domain image matching with deep feature maps. Int. J. Comput. Vis. **127**, 1738–1750 (2019)
14. Kortylewski, A., Albrecht, T., Vetter, T.: Unsupervised footwear impression analysis and retrieval from crime scene data. In: Lecture Notes in Computer Science (including Subseries Lecture Notes in Artificial Intelligence and Lecture Notes Bioinformatics), vol. 9008, pp. 644–658 (2015)
15. Lim, J.S.: Two-Dimensional Signal and Image Processing, pp. 469–476. Prentice Hall, Englewood Cliffs (1990)
16. Luostarinen, T., Lehmussola, A.: Measuring the accuracy of automatic shoeprint recognition methods. J. Forensic Sci. **59**, 1627–1634 (2014)
17. MATLAB, R2019a. The Matworks Inc. (2019). https://uk.mathworks.com/products/matlab. html. Last accessed on 29th Feb 2020
18. Mayo, M., Watson, A.: Automatic species identification of live moths. In: Ellis et al. (eds.) Proceedings of the 26th SGAI International Conference on Innovative Techniques and Applications of Artificial Intelligence, pp. 195–202 (2006)
19. Motwani, M.C., Gadiya, M.C., Motwani, R.C., Harris, F.C.: Survey of Image denoising techniques. In: Proceedings of GSPX, pp. 27–30 (2004)
20. Nibouche, O., Bouridane, A., Gueham, M., Laadjel, M.: Rotation invariant matching of partial shoeprints. In: 13th International Machine Vision and Image Processing Conference, pp. 94–98 (2009)
21. Nixon, M., Aguado, A.S.: Feature Extraction & Image Processing, 2nd edn. Academic Press, Cambridge (2008)
22. Patil, P.M., Kulkarni, J.V.: Rotation and intensity invariant shoeprint matching using Gabor transform with application to forensic science. Pattern Recogn. **42**(7), 1308–1317 (2009)
23. Rida, I., Bakshi, S., Chang, X., Proenca, H.: Forensic shoe-print identification: a brief survey. Pattern Recogn. Lett. 1901.01431 (2019). arXiv preprint arXiv
24. Sawyer, N., Monckton, C.: "Shoe-fit"-a computerized shoe print database. Eur. Convention Secur. Detect. **1995**, 86–89 (1995)
25. Silven, O., Niskanen, M., Kauppinen, H.: Wood inspection with non-supervised clustering. COST Action E10 Workshop—Wood Properties for Industrial Use, pp. 18–22, Espoo, Finland (2000)
26. SoleMate FPX. https://www.fosterfreeman.com/qdelist/trace-list/511-fpx-shoeprint-identific ation.html. Last accessed 29th Feb 2020
27. Srihari, S.N.: Analysis of Footwear Impression Evidence. US Dept. of Justice, Report (2011)
28. Tang, Y., Srihari, S.N., Kasiviswanathan, H., Corso, J.J.: Footwear print retrieval system for real crime scene marks. In: International Workshop on Computational Forensics, pp. 88–100. Springer, Berlin (2010)
29. Torr, P.H.S., Zisserman, A.: MLESAC: a new robust estimator with application to estimating image geometry. Comput. Vis. Image Underst. **78**(1), 138–156 (2000)
30. Wang, X., Wu, Y., Zhang, T.: Multi-layer feature based shoeprint verification algorithm for camera sensor images. Sensors **19**, 1–20 (2019)

Human-Centric Cyber Security
and Resilience

Cyber Resilience Using Self-Discrepancy Theory

Jassim Happa

Abstract Threats take many forms, and understanding them in order to make organisations more cyber resilient remains challenging. Many resilience management models and standards exist. They can help enterprises recover from harmful incidents. No approach today comprehensively examines perspectives of resilience concerns. In this paper, I argue it is necessary to consider a much broader spectrum of threats and harms to better understand the complex dependencies and interactions between an enterprise ('self') and the environment ('world'). This paper adapts *Self-Discrepancy Theory* from psychology to help establish and reason about multiple views of the 'self': the *actual*, *ideal* and *ought* enterprise, as viewed by the enterprise/risk owner and others (e.g. competitors). The paper investigates how changes in priorities and operations can affect the self (again, as viewed by the self and others). This framework does not compete with existing models and standards. Instead, the purpose of this work is to complement them by exhaustively considering different perspectives (views) of enterprises with the aim to re-contextualise resilience concerns. By using this framework, risk-owners can start making decisions with new insights, akin to: *what would my opponent do in my position?*; or, *if I change my mission, what effect might this have on my security (and vice versa)?* Viewing enterprise resilience from different perspectives is an underexplored topic in resilience and security research, and is a key motivation of this article. This is a position paper, and further studies will be necessary to provide empirical evidence of feasibility in real-world settings.

Keywords Resilience management · Risk management · Self-discrepancy theory · Graphs · Feedback loops

J. Happa (✉)
Information Security Group, Royal Holloway, University of London, London, UK
e-mail: Jassim.Happa@rhul.ac.uk

© The Author(s), under exclusive license to Springer Nature Switzerland AG 2021
T. Tagarev et al. (eds.), *Digital Transformation, Cyber Security and Resilience of Modern Societies*, Studies in Big Data 84,
https://doi.org/10.1007/978-3-030-65722-2_23

1   Introduction

Assessing threats posed to enterprises is challenging, and improving resilience of systems is important to limit potential harm. Conducting such assessments however, is rarely a straightforward task. This is because the threat landscape perpetually evolves, and new types of threats can emerge faster than decision makers are able to keep up with. Resilience concerns can relate to actors such as insider threats, advanced persistent threats, or external threats. Threats can also exist outside the realm of security. At the surface level, these may appear unrelated to security and mission-continuity management. However, examples of such threats may include evolving business markets, political pressures, psychological concerns, financial disruption and natural forces (e.g. natural disasters) may directly or indirectly affect operational decisions. In this paper, we examine how understanding these threats using an aggressive peer-reviewing approach may improve an enterprise's cyber resilience.

Improving resilience is a matter of identifying, knowing and managing system functions with respects to known threats and the risks they pose. It is necessary to investigate properties and functional relationships of assets, controls, configurations, vulnerabilities, incidents, services, risks, external dependencies, training and situation awareness [8], in order to *"plan and prepare for"*, *"absorb"*, *"recover from"* and *"adapt to"* potentially harmful events [37]. Managing these risks remain challenging due to their, often, qualitative nature. Regardless of resilience and/or risk management method, it is often comprised of five main steps:

1. **identify the risk**: enumerate assets and the threats posed to them;
2. **analyse the risk**: identify the likelihoods and consequences;
3. **prioritise the risk and plan a response**: by ranking their importance;
4. **act on the risk**: avoid, treat, transfer or accept the risk, and finally;
5. **learn from, and keep monitoring the risk**.

Existing literature has already pointed out the importance of enterprises through dependency modelling and effective risk management [4, 7, 8, 11, 15, 30, 33, 36, 37, 42, 47]. Instead, in this article, I am making the point that existing approaches to understanding the 'self' remain subjective, and that by recognising subjective aspects of resilience management (in particular: how one stakeholder views a resilience concern different from another), we may use this insight to provide a more complete understanding about an enterprise in order to deliver more robust suggestions for improvement. Numerous nuanced aspects may otherwise be overlooked by any single risk-owner or decision-maker. Different types of stakeholders will have different views of what constitutes a resilience concern. I postulate that enterprise operations (e.g. business model, mission etc.) and security postures are intrinsically linked, and that resilience ought to consider the complex dependencies across both the technology and enterprise estate for more well-informed decision-making. More specifically—as perceived by different perspectives (views).

Changes in enterprises can directly and indirectly affect its security and their operations. As an illustrative example: the introduction of Napster and other file sharing technologies in the late 1990s disrupted established ideas of copyright [34, 46] with the ability to share copies of media files to large audiences. In order for entertainment enterprises (particularly those in the music, film and games industry) to combat piracy, the public saw an increase in lawsuits [39] as well as increased deployment of Digital Rights Management (DRM) technologies [45]. While piracy is to this day still an issue for the entertainment industry, it is clear the threat was not an existential one. In the last decade, the public has seen a shift in how these enterprises operate. Instead of combating piracy with anti-piracy technologies and lawsuits, we instead see changes in the business models themselves (e.g. streaming services and app stores). Publishers and distributors have made legitimate options more attractive than piracy by adding: new services, convenience, competitive pricing and peace of mind for the customers (i.e. not having to worry about pirated material containing malware).

This example illustrates how a shift about the 'self' and the 'environment' in which a threat operates had to change in order to appropriately combat the threat. Specifically, a threat to an enterprise's operation had to be regarded as a competitor, as opposed to outside the realm of legitimacy. This shift in understanding of the self enable resilience concerns to be addressed at scale. With the introduction of blockchain, it is not challenging to see how a similar disruption might happen for sectors that require verifiable traceability of access logs (e.g. in the e-Government, health sector or selling of digital commodities). This is another example in which technological advancements can change enterprise operations.

Good resilience relies on well-informed understanding about an enterprise's operations and security postures. By obtaining an understanding of these relationships and perceptions of them, it is hypothesised that this can improve an enterprise's overall resilience. If we understand how security can affect missions and mission tasks, we can use this insight to improve our ability to recover from any potential future incident. Good security practices ought, also, to inform enterprises changes. It is worth noting that for the purposes of this paper, we adopt the term 'enterprise' as a synonym for: 'organisation', 'business', 'system', 'mission', 'project' or similar. This is done to be able to generically discuss resilience concerns in a manner that can apply to all of the aforementioned settings, instead of continually re-appropriating what we mean. The framework discussed in this paper is equally applicable to (e.g.) a business context as it is to (e.g.) a enterprise mission.

By considering the '*self*' (the enterprise as one unit) existing in different 'environments', all potential futures are treated as real during the resilience assessment. The purpose is for the risk-owner or enterprise-owner to make an informed choice about which path to take moving forward. In order to comprehensively cover resilience concerns posed to an enterprise, I postulate that:

- (1) enterprise operations and security postures are intrinsically linked, and will affect each other in a feedback-loop manner;
- (2) it is meaningful to regard resilience concerns from different perspectives (views of 'self'), which suggests that;

- (3) potential changes to the self (as a thought experiment) can inform risk-owners about resilience discrepancies and help them make well-informed decisions moving forward.

1.1 Paper Contributions

Several challenges become immediately apparent when examining the resilience and risk-management literature: *As a risk-owner, do you adopt one or more resilience management model across the whole enterprise estate, which one(s), and how do you benchmark your resilience management in order to validate that your approach is both comprehensive and capable enough?* This paper does not fully address any of these grand challenges. However, the paper proposes how we might be able to achieve comprehensive examination of resilience through a much broader and more nuanced spectrum (set of views). Detailed insight is necessary to better understand the complex dependencies and interactions between an enterprise ('self') and the environment ('world') in which the enterprise operates. In order to achieve this understanding, the paper explores whether concepts such as *Self Discrepancy Theory* [27] can be used to improve enterprise resilience.

Specifically, **this paper explores whether multiple views of the 'self' can re-contextualise threats posed to an enterprise**, and examine whether this re-contextualisation can help risk-owners make more well-informed decisions about how to adapt to those threats. The paper proposes the outline of a framework on how to reason about threats and the self, facilitating a peer-reviewed approach to improving enterprise resilience. By using this framework, risk-owners can begin to ask questions such as: what would my opponent do in my position?; or, if I change my enterprise's mission, what effect might this have on my security (or vice versa)? This paper is a position paper, and further studies will be necessary to provide empirical evidence that this work is both feasible and viable in real-world settings.

2 Background and Motivation

2.1 Enterprise Cyber Resilience

The concept of "resilience" as "*being able to withstand potentially damaging issues and keep operations running*" stems from Business Continuation Management [19]. Resilience involves the identification, assessment and prioritisation of uncertain effects that may or may not happen to an enterprise in order for it to recover faster in the event of harmful incidents. Many standards and bodies of work exist that examine risk management and resilience of organisations [8, 14, 18, 44]. They facilitate protection of an enterprise. More recently, the cybersecurity research community has adopted this concept and proposed a number of models to withstand threats. Common across all models is the notion that a resilience model is a dynamic effort that needs to be aware of mission contexts, including conditions, dependency relationships and time, and that a system that is not highly resilient is vulnerable. In this section, we review some of the key efforts in the enterprise resilience space.

Gibson and Tarrant [20] present a discussion on uses of conceptual models to describe resilience, outlining first the general principles of resilience before describing several models. The *principles model of resilience* shows how long an organisation is able to cope over time during disruption, highlighting that it is not only enough to be able to reactive in defences, but prepared and adaptive as the situation changes. The model is derived from common themes that emerge from comparisons of resilience and is based upon six key principles: resilience is an outcome, it is not a static trait, it is not a single trait, but multi-dimensional, exists over a range of conditions, and is founded on good risk management. The authors propose that changes in resilience capability, e.g. context (e.g. conditions, effects and time) play an important role in becoming resilient. They also describe several conceptual resilience models from business continuity management that considers resilience as a process which is more tightly integrated into risk management programs.

The *CERT Resilience Management Model* (CERT-RMM) is perhaps the most influential work in the resilience literature. It is "*a maturity model that promotes the convergence of security, business continuity, and IT operations activities to help organizations actively direct, control, and manage operational resilience and risk*" [29]. Caralli et al. [8] present the CERT-RMM as a process improvement model to help manage operational resilience in large complex environments where risk is continually evolving. It helps manage the security and survivability of the assets that ensure mission success by considering ten domains of interest: asset management, controls management, configuration and change, management, vulnerability management, incident management, service continuity management, risk management, external dependency management, training and awareness, and situational awareness.

Goldman et al. [21] present a series of proactive and reactive approaches to resilience and describes an application scenario. The authors argue organisations have to be prepared to "fight through" cyber attacks to ensure mission success even in a degraded or contested environment. The paper has a technical focus, describing at a high level the actionable architectural and operational recommendations to address the advanced cyber threat and to enable mission assurance for critical operations. The proactive approaches range from containment of emerging resilience issues to diverting attackers (among others). The reactive approaches range from dynamic reconfiguration of a system, to deception, to dynamic reconstitution and assigning alternative operations (among others).

Bodeau and Graubart [6] present how to structure discussions and analyses of cyber resiliency goals, objectives, practices, and costs. It also serves to motivate and characterise cyber resiliency metrics. The framework is intended to evolve as the discipline of cyber resiliency engineering matures. Currently, the key sources for the framework includes: (1) *Resilience Engineering* (establishment of the goals to anticipate, withstand, recover and evolve); (2) *Mission Assurance Engineering* (providing the mission assurance focus) and (3) *Cyber Security* (providing the emphasis on addressing threats). For each element within the framework, different objectives are defined. Anticipate for instance has the objectives: Understand, prepare and prevent available, while Withstand has the actions: Understand, Continue and Constrain affiliated with it etc. Cyber resiliency engineering can be characterised as engineering focused on resilience. The framework focuses on architectural strategies and practices, emphasising technical systems in which socio-technical aspects are treated as supporting components rather than central to resilience.

The European Union Agency for Cybersecurity (ENISA) [14] suggests there is a lack of a standard framework and metrics for network resilience. The report summarises key works on network resilience w.r.t. security and dependability. The report proposes a classification scheme in which incident-based classifications are related to domains-based classifications (such as dependability, security and performability) to create metrics pertaining to preparedness, service delivery and recovery. Some example metrics include for instance: Mean time to Incident Discovery; Mean time to Patch; Patch management coverage; Vulnerability scanning coverage; Operational mean time between failures; Fault report rate; Mean time to incident recovery; Incident rate; Packet loss; Mean down time; Mean time to repair; Risk assessment coverage; to list a few.

de Crespigny [12] tackles the challenge of how organisations can strike a balance between the risks and rewards, and prepare effectively to counter the growing threats from cyberspace, without losing the potential benefits. The paper outlines how an organisation has two competing forces: one that increases impact of cyber-resilience concerns (Known CIA, Known non-CIA, unknown, unpredictable, uncertain and unexpected malicious forces) that pull in one direction. The second, opposing force decreases impact of cyber-resilience concerns (governance, situational awareness, resilience assessment, responses and security). The paper postulates that a 'sweet spot' can be achieved through careful considerations of these opposing forces.

Vugrin and Turgeon [50] describe a resilience assessment methodology that combines qualitative analysis techniques with performance-based metrics. The assessment methodology enables identification of system features that limit resilience (identification of weaknesses), while quantitative data can be used to confirm the effectiveness of mitigation options. A case study in the paper demonstrates how the approach could be applied to a hypothetical HR system and shows the ramifications of those cyber systems being affected. In the example, the system is disrupted due to human error by an employee of the company.

Liu et al. [38] focuses on network resilience management, going from a "reactive" paradigm to a "proactive" one through the use of situational awareness of internal network factors and external ones of complex, dynamic and heterogeneous network environments. Their situational awareness model covers three main network resilience issues; *perception* (metrics to measure resilience of network and external factors from environment), *comprehension* (pattern identification) and *anticipation* (threat analysis) of emerging issues before doing actions to support the network resilience (in particular remediation and recovery actions).

2.2   Common Themes

Overall, the research literature has covered much ground (including many more works not covered in this paper). The resilience research literature remains fragmented as the topic spans across all sectors of any enterprise, including for instance: tasks, finance, medical, psychology, software development, security etc. Different enterprises will have different requirements to achieve resilience. All sectors need to examine at least their security posture and their operational/business model in order to improve their resilience. Common themes in the literature highlight how:

- achieving resilience is exhaustive and often not practical;
- resilience is not well-understood, with many different definitions;
- resilience is nuanced;
- most resilience models and frameworks are either conceptual, prescriptive or bespoke solutions to a particular organisation;
- there is a lack of tools and standards for straightforward integration into modern and legacy systems;
- resilience as a topic remains fragmented as systems often remain incompatible, spanning across many different sectors that appear unrelated, such as finance, medical, psychology, software development, security, risk etc.

From this literature, we see that it is necessary to consider the wider spectrum of resilience because:

- different enterprises have different requirements;
- behaviour of enterprises and systems are complex with linear and non-linear behaviour over time that resemble workflows, feedback loops and dependency functions;
- threats are often emerging faster than we can keep up with;

As mentioned, the research question we are exploring is whether adapting an understanding of the 'self' from psychology can benefit enterprise resilience? In this paper, we examine possible uses of Self-Discrepancy Theory [27].

2.3   Self-Discrepancy Theory

In social psychology, early works propose that there is a relation between discomfort and specific kinds of 'inconsistency' in a person's beliefs [1, 16, 26, 41], and that in turn these theories about the self and self-conflict and self-inconsistencies can lead to emotional problems [2, 3, 10, 28, 35, 43]. Higgins [27] introduces Self-Discrepancy Theory (SDT) as a theory to help understand how different types of discrepancies relate to different types of emotional vulnerabilities. The theory describes how an individual compares their 'actual' self (a representation of attributes that a person believes they possess) to an internalised self. This internalised self is split between

Table 1 Self-Representation in traditional Self-Discrepancy Theory

Self-representation	Actual	Ideal	Ought
Own	Self-concept	Self-guide	Self-guide
Other	Self-concept	Self-guide	Self-guide

the 'ideal' (a representation of attributes that a person believes they might be able to possess—an idealised view) and the 'ought' (a representation of attributes that a person believes they should possess) self. The inconsistencies between the actual, ideal and ought selves are associated with emotional discomfort, and the self-discrepancy comes from the discrepancy between the selves which may lead to negative emotions (e.g. anger, frustration, guilt, nervousness and fear).

The main goal of the theory is to help create an understanding of contrasting ideas that make persons feel different kinds of negative emotions. Higgins also distinguishes between two optical views of the possible self: as perceived by oneself, and as perceived by another (such as a significant other). These internal selves are often referred to as "self-guides". People compare themselves to internalised self-guides. People are also motivated remove disparity in self-guides. Comparing all guides offers six possible 'selves', as shown in Table 1.

Harm is a detrimental effect on people, data, infrastructure and operations as a consequence of attacks, accidents, natural events, undocumented knowledge gaps, or deprecated, unmaintained equipment. In this paper, we consider harm to be analogous to negative emotions as described by Higgins. Assessing cyber resilience is important to limit potential harm in enterprises. However, doing so is not a straightforward task, especially for complex and diverse organisations where not all assets may be known, and mission processes and vital knowledge go undocumented and unpreserved. While there has been a significant amount of effort related to resilience research in the areas of documentation standards, business continuity management and disaster management, many organisations today are poorly-equipped to combat resilience concerns. This may be attributed to resilience research today having investigated aspects of resilience in unrelated disciplines, and failed to identify the interdisciplinary nature of resilience itself. Irrespective of an enterprise's mission, all enterprises will still be vulnerable to harm that may come from threats and vulnerabilities.

2.4 Cyber Resilience Using Self-Discrepancy Theory

This framework, termed "*Self-Discrepancy Resilience*", is inspired by existing resilience models and aims to improve resilience management by maximising their utility by identifying discrepancies between self-guides/views. The implementation of existing models and standards remains subjective (i.e. they rely on a risk-owner's view of the risks and security concerns), too theoretical to be used in practice, or consumes too many organisational resources to be adopted straightforwardly. A key motivation of this work, is to develop a framework from which risk-owners can use as a straightforward, structured, reflective thinking-exercise when looking to improve systemic resilience. The framework allows for abstraction-based reasoning where deemed appropriate by the risk-owner. Further motivations include:

1. the qualitative nature of existing resilience management approaches means it can be a non-trivial task to meaningfully compare against a baseline and demonstrate progress and improvement;
2. identifying whether psychology models can be applied to evaluate attributes of an enterprise, and finally;
3. identifying how the technology estate has transformed the enterprise space the last few decades.

Any reader new to the topic of resilience can easily conflate topics related to risk management, resilience and robustness of enterprises. To avoid confusion, we consider risk management to be the prediction and evaluation of threats posed to an enterprise, their impact and likelihood to identify procedures to avoid, transfer, treat or accept their harm. Resilience refers to an enterprise's ability to recover from incidents smoothly, while robustness refers to an enterprise's capacity to cope with incidents (i.e. while being exposed to threats). Risk management is a crucial element to improve enterprise resilience and robustness, but the scope of resilience is wider than that of identifying and addressing risks.

Our framework consider the use of the word "threat" to cover the widest possible sense of the word: anything that can pose some form of risk or detriment to the enterprise either permanently or temporary. It can be regarded as an analogue to Higgin's idea of vulnerabilities. To the best of the author's knowledge, no approach has explored how psychology models and theories can be re-purposed for enterprise resilience. The remainder of this paper will focus on describing the framework and discussing its potential uses.

3 Framework

This framework aims to re-contextualise Self-Discrepancy Theory in the terms of an enterprise (as opposed to a person) in order to understand, reason about and, overall, improve enterprise resilience. We adapt Self-Discrepancy Theory to help establish, understand and reason about multiple views of the enterprise: the actual, ideal and ought enterprise, as viewed by the self and others. Examining threats and vulnerabilities through a multi-view representation of an enterprise might yield new insight and highlight new concerns that otherwise would go unnoticed. We hypothesise that a multi-view representation of enterprises as a 'self' can be useful to risk-owners because it enables risk-owners to reflect on the enterprise as seen in terms of where it is today, and where it might be in the future, as viewed by

Fig. 1 Identity view shows
the basic reasoning structure
in how comparing the actual
and ideal self views can lead
to an ought self, as perceived
by the self and others

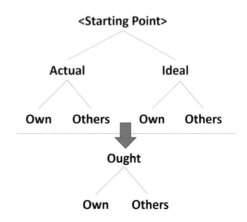

themselves and by others. In essence, this framework aims to enhance perception of risk by exhaustively considering how enterprise operations and security posture of the self and the environment these can operate in.

The proposed framework is one solution that is comprised of two constituent components, the *Identity View* and the *Convergence View*. At a high level, the framework is an iterative feedback loop that helps risk-owners converge to a refined understanding of the self in order to identify and address resilience concerns—both at the operational—and the security posture level. In turn, this help them converge towards their '*ought self*' (optimised according to their enterprise operations and security posture).

1. **Identity View**: a graph reflecting on how the enterprise "*self*" views itself as well as how others perceive it, see Fig. 1.
2. **Convergence View**: a feedback loop consisting of graphs (identity views) that that refines the priorities of the risk-owner in order to converge to their 'ought self' overall (and not simply enterprise operations or the security posture alone), see Fig. 2.

3.1   Identity View: Understanding the Self

The identity view is a graph reflecting how the enterprise self views itself and others perceive it, see Fig. 1.

The purpose of this view is to examine attributes of the self in the context of the actual, ideal and ought self. The attributes are determined and set by the risk-owner, using the risk management or resilience management model or standard of choice. The ought-self is widely regarded as the self representation of attributes that the self believes it *should* possess (i.e. a sense of duty or responsibility, rather than the 'optimal'). For the purposes of this paper, we are re-interpreting this 'ought' self to be the compromise between the actual (attributes that are factually correct) and the

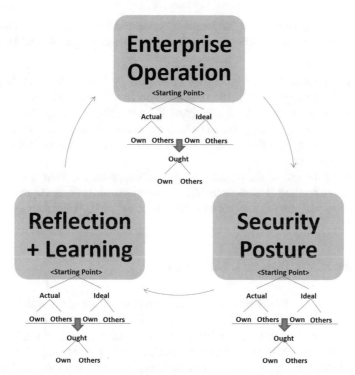

Fig. 2 Convergence view shows a feedback loop that uses the identity view as seen from an enterprise operation and security posture. Upon reflection, risk-owners might better identify how to improve attributes of different parts of the organisation, whether this be the enterprise operations or security. The final step involves comparing both selves in order to reflect and learn from the thinking-exercise

ideal (attributes in the realm of potential, hopes and aspirations—as organisations have to comply with the law in any of the selves regardless). Once a particular set of attributes related to the enterprise operation has been examined by the risk-owner (such as the enterprise operation), they may examine the security posture of the enterprise, and then both combined, as show in in Fig. 2.

3.2 Convergence View: What Can Be Refined About the Self?

Self-discrepancy describes the gap between two self-representations. People are motivated to reduce this gap to remove disparity in self-guides. Similarly, this approach might help remove disparity by establishing where the enterprise is 'today' (actual), and where the enterprise see themselves in an ideal setting (ideal being recognised as largely unachievable/unrealistic aspirations), in order to reach the 'ought' self. This ought self should help risk-owners identify steps towards a more resilient enterprise (through the part of the loop: reflection and learning), one with the capability and capacity to smoothly recover from considered threats. We might achieve this by:

- Examining the self of the enterprise operations by considering how to change the mission to respond to the threat landscape. In a business enterprise, we may consider these threats to be the evolving business market, competition or social threats (e.g. reputational, political or financial harms). In a political context, such threats might include deception or non-security related incidents that affect the mission operations.
- Examining the self of the security posture of the enterprise. Here a threat could mean not-actualised attacks, unintentional albeit harmful events and maintenance need (wear and tear). It may relate to both physical and cyber security postures.

4 Discussion: Towards Validation

In Self-Discrepancy Resilience, we see two areas in the enterprise in which we might consider applying SDT: the enterprise operation and the security posture. These two areas have been determined to be the most reasonable aspects to consider for this position paper, but there is no empirical evidence or rigorous analysis to support decision. Indeed, there may be other areas to consider as part of the analysis. The purpose of this framework is not to replace existing models and standards, but rather complement them by providing a thinking-exercise framework for risk-owners. I expect any resilience management framework or tools to work seamlessly with this approach as its primary purpose is to generate new views on existing insight of resilience concerns.

Presently, **this framework should be used as a thought-experiment to re-contextualise the mission and security of an enterprise**. It provides risk managers with the space and systemic methodology to identify risks posed to the self (enterprise), by viewing the mission, security and their relationship with different views. Each view can be the use of a resilience management framework/model/method or simply viewing the resilience of the organisation with competitors in mind, or making use of aggressive peer review to identify discrepancies in the understanding of threats posed to an organisation. Below follows a simple, generic examples detailing how the Self-Discrepancy Resilience model might be adopted.

4.1 Use Case Example: Logistics Company

A risk-manager of a logistics company has been tasked to improve the company's resilience against cyber threats and business risks. First they will need to identify

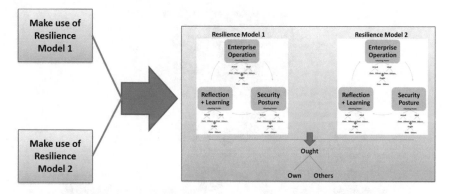

Fig. 3 Extended view 1 shows how other resilience management frameworks or models can become inputs for the Self-Discrepancy Resilience model as they provide a different perspectives on the resilience of the enterprise

the external dependencies, including supply chains, document the business processes, understand their competition, and digital and human assets used to operate the company.

After an exhaustive exercise in understanding the company's functional relationships to assets, business processes, vulnerabilities, incidents, dependencies and risks using two separate resilience management frameworks, the risk-manager can use both as inputs to the Self-Discrepancy Resilience in order to identify how the perspectives of resilience may change dependent on the model used. This is done in the interest of determining the ought self. If there is sufficient time and resources available, the risk-manager may make use of both models in tandem with 'other' views, as shown in Fig. 3, or the 'other' view can simply be an opposing resilience model, as shown in Fig. 4.

This framework leans itself to allow peer-reviewing as an approach to develop new views. Involving different stakeholders of the enterprise (as 'others') is likely to improve its resilience [23–25]. Risk-owners ought to engage with policy makers, analysts, legal team members, etc. in order to cover as many 'other' views as possible.

4.2 Future Work

The value of this peer-reviewing approach in resilience is still poorly understood, and therefore subject to future work. Presently, this framework would require further refinement, as no empirical evidence demonstrate its value.

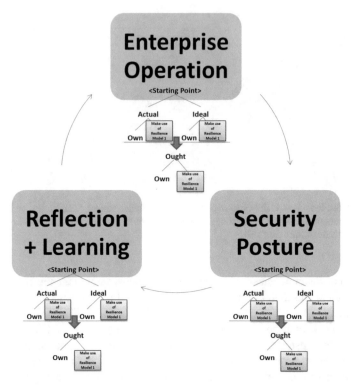

Fig. 4 Extended view 2 shows the how the 'other' self-guide has been replaced by resilience model 1

4.2.1 Validation

Future work will need to validate whether Self-Discrepancy Resilience is feasible in real-world settings. Tools might be developed to facilitate the thinking exercises of the framework and provide new ways to annotate and reason about resilience using different views. This might be achieved on a per-sector basis or by providing generic tools that apply across the board.

4.2.2 Psychology Models

Many psychology models exist with respects to the 'self'. It might be worth exploring other psychology models to provide new insight into resilience management. Examples of self-representation works include: Kohut [32] propose a bipolar self made up of two systems of narcissistic perfection: (1) a system of ambitions and, (2) a system of ideals. These poles of the self represent natural progressions in the earliest years of life. Winnicott [52] makes a distinction between a "*true self*" from the "*false self*" in the human personality. The true self is based on the individual's sense of being, rooted in the experiencing body. Berne [5] takes a transactional view of the self, and distinguishes between ego states: from a child, adulthood to parenthood. The parents state consists of behaviours and feelings from previous caregivers. The adult state is one driven by being able to judge information based on facts. Finally, the child state holds all our memories and emotions. People carry this state with them at all times. Analogously, we might be able to make use of this insight to understand how institutional memory affects enterprise 'self' decision-making processes. Cooley [9] propose a 'looking-glass self' model, and describe the 'self' as reflections of how we believe we appear to others. Other psychology works might include concepts and models of Resilience [17, 22, 40, 48] and Risk Perception [13, 31, 49, 51].

5 Conclusion

In this paper, we have explored the potential value of using social psychology theories that focus on identifying personality discrepancies, and using this in resilience management of enterprises. Instead of considering the 'self' as a person, we treat the enterprise (organisation, business, mission or operation) as the 'self'. We have adapted Self-Discrepancy Theory [27] specifically to examine whether identifying discrepancies in enterprises, with a multi-view representation in mind in order to better identify how different perspectives of the same information and priorities affect perception of threats and vulnerabilities. We have explored the actual, ideal and ought self, as well as the idea of the 'own' self and 'other' self representations. These self-guides and views might help risk-owners identify risk and resilience management concerns that would otherwise not be straightforwardly visible when using more existing models and standards. Future work will need to determine feasibility of these ideas with empirical evidence.

References

1. Abelson, R.P.: Modes of resolution of belief dilemmas. J. Conflict Resolution 3(4), 343–352 (1959)
2. Adler, A.: Problems of Neurosis (1964)
3. Allport, G.W.: Becoming; Basic Considerations for a Psychology of Personality, vol. 20. Yale University Press (1955)
4. Alpcan, T., Bambos, N.: Modeling dependencies in security risk management. In: 2009 Fourth International Conference on Risks and Security of Internet and Systems (CRiSIS 2009), pp. 113–116. IEEE (2009)
5. Berne, E.: What do You Say After You Say Hello. Random House (2010)
6. Bodeau, D., Graubart, R.: Cyber Resiliency Engineering Framework. MITRE Corporation (2011)
7. Brandt, C., Hermann, F., Groote, J.F.: Modeling and Reconfigurating Critical Business Processes for the Purpose of a Business Continuity Management Respecting Security, Risk and Compliance Requirements at Credit Suisse Using Algebraic Graph Transformation: Long version (2010)

8. Caralli, R.A., Allen, J., White, D.W.: CERT Resilience Management Model: A Maturity Model for Managing Operational Resilience. Addison-Wesley Professional (2010)
9. CMU Software Engineering Institute. Introduction to the CERT Resilience Management Model. https://www.sei.cmu.edu/education-outreach/courses/course.cfm?coursecode=P66 (2019)
10. Cooley, C.H.: Human Nature and the Social Order. Routledge (2017)
11. Cooley, C.H.: Looking-glass self. In: The Production of Reality: Essays and Readings on Social Interaction, vol. 6 (1902)
12. Creese, S., Goldsmith, M., Moffat, N., Happa, J., Agrafiotis, I.: Cybervis: visualizing the potential impact of cyber attacks on the wider enterprise. In: 2013 IEEE International Conference on Technologies for Homeland Security (HST), pp. 73–79. IEEE (2013)
13. de Crespigny, M.: Building cyber-resilience to tackle threats. Netw. Security **2012**(4), 5–8 (2012)
14. Dowling, G.R.: Perceived risk: the concept and its measurement. Psychol. Marketing **3**(3), 193–210 (1986)
15. ENISA. Measurement Frameworks and Metrics for Resilient Networks and Services: Technical report. Technical Report Published by ENISA (2011)
16. Erola, A., Agrafiotis, I., Happa, J., Goldsmith, M., Creese, S., Legg, P.A.: Richerpicture: semi-automated cyber defence using context-aware data analytics. In: 2017 International Conference on Cyber Situational Awareness, Data Analytics And Assessment (Cyber SA), pp. 1–8. IEEE (2017)
17. Festinger, L.: A Theory of Cognitive Dissonance, vol. 2. Stanford University Press (1962)
18. Fletcher, D., Sarkar, M.: Psychological resilience: a review and critique of definitions, concepts, and theory. Eur. Psychologist **18**(1), 12 (2013)
19. Gallagher, M.: Business continuity management. Accountancy Ireland **35**(4), 15–16 (2003)
20. Gibson, C.A., Tarrant, M., et al.: A 'conceptual models' approach to organisational resilience. Austr. J. Emergency Manage. **25**(2), 6 (2010)
21. Goldman, H., McQuaid, R., Picciotto, J.: Cyber resilience for mission assurance. In: 2011 IEEE International Conference on Technologies for Homeland Security (HST), pp. 236–241. IEEE (2011)
22. Goldstein, S., Brooks, R.B.: Resilience in Children. Springer, Heidelberg (2005)
23. Happa, J., Fairclough, G., Nurse, J.R.C., Agrafiotis, I., Goldsmith, M., Creese, S.: A pragmatic system-failure assessment and response model. In: International Conference on Information Systems Security and Privacy (ICISSP), pp. 503–508 (2016)
24. Happa, J., Fairclough, G.: A model to facilitate discussions about cyber attacks. In: Ethics and Policies for Cyber Operations, pp. 169–185. Springer, Heidelberg (2017)
25. Happa, J., Nurse, J.R.C., Goldsmith, M., Creese, S., Williams, R.: An Ethics Framework for Research into Heterogeneous Systems (2018)
26. Heider, F.: The Psychology of Interpersonal Relations (1958)
27. Higgins, E.T.: Self-discrepancy: a theory relating self and affect. Psychol Rev.**94**(3), 319 (1987)
28. Horney, K.: Our Inner Conflicts: A Constructive Theory of Neurosis. Routledge (1945)
29. International Organization for Standardization. ISO 31000:2018. https://www.iso.org/obp/ui#iso:std:iso:31000:ed-2:v1:en (2018)
30. Jakobson, G.: Mission cyber security situation assessment using impact dependency graphs. In: International Conference on Information Fusion, pp. 1–8. IEEE (2011)
31. Joffe, H.: Risk: from perception to social representation. Br. J. Soc. Psychol. **42**(1), 55–73 (2003)
32. Kohut, Heinz: Forms and transformations of narcissism. J. Am. Psychoanalytic Assoc. **14**(2), 243–272 (1966)
33. Kotenko, I., Chechulin, A.: Attack modeling and security evaluation in siem systems. Int. Trans. Syst. Sci. Appl. **8**, 129–147 (2012)
34. Legg, P., Moffat, N., Nurse, J.R.C., Happa, J., Agrafiotis, I., Goldsmith, M., Creese, S.: Towards a conceptual model and reasoning structure for insider threat detection. J. Wireless Mobile Netw. Ubiquitous Comput. Dependable Appl. **4**(4), 20–37 (2013)

35. Linkov, I., Eisenberg, D.A., Plourde, K., Seager, T.P., Allen, J., Kott, A.: Resilience metrics for cyber systems. Environ. Syst. Decis. **33**(4), 471–476 (2013)
36. Liu, M., Feng, T., Smith, P., Hutchison, D.: Situational awareness for improving network resilience management. In: International Conference on Information Security Practice and Experience, pp. 31–43. Springer, Heidelberg (2013)
37. McCourt, Tom, Burkart, Patrick: When creators, corporations and consumers collide: Napster and the development of on-line music distribution. Media, Culture & Society **25**(3), 333–350 (2003)
38. Meredith, L.S., Sherbourne, C.D., Gaillot, S.J., Hansell, L., Ritschard, H.V., Parker, A.M., Wrenn, G.: Promoting psychological resilience in the us military. Rand Health Quarterly **1**(2) (2011)
39. Osgood, C.E., Tannenbaum, P.H.: The principle of congruity in the prediction of attitude change. Psychol. Rev. **62**(1), 42 (1955)
40. Paul Slovic and Elke U Weber. Perception of risk posed by extreme events. *Regulation of Toxic Substances and Hazardous Waste (2nd edition), Foundation Press*, (2002)
41. Prescott Lecky. Self-consistency; A Theory of Personality (1945)
42. Ray, K., Shih, R.: The creative destruction of copyright: napster and the new economics of digital technology. U. Chi. L. Rev. **69**, 263 (2002)
43. Rodríguez, Alfonso., Fernández-Medina, Eduardo, Piattini, Mario: A bpmn extension for the modeling of security requirements in business processes. IEICE transactions on information and systems **90**(4), 745–752 (2007)
44. Rogers, C.R.: On Becoming a Person: A Therapist's View of Psychotherapy. Houghton Mifflin Harcourt (1995)
45. Ross, R., et al.: NIST SP 800-37, Revision 2. https://csrc.nist.gov/publications/detail/sp/800-37/rev-2/final (2018)
46. Samuelson, P.: Digital rights management {and, or, vs.} the law. Commun. ACM **46** (2003)
47. Saroiu, S., Gummadi, K.P., Gribble, S.D.: Measuring and analyzing the characteristics of napster and gnutella hosts. Multimedia Syst. **9**(2), 170–184 (2003)
48. Sawilla, R.E., Ou, X.: Identifying critical attack assets in dependency attack graphs. In: European Symposium on Research in Computer Security, pp. 18–34. Springer, Heidelberg (2008)
49. Seligman, M.E.P.: Building resilience. Harvard Bus. Rev. **89**(4), 100–106 (2011)
50. Vugrin, E.D., Turgeon, J.: Advancing cyber resilience analysis with performance-based metrics from infrastructure assessments. Int. J. Secure Softw. Eng. (IJSSE) **4**(1), 75–96 (2013)
51. Weber, E.U., Blais, A.-R., Betz, N.E.: A domain-specific risk-attitude scale: measuring risk perceptions and risk behaviors. J. Behav. Decis. Making **15**(4), 263–290 (2002)
52. Winnicott, D.W.: Ego distortion in terms of true and false self. In: The Person Who Is Me, pp. 7–22. Routledge (2018)

Insider Threats to IT Security of Critical Infrastructures

Ivan Gaidarski and Zlatogor Minchev

Abstract The chapter provides an outlook to contemporary innovative methods for detecting internal threats to the information security of critical infrastructure objects, mitigating these threats, as well as preventing the leakage of sensitive information. Internal threats are unpredictable and pose a major challenge to traditional IT security measures. A specific emphasis is placed on the insider threats problem emerging due to: careless behaviour of insiders, vendors and contractors, cybersecurity policies, e-identity theft, and malicious users. Methods for detecting and protecting internal threats encompass user behaviour analysis, consumer behaviour analysis, risk assessment and profiling, analysis of information flow within the organisation, and definition of sensitive information. Some useful methods for protecting sensitive data through a holistic approach that covers data both inside and outside the organization are also presented. Consumer activity monitoring systems and Data Leak Prevention (DLP) data leakage monitoring systems are finally discussed in the context of practical handling of internal threats.

Keywords Critical infrastructure · Security measures · Internal threats · Sensitive information · Holistic approach · Activity monitoring · Data leak prevention

1 Introduction

Modern life is highly dependent on the numerous achievements of modern civilisation, such as electricity, telecommunications, the Internet, water and sewerage, roads and railways, the gas transmission networks and many more. The normal life of the society depends on their regular and continuous safe work. These assets and services,

I. Gaidarski · Z. Minchev (✉)
Institute of Information and Communication Technologies, "Acad. G. Bonchev" Str., Bl. 25-A, 1113 Sofia, Bulgaria
e-mail: zlatogor@acad.bg

I. Gaidarski
e-mail: ivangaidarski@abv.bg

in turn, depend on the robust and reliable operation of facilities and large-scale infrastructures, known as critical infrastructure (CI). All critical infrastructure definitions recognize the vital and indispensable importance of the service or function provided by the asset to the society. CI also could be described as Infrastructures of significant national importance, that enable society and the economy to function [1]. Here are some definitions for CI from different organizations:

EU: According to the EU Directive 2008/114/EC, critical infrastructures can be defined as assets, systems, or parts thereof, essential for the maintenance of vital societal functions, health, safety, security, economic, or social well-being [2].

International Organization for Standardization: Organizations and facilities that are essential to the functioning of society and the economy as a whole. The standard elaborates that a failure or malfunction of such organizations or facilities would result in sustained supply shortfalls, make a significant impact on public security, and have other wide-ranging impacts [3].

International Telecommunication Union: The key systems, services, and functions whose disruption or destruction would have a debilitating impact on public health and safety, commerce, and national security, or any combination of these [4].

NATO: Physical or virtual systems and assets under the jurisdiction of a state that are so vital that their incapacitation or destruction may debilitate a state's security, economy, public health or safety, or the environment [5].

Critical infrastructure refers to various technologies, facilities, systems, assets, services and processes, which are essential to the health, safety, security or economic well-being of the society and citizens, as telecommunications, electrical power grid and plants, transportation facilities as roads and railroads, water distribution systems, healthcare services, security, banking and financial systems and many more [6].

In terms of interconnection, CI can be stand-alone or interconnected geographically across local territories or national borders. The CI elements are usually interdependent from each other, e.g.: all communications depend from electrical power, the electrical distribution himself is dependable from the communications, roads and etc. The eventual disruption or impediment to the normal operation of these infrastructures would have serious consequences and would have a critical impact on national security, economy, public health and safety, or combination of these issues [7].

There are many possible threats to CI, including:

- Natural Disasters (earthquakes, flooding, fires, etc.);
- Extreme Weather (storms, hurricanes, heavy snow, etc.);
- Pandemics (Plague, HIV, Smallpox, Chollera, etc.);
- Accidents (accidental, random, intentional, human errors, etc.);
- Technical Faults (Software errors, human factor, equipment faults, etc.);
- Malicious Attacks (direct or remote);
- Cyber Threats (External or Internal);

Some of the threats to CI are unpredictable and unavoidable as natural disasters or extreme weather, but their effect can be mitigated or even avoided, if we take proactive measures as monitoring and control of key parameters and appropriate measures afterwards to ensure that infrastructure can continue it's normal function after the failure event. A failure event can be defined as a negative incident, which can impact the normal functioning of the infrastructures or it's subsystems [7]. For example, the heavy snow can cause interruption of the communication or electrical cables, causing the malfunctioning of the communication or electrical infrastructure or part of them. If we look to the CI as a system, one of the basic requirements is that the system should continue to operate even when one or more of its components have failed. To achieve that, a systematic approach must be performed. One good approach is the holistic approach, which ensures the protection of all the system (CI) components and grants the normal operation after potential failure. The holistic approach includes risk assessment of the security (confidentiality, integrity, authenticity, availability) level of all assets (physical and cyber) in all layers of the system.

Due to deregulation purposes, one of the important requirements of CI is that they consist of various autonomous subsystems. Thus, even when one or more of subsystems have failed, the system as whole can continue to operate normally. On the other hand, the different CIs are interconnected and potential damage to one infrastructure can lead to cascading failures in others, which can have serious and unpredictable consequences [8].

To apply the holistic approach, the measures to protect the different subsystems have to include all possible vectors of attacks and take into account all of the possible threats. It is extremely important to develop methods and measures for quickly detecting, isolating, and recovering the faults. This can be achieved with the development of a common framework for modelling the behaviour of CI subsystems and for designing methods and measures for intelligent monitoring, control, and security of the CI and it's subsystems.

In this paper we will look at the protection from insider threats, as one of the components of the holistic approach for CI protection. Due to the unpredictable nature of the internal threats, the usage of innovative methods of protection are necessary, such as behaviour analysis, consumer behaviour analysis, risk assessment and profiling, analysis of information flow and definition of organisation-sensitive information. We look at the causes of insider threats, such as neglected behaviour of insiders, e-identity theft and malicious users. We also present methods for protecting sensitive data through holistic approach that covers data both inside and outside the organisation. Consumer activity monitoring systems and Data Leak Prevention (DLP) data leakage monitoring systems are finally discussed.

2 It Security of Critical Infrastructure

Each of the economic sectors has its own physical assets, such as: roads, buildings, power plants and grids, water and gas pipelines, government offices, etc. All of the critical sectors, part of CI, however depend upon information and communication technologies—the Critical Information Infrastructure (CII) to deliver the necessary services and to conduct their normal operations. A disruption to the CII could have significant impact to the ability of CI to perform its essential missions in multiple sectors. In order to provide effective protection and assurance of the resilience of CI, a security plan and policies have to be developed, part of Critical Information Infrastructure Protection (CIIP) program, intended to protect the virtual components of the CII. These policies are part of the holistic cybersecurity idea. The cybersecurity in fact is expected to guarantee protection against all forms of cyber threats and cyber incidents by strengthening the safety of the CII and securing the networks and services which serve the everyday needs of the users from the critical sectors.

Cyber incidents can affect the CII and can take many forms of malicious activity, like: using botnets to perform DoS (Denial-of-Service) attacks or malware infections, which can affect the ability of networks to operate.

Other threats as phishing, ransomware attacks or identity theft can cause the ability of the affected organization to perform it's normal operations and even bankruptcy [4]. The field of security deals with the protection of assets, taking associated risk into account. The protection of information in all its forms is the main goal of the Information Security, part of which is Cybersecurity. Each asset must be considered in the context of its value and the associated risk of it's loss. Information security is based on CIA triad: "Confidentiality", "Integrity", and "Availability". "Confidentiality" restricts access to the information only to authorized systems, devices, application or persons. "Integrity" assures that the data in the system has not been altered in an unauthorized way. "Availability" refers to the assurance that the relevant service will be available always when it's needed [9].

Cyber threats are one of the major threats to the CI. The threats can be both "External", with the vector of attack being outside to inside direction and "Internal", with vector of attack inside to outside direction. For the "External" threats it is assumed that the attack to the assets comes from an external source—like hacker attacks. The goal is to stop the attack from the outside and not to allow penetration to the resources of the organization. An example of such a solution is the firewalls. For protection from the "Internal" threats, the assets are protected in the direction Inside-Out. The presumption is that the attack is carried out by a person with access to the resources of the organization or an insider. The aim is to protect the assets so that it is impossible to be exported out even with a direct access to them. A typical solution is Data Loss Prevention (DLP). There are many commercial products that implement this method, to mention, e.g.: Device Lock [10] and CoSoSys Endpoint Protector [11].

Information security (IS) activities applied within an organization concern compliance with certain legislative regulations, necessary for its normal operation. There are IS standards such as ISO 27000 [12, 13], ISACA's COBIT [14], NIST "800 series" [15], sector-specific regulations—the Gramm-Leach-Bliley Act (GLBA) [16] for the financial sector, Sarbanes-Oxley Act (SOX) [17, 18] for US public companies, Health Insurance Portability and Accountability Act (HIPAA) [19] and Payment Security Industry (PCI) Data Security Standard (DSS) [20] for credit card operators.

While these standards and regulations incorporate the most important aspects of IS, they are rather a set of good practices. The cases, in which the IS is approached methodically and all requirements of the standards are satisfied, are seldom. Thus, the most common practice is single actions to solve certain tasks as incidents (leakage, attack on infrastructure, loss of information, etc.) or new challenges—for example recently adopted regulation EU GDPR [21] for protection and processing of personal data for EU Citizens.

3 Internal Threats to Critical Infrastructure

An incident caused by an internal threat occurs when an insider, employee, partner or provider with authorized access to an organization's sensitive information or system, intentionally or accidentally misuses that access, resulting in negative consequences for the organization. Due to their nature, internal threats are virtually unpredictable and pose a major challenge to traditional IT security measures. Although these types of incidents are becoming more common, many organizations do not understand the causes of these incidents, nor how to detect and prevent them.

There are numerous causes for the rise in incidents provoked by internal threats:

1. Careless behaviour of insiders—accidentally sharing sensitive data, opening malicious phishing e-mails, using illegitimate software, and other activities that can put the organization at risk;
2. Third parties—external consultants, agencies and suppliers pose a huge security risk to the organization. These types of insiders have authorized access to some of the organization's resources, but often they do not follow the internal security policies. This can lead to potential security breaches. An effective counteraction measure is the monitoring and recording of all external actions;
3. Cybersecurity Policies—Too many and too strict IT security policies, implemented in an organization may be the root cause of insider-related incidents, creating too many barriers to the usual daily work of the employees, and can lead to so-called "Security Fatigue Effect". An effective counteraction measure is continuous training, maintaining open dialogue with seeking feedback and proactive sharing of good practices with employees. It's an important element adopting policies and implementing the solutions they provide transparency of daily activities—the principle of non-intervention;
4. Electronic Identity Theft—This is the type of internal threat can be with the highest cost of an incident. The reasons can be using of weak passwords, lack of a policy for changing them periodically, the lack of strict rules for their formation, as well as the use of the same passwords to access different resources, services and systems. This facilitates attempts by unauthorized persons to access sensitive information of the organization and its expiration. Countermeasures against this kind have entailed the adoption of a strict policy for the formation, replacement and use of passwords, two-factor identification (2FA), as well as additional biometric identification through fingerprints, scan of the iris, retina or face;
5. Malicious users—Understanding the true intentions of the users is very important to combat internal threats effectively. Persons with privilege access, as well as

technicians could use their level of access to the organization's resources to steal and delivering sensitive information unnoticed, bypassing security measures in the organization. Combating these threats requires unconventional approaches;

Internal threats can be divided into several main groups, depending from their source:

1. Human threat—In their day-to-day work, employees can perform unintentional mistakes, become a target of hackers, or completely intentionally cause harm. Because some of these users have access to sensitive data and security-critical systems, in practice humans are the most serious internal threat. It's very important to classify different types of users and profile them according to the risk they carry to the organization.
   The following categories of users may be distinguished:

   • External vendors—Many of the biggest security breaches, caused the leakage of sensitive information recently have been committed by stealing electronic identification by external suppliers of the company with access to some of its resources;
   • Privileged users—a typical example is system administrators with full access to the resources of the organization;
   • Employees with access to sensitive data—Their number and the lack of knowledge and skills to prevent security incidents can cause the leak of sensitive information.
     It is essential to profile users by grade risk to the organization and implementation of solutions to reduce these risks [22–24].

2. Employees activities—It is the most common vector of internal threat. Due to negligence, carelessness or malicious intentions, individuals may take actions that threaten sensitive data and critical systems of the organization. Due to the huge number of actions performed by employees in the normal work process it is extremely difficult to identify unauthorized or risky activities. So, it is of vital importance that the organization is equipped with a IT security tools that record and block these activities for reducing the risk of leaking sensitive information through internal or external attacks;

3. Business applications—The business applications used by employees on a daily basis are serious source of risk (e.g. accounting systems, invoicing systems, CRM systems, CAD applications). Even used as intended, they present a potential risk, because of their access to sensitive data. Users can abuse them intentionally or completely by accident. Apart from specialized applications, it is already a daily routine to use modern services and tools as cloud services, sharing a desktop for training purposes (webinars), file transfer (FTP), peer-to-peer file sharing (torrents), and many more. They are a direct source of high risk for the organization data leaks.

4 Detecting the Internal Threats

Organizations that pay serious attention to their IT security, have serious investments such as capital and human resources in various systems designed to ensure the security of their infrastructure: firewalls, Virtual Private Networks (VPN), Intrusion Detection System—IDS, Database Activity Monitoring—DAM. These solutions collect a huge amount of data about the system and the infrastructure normally collected by Monitoring Systems and Event Management—SIEM (see Fig. 1, after [25]).

The problem with traditional IT security systems is that they are targeted primarily at external sources of threats, ignoring the fact that insiders—employees, administrators, vendors, etc., have direct and legal access to sensitive data and critical systems, so they have already overcome the traditional IT security systems and the security of information is completely dependent on their intentions or goodwill. In practice, none of the authorized persons is aware of employee activities. This is a huge flaw in IT security organizations that require new and unconventional methods of protection. As to mark here the consumer activity monitoring systems and data leakage prevention (DLP) systems, both based on new methods for detecting threats.

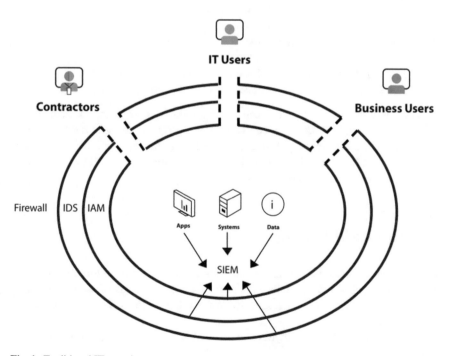

Fig. 1 Traditional IT security systems

4.1 User Behaviour Analysis

By constantly monitoring the actions of individual users, it is possible to create a profile of the user according to the prognostic risk he represents for the organization of interest. This type of profiling requires preliminary classification of the various actions during the daily work. The information could be gathered from the policies adopted by the organization for IT security, compliance with different standards, regulatory mechanisms, business documentation, etc., taking into account the specifics of the different roles that employees play in the organization. The involvement of IT professionals and management is needed, as well as more experienced employees keeping also the specifics of the organization.

Analyzing user behaviour enables automatic detecting of behavioural anomalies of users or groups of users showing negligence or targeted hostile action. For example, the actions of a user suddenly received access to new resources, hacker, stolen electronic identity, etc. Numerous types of anomalous behaviours can be distinguished in this context:

- Organization-specific system applications;
- Atypical access to systems, files or other IT resources;
- Perform unusual operations or infrequently used commands.
- Generate larger than usual reports;
- Performing an unusual number of actions compared to normal;
- Access to systems from unusual client machines or from outside;
- Access to systems outside of normal hours of the day or days of week.

The analyses of the consumer behaviour provide opportunities for detection of irregularities in behaviour and identification of a potential threat to the IT security. The suspected users can be put on closer monitoring for their actions. If necessary the user may be alerted to the potential or real violation, and even to be excluded from an appropriate resources access.

4.2 User Activities Analysis

Like behaviour analysis, consumer activities analysis enables automatic detection of abnormalities in normal daily activities of users or groups of users, indicating possible hostile acts toward the security of the organization—due to negligence or targeted attacks.

Typical examples of Employees activities, affecting the security of the organization are:

- Working with applications that process sensitive data;
- Upload sensitive data to cloud services;
- Deliberately or unintentionally sharing sensitive data with other people via e-mail, cloud apps, flash drives, etc.;

- Installing remote access applications for work from home;
- Open phishing e-mails, giving access to the internal network of people with hostile intentions;
- Visiting unauthorized websites that may install malware on systems.

Typical examples of privileged user actions, influencing the security of the organization are:

- Making changes to configuration files that can cause system fail;
- Creating unauthorized local or remote accounts (VPN or SSH);
- Escalating privileges for Unix or Linux users' workstations;
- Changing admin passwords;
- Installation of "backdoors", allowing later unauthorized access to the systems;
- Execution of malicious code leading to denial of critical services (DOS).

4.3 Risk Assessment and Profiling

With risk assessment and profiling, can be created risk profiles of users (employees) and the risk they carry for the IT Security of the organization. It is based on the real-time monitoring of users and the relevant information about their actions and behaviour in their daily activities, good business practices, accepted standards, compliance with regulations, and regulations. In this way special measures can be carried out to the high-risk users for achieving a prevented leakage or stealing of sensitive information, and even stopping the access to critical systems to which they are not authorized to have one.

4.4 Analysis of Information Flow and Definition of Organization-Sensitive Information

To define the sensitive information in an organization we need to take a holistic approach to information that the organization owns, uses, processes or generates. The holistic approach captures data both inside and outside the organization [26, 27].

The process of analyzing the flow of information within an organization and the definition of sensitive information includes the following phases:

1. Creating a model of information flow in the organization;
2. Classification of information in the organization;
3. Identification of critical activities and starting points;
4. Creating a policy to prevent leakage sensitive information.

A model of the information flow within the organization can be created through specialized tools for monitoring the information flow via endpoints, servers, databases and networks. Based on this model, the regulatory requirements for the organization, the legal framework, regulatory activity and, as well as good practices, classification of the data is performed and determining of the sensitive information for the organization is accomplished as a result.

On this basis, it is possible to determine the degree of risk of the activities with sensitive data, performed by the different users and the critical points via which this kind of data can be leaked outside the organization. This also enables management to create and adopt a common policy to prevent leakage of sensitive information [25, 26].

5 Protecting from Internal Threats

The next stage is the implementation of innovative security solutions for internal threats protection using the above methods, such as Consumer activity monitoring systems and Data Leak Prevention (DLP).

5.1 Consumer Activity Monitoring and Analysis Systems

Consumer activity monitoring and analysis systems provide an insider look at all user actions. This type of solutions gives IT security teams the opportunity for immediate detection of unsafe or unauthorized activities and the ability to block the relevant users. These solutions allow quick identification on "Who", "What", "When" and "How" accesses sensitive data, systems and applications. An example of such a solution is ObserveIT [22]. It greatly reduces the incidents related to internal threats and exposure to risks:

- Behavioural analysis and automatic risk assessment for the individual users, depending on their actions;
- Monitor user actions for suspicious actions or violations of the established security rules;
- Monitor user activity in applications, web pages visited and used by the IT organization systems, regardless of how they access them (RDP, SSH, Telnet, Citrix, Access Console, etc.);
- Alert via custom real-time messaging for risk activities or violations of established rules and stopping them;
- Compliance with regulatory requirements—GDPR, PCI, SOX, HIPPA, NERC, FFIEC, FISMA and FERPA.

5.2 Data Leak Prevention Systems

Data Leak Prevention systems (DLP) are Data-Centric based—they are focused on the data. They are able to monitor and control the data in all three dynamic states:

- Static data—Data-At-Rest;
- Data on the move—Data-In-Motion;
- Data used—Data-In-Use.

Data Leak Prevention systems are designed to prevent attempts to steal, modify, prohibit, destroy or obtain unauthorized access and usage of the data, without interrupting normal business processes. DLP can significantly reduce the leak of sensitive information, resulting from internal threats like human error, intentional action or outside breach. The main goal of DLP is to stop the data before it leaves protected environment of the organization [10].

DLP solutions can provide very useful information for process of protection of the sensitive data:

- Identification of the violations, threats, risks and vulnerabilities to the data;
- Violations of security policies and procedures;
- Discovering and identifying all sensitive information in the organization.

DLP systems can be different types, depending on their focus area:

- With focus on servers, global communications and data channels of the organization. The DLP can control e-mail servers, file transfers from file servers, and Internet traffic filtering;
- Focused on endpoints and local data channels—workstations, laptops, mobile devices (e.g. tablets and phones). Controlled channels include all possible physical ports, personal e-mails, file transfer to cloud services and more—as DeviceLock DLP Suite, shown on Fig. 2, after [10].

The implementation of DLP in organization can effectively protect the sensitive information from Internal and External threats and additionally brings the following results, after [25–30] (see Fig. 3):

- Reducing the sensitive information leak incidents;
- Limiting data leak channels;
- Increasing the visibility of sensitive information, by the discovery function of the DLP (Data-At-Rest);
- Improving compliance with the internal security policies, legal regulations and privacy directives.

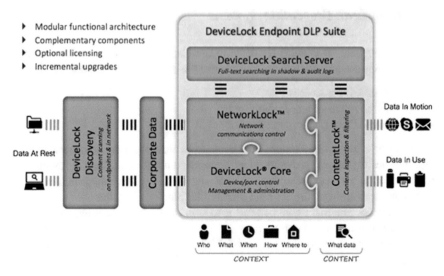

Fig. 2 DeviceLock endpoint DLP suite

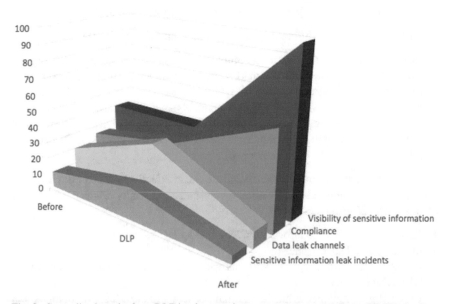

Fig. 3 Generalized results from DLP implementation

6 Conclusion

Today's IT security challenges to the Critical Infrastructure require new and non-standard solutions based on entirely new principles. Organizations need to develop a comprehensive, risk-based information security strategy to protect their sensitive information. For this purpose, in addition to new technical measures, a focus on the human factor in the organization is needed through proactive activities, involving continuous training and enhancing security knowledge—Security Awareness Training. The presented holistic approach addresses not only the organizations data. It includes profiling of the participants in the information process, based on "What" data they are using or processing, "How" and "Where" they are using or processing it. It allows us to identify the critical places, staff and activities, thus discovering the weak points in the organization. After risk assessment is done, suitable measures may be taken, by deploying appropriate security controls as Data Leak Protection (DLP) systems, which are designed to stop the data leakage from the inside to the outside. It can not only limit the data leaks, but improve the compliance with the internal security policies, legal regulations and privacy directives.

Acknowledgements The research is partially supported by the KoMEIN Project (Conceptual Modeling and Simulation of Internet of Things Ecosystems) funded by the Bulgarian National Science Foundation, Competition for Financial Support of Fundamental Research (2016) under the thematic priority: Mathematical Sciences and Informatics, contract № DN02/1/13.12.2016. Additional gratitude is also given to the National Scientific Program "Information and Communication Technologies for a Single Digital Market in Science, Education and Security (ICTinSES) 2018–2020", financed by the Ministry of Education and Science, Republic of Bulgaria.

References

1. Zaballos, A., Jeun, I.: Best Practices for Critical Information Infrastructure Protection (CIIP). Inter-American Development Bank (IDB) and Korea Internet & Security Agency (KISA) (2016)
2. EU Directive 2008/114/EC: Identification and designation of European critical infrastructures (2008)
3. ISO (International Organization for Standardization): Information Technology—Security Techniques—Information Security Management Guidelines Based on ISO/IEC 27002 for Process Control Systems Specific to the Energy Utility Industry. ISO/IEC TR 27019:2013 (2013)
4. ITU (International Telecommunication Union): Report on Best Practices for a National Approach to Cybersecurity: A Management Framework for Organizing National Cybersecurity Efforts. ITU Study Group Q.22/1, Geneva (2008)
5. Schmitt, M.N.: Tallinn Manual on the International Law Applicable to Cyber Warfare. Prepared for the NATO Cooperative Cyber Defense Center of Excellence. Cambridge University Press, Cambridge (2013)
6. USA Patriot Act. Public Law 107-56 (2001) [Online]. Available at: https://epic.org/privacy/terrorism/hr3162.html. Accessed: Dec 2019
7. Ellinas, G., Panayiotou, C., Kyriakides, E., Polycarpou, M.: Critical infrastructure systems: basic principles of monitoring, control, and security. In: Kyriakides, E., Polycarpou, M. (eds.) Intelligent Monitoring, Control, and Security of Critical Infrastructure Systems. Studies in Computational Intelligence, vol. 565, pp. 1–30. Springer, Berlin (2015)
8. Rinaldi, S.: Modeling and simulating critical infrastructures and their interdependencies. In: Proceedings of the 37th International Conference on System Sciences 2004, pp. 1–8 (2004)

9. Rhodes-Ousley, M.: Information Security: The Complete Reference, 2nd edn. McGraw-Hill, New York (2013)
10. DeviceLock Web Page. Available at: www.endpointprotector.com. Accessed: Dec 2019
11. Cososys Endpoint Protector. Available at: www.endpointprotector.com. Accessed: Dec 2019
12. Hintzbergen, J., Hintzbergen, K., Smulders, A., Baars, H.: Foundations of Information Security Based on ISO27001 and ISO27002, 3rd edn. Van Haren Publishing (2010)
13. ISO 27001. Official Web Page. Available at: https://www.iso.org/isoiec-27001-information-security.html. Accessed: Dec 2019
14. IT Governance Institute: COBIT Security Baseline: An Information Survival Kit, 2nd edn. IT Governance Institute (2007)
15. NIST Special Publications (800 Series). Available at: https://www.nist.gov/itl/publications-0/nist-special-publication-800-series-general-information. Accessed: Dec 2019
16. Gramm-Leach-Bliley Act (GLBA) resources. Available at: https://www.ftc.gov/tips-advice/business-center/privacy-and-security/gramm-leach-bliley-act. Accessed: Dec 2019
17. Anand, S.: Sarbanes-Oxley Guide for Finance and Information Technology Professionals. Wiley, Hoboken (2006)
18. Sarbanes-Oxley Act (SOX) Resources. Available at: https://legcounsel.house.gov/Comps/Sarbanes-oxley%20Act%20Of%202002.pdf. Accessed: Dec 2019
19. Herold, R., Beaver, K.: The Practical Guide to HIPAA Privacy and Security Compliance, 2nd edn. Auerbach (2011)
20. PCI Security Standards. Available at: https://www.pcisecuritystandards.org/pci_security/. Accessed: Dec 2019
21. EU General Data Protection Regulation Official Page. Available at: https://ec.europa.eu/info/law/law-topic/data-protection_en. Accessed: Dec 2019
22. ObserveIT Web Page. Available at: www.observeit.com. Accessed: Dec 2019
23. Dimitrov, W., Siarova, S., Petkova, L.: Types of dark data and hidden cybersecurity risks. Project Conceptual and Simulation Modeling of Ecosystems for the Internet of Things (CoMein) (2018). https://doi.org/10.13140/RG.2.2.31695.43681
24. Dimitrov, W.: Analysis of the need for cyber security components in the study of advanced technologies. In: INTED2020 Proceedings, 114th Annual International Technology, Education and Development Conference, INTED, 3–5 Mar 2020. ISBN: 978-84-09-17939-8. Available at: https://doi.org/10.21125/inted.2020.1423. Accessed Mar 2020
25. Gaydarski, I., Minchev, Z.: Conceptual modelling of information security system and its validation through DLP systems. In: 9th International Conference on Business Information Security (BISEC-2017), 18th Oct 2017, pp. 36–40, Belgrade, Serbia (2017)
26. Gaydarski, I., Kutinchev, P.: Holistic approach to data protection—identifying the weak points in the organization. In: International Conference "Big Data, Knowledge and Control Systems Engineering" BdKCSE'2017, 7–8 Dec 2017, pp. 125–135, Sofia, Bulgaria (2017)
27. Gaidarski, I.: Challenges to Data Protection in Corporate Environment, 30 Mar–5 Apr 2018, Sofia–Borovets (2018). Available at: https://it4sec.org/news/forum-future-digital-society-resilience-new-digital-age. Accessed: Dec 2019
28. CYREX 2018 Web Page. Available at: https://securedfuture21.org/cyrex_2018/cyrex_2018.html. Accessed: Dec 2019
29. Dimitrov, W.: Operational Cybersecurity, p. 122. Avangard Prima, Sofia (2019). ISBN 978-619-219-209-3
30. Polemi, N.: Port cybersecurity: securing critical information infrastructures and supply chains. Elsevier, Amsterdam (2017) ISBN: 9780128118184

Intelligent Decision Support—Cognitive Aspects

Valeriy P. Mygal, Galyna V. Mygal, and Oleg Illiashenko

Abstract This chapter is dedicated to the problem of dynamic systems operating in harsh conditions where informational cognitive distortions and the psychophysiological state of a human could influence the decision-making process. The main information source about cognitive distortions of information is the spatio-temporal features of periodic signals of different nature. Their consideration has led to the emergence of many models, patterns, indicators, criteria and methods for their processing, which limit the possibilities for intelligent decision support. The chapter presents the results of developing a technology for constructing a topographic three-dimensional model of the cycle of functioning of an information source from a digitized time series. It is proposed to expand the knowledge base using structural and other patterns of the cycle of functioning of self-organizing objects. The application of the proposed technology to information sources of various nature demonstrates the advantages and new opportunities for identifying hidden spatio-temporal relationships that determine the features of functioning the dynamic systems in harsh conditions. The authors outline the prospects of utilising the developed technology for conducting intelligent trainings and individual approach to training of operators for critical application systems, as well as in the transition from Industry 3.0 to Industry 4.0.

Keywords Decision making system · Cognitive distortions · Digitalization · Human–machine interaction · Safety · Security

V. P. Mygal (✉)
Department of Physics, National Aerospace University "Kharkiv Aviation Institute", Chkalova Street, 17, Kharkov 61074, Ukraine
e-mail: valeriymygal@gmail.com

G. V. Mygal
Department of Automobile and Transport Infrastructure, National Aerospace University "Kharkiv Aviation Institute", Chkalova Street, 17, Kharkov 61070, Ukraine
e-mail: mygal.galina@gmail.com

O. Illiashenko
Department of Computer Systems Networks and Cybersecurity, National Aerospace University "Kharkiv Aviation Institute", Chkalova Street, 17, Kharkov 61070, Ukraine
e-mail: o.illiashenko@csn.khai.edu

1   Introduction

Ensuring the reliability and cybersecurity of Instrumentation and Control systems (I&Cs), used to automate process control in various critical areas (e.g., critical information infrastructure facilities, Smart Grids, etc.), is one of the most important tasks, since efficiency and failure-free operation of complex technical objects depends on their safe continuous operation. The above task includes the assessment and assurance of cybersecurity, the identification of factors affecting the given risk level, as well as the improvement of methods for its ensuring during the design, operation and modernization of I&Cs.

The safety of the dynamic systems (DS) functioning imposes increasingly strict requirements for scientific and technological development, which, in its turn, prompts scientists and engineers to develop intelligent decision support systems. The widespread use of intelligent technologies (Knowledge Discovery Data-bases, Data mining, etc.) is associated with the need for intelligent analysis of the functioning of dynamic systems [1, 2]. However, their development did not pay enough attention to the problem of the human factor [3–5]. Therefore, latent security problems of DS appear in complex conditions. They are caused by cognitive aspects of perception, representation and developed mainly divergent thinking of a person [6–8]. Therefore, the effectiveness of intellectual support cannot be achieved without taking into account the cognitive aspects of human activity.

It turned out that the digitalization of technology and human–machine interaction itself does not guarantee the safety of the DS functioning [9]. There are features of cognitive perception of information by a person that determine the effectiveness of human–computer interaction. The intersection of problems is due to the relationship of the psychological characteristics of a person, his (or her) psychophysiological limitations and capabilities, as well as awareness and desire to use their resources and knowledge [10, 11]. If the peculiarities of the nature of the human factor in the design of DS is not taken into account, their maintenance and operation as well as the likelihood of risk realization significantly increases. The safety, reliability and effectiveness of complex DS and technologies under the influence of external and internal stress factors depend on their prediction. Often, the effectiveness of human–machine interaction, which determines the viability of complex DS, as well as human-dependent technologies and processes, depends on a person's cognitive perception of information flows of a different nature.

The emergence and diffusion of advanced digital production and processing technologies in the framework of the fourth industrial revolution radically changes approaches to the automation of production and industry, more and more blurring the boundaries between physical and digital production systems. The industry and the academy require the integration of various areas of security (functional, informational, cybersecurity, cybersafety) into an integral concept that will take into account many internal and external factors [12].

The aim of the work is to develop means of cognitive visualization of the signals structure of the self-organized objects functioning of, which allows to apply interdisciplinary principles, ideas and concepts when analyzing their functioning.

2 Problems of Human–computer Interaction in the Complex Conditions of the Functioning of Dynamic Systems

The problem of increasing the complexity of human–computer interaction. The need to apply interdisciplinary principles, ideas and concepts for the intellectual support of decision-making is caused by a constant increase in the complexity of DS and the volume of poorly structured information. Moreover, the inevitable increase in the number of sensors and other sources of information in DS leads to new problems that limit the ability to effectively predict their safety, reliability and stability in the context of the sustainable development paradigm [13]. In difficult conditions of DS functioning, the need arose for technologies of KDD, Data mining, and others.

Their use for searching in large volumes of data of unobvious, objective and practical patterns has increased the number of:

- methods for processing complex signals,
- methods and types of visualization,
- modeling methods and analysis tools.

Under normal operating conditions of the DS, the use of intellectual support is quite effective. However, in extreme conditions of DS functioning, the new problems are manifested (see Fig. 1). These are interrelated problems that are caused by limited human capabilities and hidden human errors. Their complex nature determines the individual characteristics of human–computer interaction, which greatly complicate the extraction of new knowledge by intelligent technologies (Data mining, etc.).

System problems of human–computer interaction. Complex operating conditions induce hidden relationships between the elements of the DS. At the same time, the growth of non-linearity of processes, their ambiguity and uncertainty are accompanied by loss of information at the main stages of the implementation of Data mining technology, etc., and also complicate the modelling and prediction of the functioning of DS. Extension of methods for solving descriptive and predictive problems within the framework of existing approaches and tools limits the possibilities of human–computer interaction. The reasons in such case are as follows:

- the complexity and uncertainty of the information;
- the variety of sources of information;
- multidimensionality and fractality of signals.

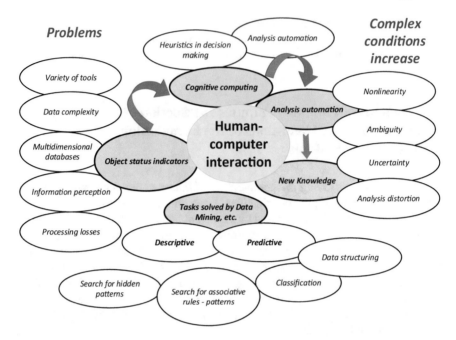

Fig. 1 Interconnection of human–computer interaction problems that manifest themselves in difficult conditions

All abovementioned factors complicate the perception, presentation and thinking when choosing relevant digitized information. Therefore, in cases where the multi-variability of solutions is not amenable to classical methods of analysis, in order to select alternatives, it is necessary to supplement divergent thinking with convergent visualization tools.

Human–machine interaction is limited by the human capabilities of simultaneous accounting (no more than 5–7 different factors). This fact must be taken into account when expanding the methods and means of intellectual support (Data mining, etc.), which increases the complexity of information processing and decision making.

For effective human–computer interaction the following factors are extremely necessary:

1. the criteria of human selection in order to work in a complex information environment;
2. monitoring of the psychophysiological state of a person (PPSP) for its forecasting and assessment.

Without visualization of the individual characteristics of electrophysiological signals, the problem of identifying PPSP in real time is not solved. The complexity of the identification of the PPSP is shown in the form of the Ishikawa diagram in Fig. 2.

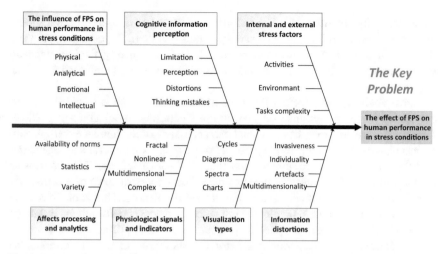

Fig. 2 Relationships of problems in the identification of PPSP in difficult conditions

As it could be seen, the problem of identifying the PPSP is due to the influence of the dynamics of processes of different nature on cognitive capabilities. However, within the framework of the information theory, cybernetics, and synergetics, it is impossible to separately explain and generalize all the data presented in Fig. 2 hidden relationships of information processes and phenomena. Indeed, different types of information differ not only in quantity and complexity, but also in quality and multi-dimensionality, which complicates the generalization and formalization. At the same time, the indicated sciences examine the dynamics of processes, phenomena, etc., which are described by fundamental principles and equations. However, in order to identify the induced information and evaluate its quantity and quality in these sciences, there are not enough funds, criteria and methods.

Local and integral distortions as sources of hidden information. One part of the hidden information about the PPSP is due to external environmental stress factors that distort the dynamics of information exchange between its elements. The second part is due to internal stress factors that induce counteraction and affect human PPSP. In extreme conditions all factors are interconnected, which distorts the human perception of information and complicates the application of mathematical modelling methods. However, the widespread use of the variational principles of dynamics made it possible to overcome mathematical difficulties in modelling the functioning of DCs under ordinary conditions. These principles contributed to the development of systemic and symbolic dynamics, as well as infodynamics and information physics. Everyone is looking for the most common patterns in the processes of transmission, transformation, processing and visualization of information. Therefore, based on extreme dynamic principles, it is possible to summarize the results of studies of objects of animate and inanimate nature and find ways to solve systemic problems. For example, in semiconductor crystalline sensors, response distortion and insta-bility of characteristics indicate the presence of structural inhomogeneities that are technologically inherited [14]. The distortions of the temporal and spectral response are more sensitive to influences of various nature—static (field, radiation, etc.) and dynamic (thermal and acoustic shocks, etc.) [15]. Consequently, the heterogeneity of the structure of the sensors is manifested in the spatio-temporal features of the response.

Stress environmental factors affect the human PPSP, which is manifested in the connection of the rough and fine structure of electrophysiological signals. Therefore, we can assume that the individual characteristics of the human electrophysiological signals are also due to the presence of local spatio-temporal heterogeneities of different nature and "communication channels" between them. This confirms an increase in the individuality of the dynamics (form and structure) of electrophysiological signals when a person is exposed to stress factors. That is obviously the reason why it is difficult to predict the influence of the psychophysiological state on the adoption by the operator (pilot, dispatcher, etc.) of the optimal solution.

The lack of effective selection criteria for relevant sources of information about the SFC of the operator and the criteria for its assessment are due to systemic problems. They are associated with:

- the presence of different definitions of information and its measures;
- loss of information during processing and visualization;
- the fractality of electrophysiological signals.

In addition, both the complexity of their structure, and multidimensionality and non-linearity, have common sources and causes that lead to distortions in information flows (see Fig. 3). It is especially difficult to separate the influence of external

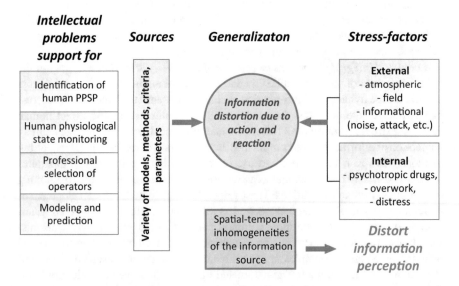

Fig. 3 Problems, sources and causes of distortion of information on the PPSP

and internal stress factors on a person's perception of information, which complicates the prediction of behaviour and errors in decision making. Their influence is studied by cognitive sciences (ergonomics, linguistics, psychology, etc.), as well as neuroscience (neuropsychology, neuroergonomics, neurobiology, neurophysiology, neuroeconomics, etc.) [16]. Little is known about the influence of the PPSP on its perception of complex dynamics, instability, individuality and nonlinearity of signals.

When making a decision, cognitive distortions (e.g. the illusion of truth, distortions of probabilistic events, overestimating or underestimating the significance of one's own opinion, etc.) have a great influence. Thus, the main source of systematic problems of intellectual support methods (Data mining, etc.) is the distortion of the perception of information, which is due to the variety of information sources, their processing tools and the de-synchronization of information flows of different nature. Cognitive problems are often the result of predominantly divergent learning and lack of attention to the development of convergent thinking in people who make decisions in harsh conditions. Therefore, the main reason for the growing safety problems of dynamic systems is an increase in the diversity and complexity of information sources.

3 Interdisciplinary Principles, Ideas and Concepts in the Analysis of the Individual Functioning of Dynamic Systems

The widespread use of new technologies (KDD and Data mining, etc.) limits the human perception of the totality of information flows, types of visualization and analytics. In addition, the psychophysiological state of a person, as well as his patterns of successful experience or anti-patterns of unsuccessful activity, affect the effectiveness of intellectual decision support. Therefore, the requirements for the competencies and capabilities of engineers (pilots, controllers, etc.) are increasing. In turn, all this indicates the need for their intellectual training and education in the field of human factor engineering [10, 11].

Intelligent analysis of the totality of information flows can be carried out on the basis of an interdisciplinary synthesis of several ideas, principles and new concepts.

The first idea is to consider information interaction in terms and methods of field theory. This idea connects information flows with sources, which allows:

- to consider the influence of internal and external stress factors as an "information attack";
- to evaluate the interaction of information sources by the degree of overlap of the power of subsets of their arrays using the Venn diagram.

The second idea is about the most general form of signal organization (its structure) as its spatio-temporal ordering (linear invariant) [17]. N. Wiener also introduced new fundamental concepts of the topological, projective, metric invariant of the space–time structure, which made it possible to reveal various forms of signal organization. These ideas contributed to the development of informational psychology. So, the structural unity of the world, physical, physiological and psychological, allowed K. Levin, L. Vekker [18] and others to establish that:

- the structure of all codes (technical, mental, etc.) can be described in terms of space–time isomorphism and reflects the principle of mutual ordering of two sets.
- preservation of the topological properties of time does not exclude violations of the homogeneity and scale of time (changes in tempo and rhythm).
- the common component of space and time is a one-dimensional series, since space is three-dimensional, and time is one-dimensional. This allows us to make a generalization that in objects of animate and inanimate nature the control structure is used, which is a flexible closed loop (cycle) with the feedback. In particular, the morphology of the neuron gives the same contours. Therefore, in the structure of the dynamics of the information flow, information is hidden about its connection with the functionality of the system.

The idea of reconstructing the functioning model from the measured signals (scalar time series) turned out to be fruitful. It was proposed by Packard et al. [19] and then formalized by Takens [20] in the form of a theorem on the embedding of a time series in R_Π [20]. With its help, the cardiac signal was transformed into a 3D model. However, only rough features of the signal structure appear in it, which does not allow us to analyse the influence of external and internal stress factors. The individual characteristics of the cardiocycle are due to the structure of the bonds between the spatio-temporal heterogeneities of different nature.

In general, the intellectual potential of a human and its ability to make effective decisions depend on the complementarity of convergent and divergent thinking. Therefore, to solve the problem of the viability of DS in difficult conditions, it is necessary to strive to erase the boundaries between scientific and technological knowledge, between theory and experience. To do this is necessary to search for visualization tools for transitions from analysis to synthesis, from the whole to the particular, from macrostates to microstates on the basis of interdisciplinary generalization.

4 Generalization of Experimental Results Based on the Extreme Principles of Physics and the Development of the Signature Approach

Complementarity of the extreme principles of physics. The use of the variational principles of dynamics in various sciences (continuum mechanics, thermodynamics, electrodynamics, field theory, synergetics, etc.) made it possible to decrease the mathematical constraints as much as possible. In particular, these principles underlie the theory of optimal control. The possibility of various approaches to the principle of least action (Hamilton, Lagrange, Jacobi and others) testifies to their fundamental nature. The extreme principles of dynamics are interconnected and have a geometric interpretation—the principle of least curvature or length of G. Hertz. The principle of least coercion of Gauss has an energy interpretation. In turn, the relationship between the symmetry of the physical system and conservation laws (E. Noether's theorem) explains the dynamic similarity of processes of different nature.

Non-equilibrium processes form the fractal properties of objects during self-organization, which features are due to the manifestation of the Le Chatelier-Brown thermodynamic principle (opposition to external influences). The obvious consequence of the interconnections of the extreme principles of physics is the dynamic similarity of the processes of action and reaction. These processes form the fractal properties during the self-organization of objects of various nature. Therefore, to identify the functional state of the information source, it is necessary to visualize the spatio-temporal structure of the information stream. Identification of individual features of the spatio-temporal structure of signals of various nature is possible due to their visualization in the space of dynamic events (state-speed-acceleration) [21, 22]. This space is formed by the state $X(t)$, its rate of change $dX(t)/dt$ and the acceleration $d^2X(t)/dt^2$, which are included in different formulations of the extreme principles of mechanics. Therefore, the extreme principles of dynamics are complementary. Therefore, the application of the extreme principles of mechanics to the information flow allows us to move from a deterministic description of the sensor signal to probabilistic consideration.

Experiment. Transformation of biosignals into dynamic signatures. We illustrate the technology of transformation of information flows by the example of a signature analysis of ECG, EEG, EOG and rheograms that are in the PhysioNet database [23]. Signal processing was carried out using well-known software packages (Matlab, Origin, Mathematics, etc.) and original algorithms. Processing of the cardiac signal $X(t)$ made it possible to obtain time series $dX(t)/dt$ and $d^2X(t)/dt^2$. In the space of dynamic events, each point with coordinates $\{X, dX/dt, d2X/dt^2\}$ represents the probability of the ith event. In addition, each event is determined by calculating a dynamic microcycle (state-small change in speed-small change in acceleration-small change in state).

For a probabilistic description of the evolution of the dynamic state $X(t)$, we applied extreme principles to its small changes, as well as to small changes in the speed $d(dX(t)/dt)$ and acceleration $d(d^2X(t)/dt_2)$. In this case, the scalar time series $X(t)$ (the digitized cardiac signal, the characteristic of the sensor or biosensor, the information flow) is converted into three time dependences $X_i(t)$, $dX_i(t)/dt$ and $d^2X_i(t)/dt^2$, which are natural way interconnected. Therefore, the display of a cardiac signal in space (state-speed-acceleration) turns a scalar time series into a discrete sequence of dynamic events in the form of a topological three-dimensional model of a functioning cycle (see Fig. 4). In fact, the application of the variational principles to the state $X(t)$, speed $dX_i(t)/dt$ and acceleration $d^2X_i(t)/dt^2$ of the cardiac signal made it possible to reconstruct the 3D model of the cardiocycle. The phase

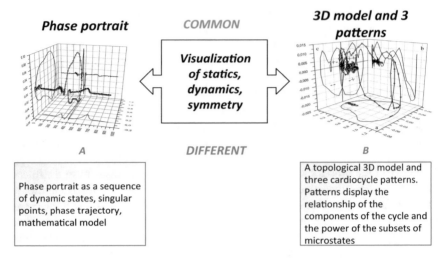

Fig. 4 Human cardiac signal as a sequence of dynamic states in space (state-speed-time) (A), as well as the trajectory of dynamic events in space (state-speed-acceleration) (B)

portrait of a human cardiac signal as a sequence of dynamic states in space (state-speed-time) (Fig. 4, A) is transformed into the trajectory of dynamic events in space (state-speed-acceleration) (Fig. 4, B).

The cardio signal is transformed into a 3D trajectory of dynamic events, which is a geometric interpretation of the principle of least curvature or length of G. Hertz. The trajectory can also serve as a topological three-dimensional model of the cardiocycle in accordance with the Jacobi principle of least action. Its geometrical interpretation is the geodesic curve [11] (see Fig. 4, B). As can be seen from the figure, the geometrization of the dynamics of the cardiocycle is accompanied by its decomposition into natural components. They differ in spatio-temporal ordering and the time interval between events.

5 Visualization of Hidden Informational—Signature Approach

Natural decomposition of a time series. Hidden individual features of the cardiocycle are most evident in the orthogonal projections of its three-dimensional model, which are first and second order signatures [24, 25]. The configurations of these signatures reflect the relationship of dynamic parameters. This relationship is manifested in the form of a closed sequence of geometrically ordered components that differ in length, steepness or curvature. Such a decomposition distinguishes processes (physical, biological, informational, etc.), which are characterized by the constancy

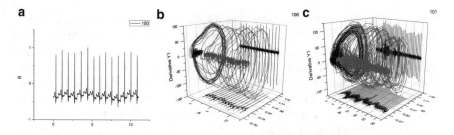

Fig. 5 Cardiocycle sequence and batch representation of their signatures

of speed or acceleration, respectively. Therefore, in signature configurations, hidden cardiac cycle features are manifested.

Analysis of the DS functioning using the configurations of the 1st and 2nd orders signatures. The markers of the functional state are the symmetry/asymmetry of the configurations, the number of geometrically ordered components n and their partial contributions Pn, the ratio of the partial contributions of antiphase components. Together, they will allow us to evaluate the dynamic complexity of the cardiocycle, which is due to dynamic, energy and structural features. They determine the individuality of dynamic behaviour. The configuration $X(t) - dX(t)/dt$ is traditionally analysed as a phase portrait, which is perceived as a sequence of dynamic states (Fig. 5, A). However, the configuration of the signature $X(t) - dX(t)/dt$ is perceived as a trajectory of dynamic events that are causally related. It reflects their spatio-temporal relationships, i.e. is a dynamic structure pattern. Therefore, in the developed approach, the fundamental principle of correspondence is fulfilled—the configurations of the first-order signature and phase portrait are the same, but the method of obtaining them is different.

New opportunities for cognitive perception and analysis are provided by the second-order signature $X(t) - d^2X(t)/dt^2$ of the information source, the area of which has an energy dimension. In the configuration of the signature, the decomposition of the information source into antiphase energy components is most clearly manifested. Therefore, the configuration of the signature $X(t) - d^2X(t)/dt^2$ represents the spatio-temporal distribution of the energy of the antiphase components.

The nature of the relationship of the dynamic variables $dX(t)/dt$ and d^2X/dt^2 is of particular interest, which displays the configuration of the signature $dX(t)/ - d^2X(t)/dt^2$. It visualizes target control functions based on a feedback loop. The signature configuration is perceived as a spatio-temporal structure of the relationships between the dynamic parameters of the cardiocycle. The advantage of a signature analysis of a cardiocycle is the identification at a glance of its dynamic, energy and structural features, as well as the assessment of the nature of changes when compared with previous signatures or models that are in the knowledge base.

Identification of statistical markers of the functioning cycle in the 1st and 2nd order signatures. Significant statistical parameters of signatures are the powers of subsets of information arrays, the value of which is proportional to the area of configurations of signatures of the first and second orders. These areas have a physical meaning, which allows one to analyse statistically the dynamic, energy and informational features of the information source (sensor, etc.) functioning. The area covered by the configuration of first-order signatures is proportional to the power of a subset of microstates W, and the natural logarithm of W is proportional to the entropy H [26]. The entropy of a signature is a universal measure of ordering the distribution of W. Informative is the ratio of the areas of the antiphase components of the H_b/H_n signature. For example, for semiconductor sensors, this ratio reflects the features of the generation—recombination processes [22]. Therefore, comparing these signatures simplifies the search for relevant information sources.

The area of the configuration of the second-order signature $X(t) - d^2X(t)/dt^2$ visualizes the energy balance of the information source. This signature is characterized by the following indicators:

- energy intensity, which is proportional to the area of the signature,
- energy balance, which is equal to the ratios of the antiphase regions of the signature.

The area of the configuration of the second-order signature is of particular interest, since it displays the energy exchange in the information source. Therefore, the second-order signature $X(t) - d^2X(t)/dt^2$ is the energy pattern of the information source.

The area of the configuration of the second-order signature $dX(t)/dt - d^2X(t)/dt^2$ is proportional to the degree of the information cycle, which displays the evolution of the action–reaction. The configuration is perceived as a spatio-temporal structure of the power distribution of microinformation sources between the main phases of the cycle. Therefore, the nature of the distribution of the area over the quadrants of the signal signature can be analysed statically using the matrix of integral balance indicators [21]. Consequently, the signature $dX(t)/dt - d^2X(t)/dt^2$ of a self-organizing object can serve as a structural model of sources of information about its state.

Evolution patterns of the cycle of information source functioning. Analysis of the evolution of DS is carried out in a generalized phase space (state-speed-time). Visualization of the sequence of cardiocycles (see Fig. 5, A) in the form of a batch representation of the evolution of their phase portrait is quite informative (Fig. 5, B). The influence of information noise leads to a spread of phase trajectories and cardiocycle instability (Fig. 5, C).

The nature of local changes in various quasiperiodic signals and sensor responses (sensors, biosensors, etc.) under external influence is most fully displayed in the packet representation of their signatures in the form of a restructuring of the spatiotemporal structure of the cardiocycle [27]. Therefore, packages of signatures of electrophysiological signals of a person are cognitive patterns of restructuring of their structure, which reflect the nature of changes in the psychophysiological state of a person. The sequence of stages of the visualization of the cardiac signal is shown in Fig. 6.

Monitoring signs of evolutionary patterns. The nature of the rearrangement can be analysed by: antiphase components and subsets of microstates that reflect their contribution to the main phases of the cycle. The configuration structure is perceived

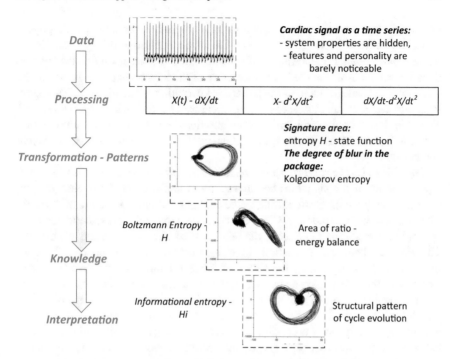

Fig. 6 Algorithm for transforming a cardiosignal into patterns

holistically as a combination of ordered components and subsets that can be analysed by deterministic and statistical methods. In particular, the evolution of the entropy H of the cardiac signal and the generation of the entropy $dH(t)/dt$ reflect the individual characteristics of adaptation under the influence of stress factors.

6 Cognitive Perception of Time Series Patterns in the Space of Dynamic Events

The multidimensionality of cognitive space. In the ontological sense, the concepts of "*space*" and "*time*" constitute *a single space–time continuum*, which most fully reflects the space of dynamic events. Thinking about time, human brain usually refers to spatial images, as well as to relationships and concepts. In the end, when we talk about "*time*" we mean "*duration*", "*change*", "*dynamism*", "*process*". At the same time, we present time as a *change in spatial properties and relationships*. The structural features of the spatio-temporal relationships of the time series at different angles of view can be analysed. Therefore, such structures can serve as patterns of cognition of reality, a generalization of which is the concept of cognitive space. It is considered as a multi-structured education, which includes cognitive, semiotic, psychophysiological and other aspects [28]. The multidimensional nature of the manifestation of hidden spatio-temporal features of information flows is most evident in the space of dynamic events in which the hidden structure of the interconnections is visualized. Therefore, we can assume that the patterns of evolution of the electrophysiological signals of the brain are closest to psychological patterns (behavior, experience, etc.). This is indicated by the similarity of configurations of patterns of evolution of electrophysiological signals in one space, which confirms the connection of the psyche and physiology.

Cognitive visualization of spatio-temporal relationships. In fact, in the batch representation of the signature configurations of self-organized DS elements, we simultaneously see the restructuring of the configuration and the nature of the redistribution of areas by quadrants. Consequently, latent dynamic and statistical features appear in the evolution of patterns, the relationship between which can be analysed. Therefore, the batch representation of the signatures of a self-organized DS is quite informative. In particular, by the nature of the restructuring of the spatio-temporal structure of the information source under the influence of stress factors, a change in its functional state can be predicted. The use of evolutionary patterns simplifies the search for relevant information sources that determine the functioning of complex dynamic systems in extreme conditions. Therefore, using integrative indicators of the balance of antiphase processes, it is possible to assess the degree of influence of stress factors (intrusion, information attack, etc.). And by the nature of the restructuring of the structure of the source of information, the consequences are predicted. Using integrative indicators of the degree of adjustment (dynamic, energy, and entropy) of antiphase processes, changes in the sources of information can be estimated. The unification of the means of electrophysiological signals processing and their cognitive visualization will help to solve the qualitative problem of professional selection of pilots and human operators. The inclusion of patterns of evolution of the functioning of the elements of the DS in the knowledge base will increase the effectiveness of their learning.

Flexible logic of antonyms and patterns of cognition. Pattern configurations of different sources of information are perceived as a combination of highs and lows. Therefore, a flexible logic of antonyms is effective for their analysis. On the one hand, the features of pattern configurations can be described by mathematical terms-antonyms (maximum–minimum, convexity–concavity, composition–decomposition, continuous–discrete, analysis–synthesis, etc.). On the other hand, it is possible to use physical antonyms (action–reaction, source–stock, order–disorder, balance–imbalance, symmetry–asymmetry, etc.). The antiphase components of the configurations of patterns of self-organized elements of the DS are distributed across different quadrants. Therefore, the signature configurations of the information source can be a kind of cognitive patterns of cognition. Indeed, visualization of hidden relationships stimulates the activity of both hemispheres of the brain, which contributes to the synthesis of new knowledge. The presence in the configurations of patterns of elements of geometric, kinematic and dynamic similarities expands their perception, and the use of flexible logic of antonyms contributes to the development of intuition in the learning process [11].

The innovative potential of cognitive visualization. A variety of complementary tools for analysing and synthesizing pattern configurations has innovative potential. The proposed patterns of the cycle of functioning and its evolution allow the dynamic, energy and informational features of the elements of complex DS to be revealed. In the symmetry of the configurations of patterns and in different types of similarity of configurations (geometric, kinematic, dynamic, etc.) conservation laws and extreme principles of physics are displayed. This makes it possible to use thermodynamic criteria of stability, reversibility, etc. in an intelligent analysis. It is very important that visualization of the restructuring of the spatio-temporal structure of fractal signals of various nature in the space of dynamic events is accompanied by natural decomposition into antiphase components. Thus, the space of dynamic events is an analogue of cognitive space. It is multifaceted and allows using one toolkit to compare the functioning cycles of all elements of the DS. Visualization of the restructuring of the space–time structure in the patterns of evolution of cyclic processes contributes to the "*cognitive effect*" and the development of system thinking.

Application in intelligent support and predictive maintenance/analytics in Industry 4.0. Digitized Industry 4.0 uses the Internet of Things and cyberphysical systems, such as sensors (smart sensors), to collect huge amounts of data that can be used in factories to analyse and improve their performance. Recent advances in big data and analytic platforms mean that systems can process huge datasets and receive information that needs to be quickly responded to. It is assumed that the Smart Factory, which forms the basis of Industry 4.0, will use information and communication technologies to develop the supply chain and production line, which will provide a much higher level of automation and digitization. It is assumed that industrial automation systems will use self-optimization, self-configuration and even artificial intelligence to perform complex tasks in order to ensure significantly higher cost-effectiveness and improve the quality of goods or services.

At the same time, during the transition phase from Industry 3.0 to Industry 4.0, as well as for conducting trainings and preparations for human to work as an operator of critical application systems (where the reliability (and safety) of the entire system is significantly reduced due to the presence of a human itself), the cognitive distortions in the human perception of information should be eliminated, or alternative methods and means based on a risk-based approach must be provided to ensure the safety and cybersecurity of a complex human–machine system.

7 Conclusions

Harsh conditions of operating dynamic systems induce distortions in the information flows varied in nature, and affect the psychophysiological state of a human who makes responsible decisions. The analysis is also complicated by the lack of complementarity of information visualization types, which does not allow revealing hidden spatio-temporal features of the functional state of dynamic systems' elements, including human psychophysiological state. The intellectual support of decision-making is limited by the variety of information sources different in its nature, methods of their processing, as well as models, patterns, parameters and criteria. The article proposes a technology for cognitive visualization of the spatio-temporal structure of information sources of various nature, which is based on an interdisciplinary generalization of approaches, methods and concepts. This allowed to transform a one-dimensional time series (electrophysiological signal, sensor response, etc.) into patterns which configurations reflect the structure of spatio-temporal relationships of antiphase processes. Visualization of the coarse (dynamic) and thin (information) structure of information flows, varied in nature, helps to identify the hidden patterns of information sources functioning. Their relationship determines the dynamic and statistical features of information sources functioning. Signature analysis of such patterns allows probabilistic and deterministic methods for studying the dynamic systems to be balanced.

Application of the technology to the information sources varied in nature (EMR, radiation and acoustic radiation sensors), as well as to human electrophysiological signals (cardiograms, encephalograms, oculograms, rheograms, etc.) demonstrate the advantages and new possibilities of revealing the hidden spatio-temporal relationships, which determine the features of functioning the dynamic systems in complex conditions.

The unification of the processing of digitized information flows of various nature and their cognitive visualization in the form of structural and other patterns widens opportunities of intellectual support. This will simplify human–machine interaction, and therefore the cybersafety of the dynamic systems functioning in the complex conditions of industrial automation and control systems.

References

1. Newby, G.: The strong cognitive stance as a conceptual basis for the role of information in informatics and information system design. Cognitive space and information space. NY J. Am. Soc. Inf. Sci. Technol. Arch. **52**(12), 1026–1048 (2001). https://doi.org/10.1002/asi.1172
2. Zhou, Y.L., Wahab, M.A., Maia, N.M.M., Liu, L., Figueiredo, E.J.F.: Data Mining in Structural Dynamic Analysis. A Signal Processing Perspective. Springer Nature Singapore Pte Ltd. (2019) https://doi.org/10.1007/978-981-15-0501-0
3. Wickens, C.D., Lee, J.D., Liu, Y., Gorden, B.S.E.: An Introduction to Human Factors Engineering. Prentice Hall (2003)
4. Dul, J., Bruder, R., et al.: A strategy for human factors/ergonomics: developing the discipline and profession. Ergonomics (2012). https://doi.org/10.1080/00140139.2012.661087
5. Fedota, J.R., Parasuraman, R.: Neuroergonomics and human error. Theor. Issues Ergon. Sci. **11**(5), 402–421 (2009). https://doi.org/10.1080/14639220902853104
6. Young, M.S., Brookhuis, K.A., Wickens, C.D., Hancock, P.A.: State of science: mental workload in ergonomics. Ergonomics **58**(1), 1–17 (2014). https://doi.org/10.1080/00140139.2014.956151

7. Gevins, A., Smith, M.: Neurophysiological measures of cognitive workload during human-computer interaction. Theor. Issues Ergon. Sci. **4**(1–2), 113–131 (2010). https://doi.org/10.1080/14639220210159717
8. Runco, M.A., Yoruk, S.: The neuroscience of divergent thinking. Activitas Nervosa Super. **56**(1–2), 1–16 (2014)
9. Mygal, V.P., Mygal, G.V.: Problems of digitized information flow analysis: cognitive aspects. Inf. Secur. Int. J. **43**, 134–144 (2019). https://doi.org/10.11610/isij.4312
10. Mygal, V., Mygal, G.: Interdisciplinary approach to informational teaching environment formation. Odes'kyi Politechnichnyi Universytet. Pratsi (2018). https://doi.org/10.15276/opu.1.54.2018.1
11. Mygal, V.P., Mygal, G.V., Balabanova, L.M.: Visualization of signal structure showing—cognitive aspects. J. Nano Electron. Phys. **11**(2) (2019). https://doi.org/10.21272/jnep.11(2).02013
12. Kharchenko, V., Dotsenko, S., Illiashenko, O., Kamenskyi, S.: Integrated cyber safety & security management system: industry 4.0 issue. In: 10th International Conference on Dependable Systems, Services and Technologies (DESSERT), Leeds, United Kingdom (2019). https://doi.org/10.1109/DESSERT.2019.8770010
13. Illiashenko, O., Kharchenko, V.: Concepts of green IT engineering: taxonomy, principles and implementation. In: Green IT Engineering: Concepts, Models, Complex Systems Architectures. Studies in Systems, Decision and Control, 74, pp. 3–19. Springer, Berlin
14. Mygal, V., But, A., Phomin, A., Klimenko, I.: Hereditary functional individuality of semiconductor sensors. Funct. Mater. **22**(3), 387–391 (2015). https://doi.org/10.15407/fm22.03.387
15. Komar, V., Gektin, A., Nalivaiko, D., Klimenko, I.: Characterization of CdZnTe crystals grown by HPB method. Nucl. Instrum. Methods Phys. Res. Sect. A **458**(1–2), 113–122 (2001). https://doi.org/10.1016/S0168-9002(00)00856-1
16. Parasuraman, R., Mehta, R.: Neuroergonomics: a review of applications to physical and cognitive work. Front. Hum. Neurosci. 23 Dec 2013. https://doi.org/10.3389/fnhum.2013.00889.
17. Viner, N.: Kibernetika, ili Upravlenie i svyaz v jivotnom i mashine (Wiener, N., Cybernetics, or Control and Communication in the Animal and Machine). Nauka, Moskva (1983) (in Russian)
18. Vekker, L.M.: Psihika i realnost. Edinaya teoriya psihicheskih protsessov (Vecker, L.M., Mind and Reality. Unified Theory of Mental Processes). Smyisl, Moskva (1998) (in Russian)
19. Packard, N., Crutchfield J., Farmer, J., Shaw, R.: Geometry from a time series. Phys. Rev. Lett. **45**, 712–716 (1980)
20. Takens, F.: Nonlinear Dynamics and Turbulence, pp. 314–333. N.Y. Pitman (1983)
21. Mygal, V., But, A., Mygal, G., Klimenko, I.: An interdisciplinary approach to study individuality in biological and physical systems functioning. Sci. Rep. Nature Publishing Group (2016). https://doi.org/10.1038/srep29512
22. Mygal, V., But, A., Phomin, A., Klimenko, I.: Geometrization of the dynamic structure of the transient photoresponse from zinc chalcogenides. Semiconductors **49**, 634–637 (2015). https://doi.org/10.1134/S1063782615050152
23. PhysioNet: The Research Resource for Complex Physiologic Signals (2019). Available at: https://physionet.org
24. But, A.V., Migal, V.P., Fomin, A.S.: Structure of a time variable photoresponse from semiconductor sensors. Tech. Phys. **57**, 575–577 (2012). https://doi.org/10.1134/S1063784212040044
25. Mygal, V., Klimenko, I., Mygal, G.: Influence of radiation heat transfer dynamics on crystal growth. Funct. Mater. **25**(3), 574–580 (2018). https://doi.org/10.15407/fm25.03.574
26. Keith, A.: Entropy. Am. J. Phys. **52**, 492–496 (1984)
27. Mygal, V., Mygal, G.: Method of electrocardiographic data evaluation for diagnostic purposes. Ukrainian patent № 77203 (2006)
28. Hettinger, L.J., Branco, P., Encarnacao, L.M., Bonato, P.: Neuroadaptive technologies: applying neuroergonomics to the design of advanced interfaces (2010). Published online. https://doi.org/10.1080/1463922021000020918

Empirical Study on Cyber Range Capabilities, Interactions and Learning Features

Kirsi Aaltola

Abstract Emerging technologies and the globalization require constant investment in people and their performance in actual and virtual environments. New technologies such as autonomous systems, machine learning and artificial intelligence (AI) radically re-contextualize the human dimension of the organization. Technological developments are changing the ways people experience the physical and the virtual environments. Strategic changes have revealed new critical vulnerabilities such as social media-based disinformation campaigning with impact on the human aspects at state, societal, organizational and individual levels. Scenarios of gathering information, committing fraud or getting access to critical systems are often used for follow-up actions. Cybersecurity education and training aim to raise the level of expertise, skills and competences and ensure better performance in complex situations in cyber space. Researchers have addressed assumptions, models, concepts and cognitive aspects of humans performing in the cyber domain. However, the human cognitive learning and human performance approaches in Virtual Reality (VR), Augmented Reality (AR) and Mixed Reality (XM) and Cyber Range (CR) learning platform design are only partly touched. CRs are becoming crucial means in acquisition of skills and knowledge but also to augment and mimic human cognitive behaviour for cognitive agents. Empirical studies and evaluations of the capabilities, tools and techniques for enhancing organizational cyber resilience by the human performance to better face cyber-attacks are needed. The purpose of this paper is to provide a literature review, suggest viewpoints on cybersecurity training and education and to study the current capabilities of CRs. This paper includes a literature review to support human performance and provides empirical findings on CR capabilities, interactions and learning features. The results of this study can be used as a baseline for future initiatives towards the development of CRs in accordance with human cognitive learning and future improvements in design. Furthermore, the intention is

K. Aaltola (✉)
VTT Technical Research Centre of Finland , Espoo, Finland
e-mail: Kirsi.Aaltola@vtt.fi

University of Jyväskylä, Jyväskylä, Finland

© The Author(s), under exclusive license to Springer Nature Switzerland AG 2021
T. Tagarev et al. (eds.), *Digital Transformation, Cyber Security and Resilience of Modern Societies*, Studies in Big Data 84,
https://doi.org/10.1007/978-3-030-65722-2_26

to constructively promote discussion on current issues about humans in cyber physical systems and cybersecurity domain and thus boost multidisciplinary studies to
enhance cyber awareness in different sectors.

Keywords Human Performance · Acquisition of skills · Cognitive behaviour ·
Learning · Cybersecurity · Human factors · Cyber range

1 Introduction

Emerging technologies promise benefits for humanity and reality environment.
Market growth for personal technological equipment and computing technologies
add to unforeseen security challenges of network technologies [3, p. 3]. The consequences of deploying certain applications present uncertain risks, untested mechanisms and involve challenging issues of values and ethics [19]. Core components of physical things are connected to Internet or other connected data such as
Cyber-Physical Systems (CPSs). This has led the safety-critical infrastructures to
become potential for security risks and attacks [47]. The lack of experience in dealing
with technological issues and the recent increase in adoption of digital equipment and
personal devices opens to several vulnerabilities in different sectors. As an example,
it is still quite common to have general human errors such as passwords written on top
of devices or down on paper close to PC the password should protect. Everybody can
easily peek a password stored this way with high effectiveness beyond sophisticated
phishing campaigns or malware attacks. Moreover, social engineering could lead to
the compromise of sensitive and financial information. At the same time, there is a
need to improve human performance to avoid human errors in complex CPSs and
conduct research of preventing cyber indidents [47]. Training and education play key
role in improvement of human performance and also cybersecurity industry [6] and
business have recognized the need of sector-specific learning opportunities.

Recently, there has been indicated an interest to augment and mimic reality and
expertise in AR/VR learning platform focusing on facilitating cyber threat scenarios
to increase human performance and preparedness towards cyber threats. The use of
cyber virtual environments raises the situational awareness [16] and VR simulations
can simulate and augment decision-makers' understanding of cyber-physical situational awareness [25]. This kind of work focuses on enhance security and cyber
security capabilities, where cyber ranges are acting a crucial role. At the same time,
even this study does not focus on it, this raises the interests to discuss and research
the security of devices and environments or platforms themselves. When writing
this article, there was not a comprehensive systematic literature review found by
the author on the topic of Cyber Ranges (CR). Nevertheless, several articles were
identified with focus on Cyber-Physical Systems (CPSs), computer network operations (CNO) and cyber security training. Several reviews might been completed
under the classified domain since CR work has been covered and funded strongly by
the US military (e.g. [5, p. 1]). Moreover, there are several pilots and development

initiatives on CRs, such as European Commission funded projects, launched during the year 2019. An example project is called ECHO (European Network of Cybersecurity Centres for Innovation and Operations) and it consists of 30 partners from different fields and sectors including health, transport, manufacturing, ICT, education, research, telecom, energy, space, healthcare, defence & civil protection [34]. It aims to further develop cyber security infrastructures to boost European cybersecurity industry.

CR is a system to be used or it can include learning platform to be integrated in the training and education around cyber domains, information technology or security topics. Using technology as a training mean and in acquisition of knowledge and skills in simulated environment can have positive and beneficial experiences for learning. Due to a gap of conceptual background of the definition "cyber range" among academic practice, this study points out the literature of cybersecurity learning platforms and AR or immersive learning as well as studies the current capabilities of CRs. Moreover, this study contributes to examine the role of cybersecurity learning platforms as reality-virtuality technologies and their applications for skills acquisition and learning experience. Firstly, this research reviews previous literature and clarifies some terminological concepts in an attempt to establish limits and standardize the use of the terms describing CR capabilities in the different realities. With this goal in mind, this study extends the "Reality-Virtuality continuum" [21] and Human-Technology Interaction (HTI) to development and design of CRs as training and education means or methods. Secondly, this article presents the findings of qualitative survey responses on CR capabilities. The overall aim of the study is to improve the understanding of CRs and their learning platform capabilities from the learning and human performance perspective and to contribute to the enhancement of the design and impact.

2   Literature Review

Originally or traditionally, the word or definition "cyber" can be traced back to Ancient Greece and Kybernetes, when it meant "the art of steering" [28, p. 76]. Wiener [32] published Cybernetics as a study of the importance of systems in both, living and artificial machines. In the 1980s there was at the same time, a cyberpunk movement and a spread of sophisticated computers for military operations by U.S. In the 1995, the investigation revealed some vulnerabilities in U.S. critical national infrastructure, and a phrase "cyber" was chosen to capture the challenges posed by computer vulnerabilities [17, pp. 45–46]. The definition of cyberwar grows out of Information Warfare (IW), Information Operations (IO) and revolution in military affairs [2]. The language of IW and IO begun to disappear from academic debate in early 2000s. Cybersecurity is often understood as national security issue by the military or intelligence [37, 38].

The definition "cyber" had replaced these, as an overall concept to discuss and analyse security related challenges in the information age. "Cyber" also often refers to real-time context, digital information as well as to virtual domains. It also includes both people and machines, which lead to concepts of interfaces and interactions between human, computers and machines. Cybersecurity has emerged from precise langue and the concept already has and will have even more complexity, activities, phenomena and dynamics to label. In the context of cybersecurity, "cyber" connotes with a strong relationship with information technology and relates to characteristics of security and virtual reality culture in different sectors and contexts. The definition varies naturally from narrow conceptions to broader framework.

The conceptualization to physical/mechanical, logical, informational and human/cognitive (e.g. [11]), is not enough. Beyond them, there is a need to analyse more in details physical infrastructure and hardware, command of control, collection/store/generate/rely of system data or information, and human beings and their interactions with computers, machines and technology. The semantics of cybersecurity relate with rather challenging topics and activities, such as threat, crime, and attack, variety of physical sectoral scenarios and specific contexts, and there is an ambition to find and define transversal and inter-sectoral requirements and needs. In order to specify how comprehensive the concepts are, the definition cyberattack is commonly used in tactical, operational, academic and political discourses. Often, the original purpose of the defined cyber-attack in practice can be to indirectly mislead or spark to crisis rather than aiming to damage computers, technology or machine. The semantics of phrases "cyber" and "cybersecurity", reveal the complexity and comprehensiveness of the vocabulary and taxonomy used. The challenges are only increasing when the aim is to conceptualize them in several sectors and link their inter-connectedness. The very original purposes of these meanings are based on leadership and importance of both, physical and virtual technology domains. "Resilience" provides a meaning for communicating about system performance, organizational operational risks, organize the challenges, measure and quantify organizational cyberattack surface [3]. Educating users is a proactive measure to detect attacks and education and training increase the potential victims' threshold for malicious behaviour [13].

2.1   Cyber Range (CR) Capabilities

Cybersecurity and assessment frameworks offer to look more comprehensively at the reality of cybersecurity concept and practices that are highly nuanced. The taxonomy and vocabulary provide reality-based definitions for strategical and political levels of responsibilities. Semantics, taxonomies and vocabularies offer better and more precise language for entire domain and frameworks. Lumping all terms under the framework draws the direction. The pace of technological change will only continue faster development, and the understanding of differences (through analysing activities, means, methods, capabilities, threats and responsibilities) is required especially from academic and political actors to provide more precisely and clarity. The definition *"range"* implies an environment for offensive target practice, much like a shooting range for soldiers [5].

A Cyber Range (CR) would therefore be an environment where staff can practice their skills and competences towards the challenges posed by computer vulnerabilities. While cyber range tries to replicate a particular scenario, the lack of empirical studies on cyber ranges remains a critical drawback to a systematic evaluation. CR is a technological system or tool, which includes often scenarios and simulations, and it is used an exercise learning platform. Davis and Magrath [5] defined that *"a CR provides an environment to practice CNO skills such as penetration testing, defending networks, hardening critical infrastructure and responding to attacks"*.

CR, which includes a learning platform to learn cyber skills and improve performance in different scenarios and potential simulated exercises, can be understood and defined as technological VR/AR immersive learning platform. The cyber range proof of concept can address learning objectives, exercises in e.g. network forensics, reverse engineering, social engineering, or penetration testing and learning is facilitated problem-based and self-directed [27]. Simulations and game theories are acknowledged as methods of CRs (e.g. [31]). Game theory models the interaction between to acting teams (often called red and blue teams). Cyber-attack simulations actually implement hackers to attack to network to define potential vulnerabilities with the aim to increase cyber defence preparedness. Agent-based simulation platforms on simulating the effects of attacks and analyse the impacts [12, 15]. Even the simulations are seen beneficial at providing insights, the impracticality has been raised to the technological and policy changes [22].

Depending of CR purposes the capabilities vary widely. The purposes of CRs can aim to providing so called cyber warriors, develop cyber experts' competences, developing and practicing skills as well as testing organisations' environments in virtual facility, running threat scenarios or network simulations and identifying critical processes. The CR capabilities include simulations of real-world network environment, data traffic generation and capture, penetration testing, incident response, thread injection, patch levels and network services to testing, evaluation, interoperability assessment of devices and applications, operations-based models to respond to threat scenarios, platform based security tools, simulations for real applications, continuous updating and upgradation and a cloud hosted environment. CR can also implement electronic warfare, test and rehearsal, mission refinement capabilities with use of tools, techniques and procedures [26, pp. 1–28].

CR setting provides opportunities review and analyse the benefits of capabilities for human improvement. It also raises the importance of human learning research and principles in design and use of cyber ranges in training and education. A simulation has been defined as an artificial or augmented scenario or environment that is designed to represent or simulate some aspects of reality. Simulations can incorporate different degrees and types of fidelity, or realism, namely, physical, functional and psychological. AR game simulations immerse the player in the game and immersion in the decision-making process of the game requires the player to learn the consequences of their decisions, and thus being part of an active learning process [33].

2.2 Interactions and Reality-Virtuality Continuum

As a starting point, commonly known theory of research, Human–Computer Interaction (HCI), possess the importance to study and understand interaction between human and technology. Dix [9] stated the extended concept of HCI to Human-Technology Interaction (HTI). The presented boundaries of realities and systems raise the interest towards the knowledge area on the processes between human and technologies, actions and interactions. Recent technological developments integrate technology to human body and provide opportunities for mixed reality experiences with 2D or 3D visualizations [40]. This raises the users' experiences and extend the processes to sensory, cognitive and motor functions [14]. When this kind of technological embodiment increases, the technology becomes part of human actions which changes also the HTI processes.

Among all revised taxonomies, for researchers to classify the wide variety of realities has been "Reality-Virtuality Continuum" by Milgram and Kishino [21]. Real Environments (RE) encompass the reality, and it also includes direct or indirect views of real scene such as video display. Virtual Environments (VE) are entirely computer-generated, the real objects do not exist, and users interact through technological interface in real time. Virtual Reality (VR) is also computer-generated environment and user can navigate and interact in real-time simulation with own senses [35]. VR provides a sensory immersive experience. Augmented Reality (AR) *"modifies the user's actual physical surroundings by overlaying virtual elements"*, such as images, videos, virtual and other immersive items. In addition to this approach of reality-virtuality continuum, the extended approach of realities has been defined, with the independent dimension called "Pure Mixed Reality" (PMR). In PMR the *"users can interact with both virtual and reality objects in real-time and, simultaneously, these objects can interact with each other"* [10].

The different realities of the continuum provide the factors for human to interact in physical and cyber or virtual worlds. Online games, VR and AR, in particular, are addressed under various different names and definitions, e.g. immersive learning simulation, digital game-based learning, gaming simulations, and cyber range. The merits of incorporating VR and AR design not only into education and training but should be used for solving real world challenges, from societal problems to specific issues of conflict or climate change [20].

2.3   Learning with Use of AR/VR Technologies

"After behaviourism, constructivism, the "connectivism"is a new concept of learning for the digital era. Learning is no longer an individual process. When talking about online learning groups, for example, there are different and changing tools used: formal (dedicated e-learning platforms) and informal (through chat platforms)" [36; p. 97]. When digitalization started to strongly influence to education and training sector, practitioners pointed out that traditional methods should move beyond LMS (Learning Management Systems) [41] and that online learning components are often combined or blended, such as hybrid methods which include face-to-face instruction, in order to provide more learning outcomes [42]. In practice, face-to-face learning still represents the standard in education [7: 16], and we have to consider how new tools and emerging technologies allow us to utilise and improve methods for enhancing human improvement. Siemens [43] posits that the digitization of training and learning methods should be aligned to the connectivism of the training subjects, and Friedrich [44] stresses that digital tools complement traditional methods enabling more constant interaction with users and lowering users' participation.

CRs provide platforms for learning with immersive simulations and scenario-oriented capacities in specified sectors. The use of reality-virtuality technologies in learning allows to have a more dynamic and autonomous role in experiences [23] leading to higher perceptions of value [24]. It has been recognized by the empirical studies that AR and gamification have raising potentials to model and teach complex cognitive competences in an engaging way [4] and they are practical and functional methods to train and educate. Beyond technological embodiment raises the immersive experiences, extends the human perception and enhances the motor and sense perceptual skills [10, 29]. The typical critique of CRs is that they lack an active opposition which lead to the situation that users in CRs learns to cope with an attack but do not have an tactical or strategic understanding [15].

Interestingly, there are not too many empirical findings on learning impacts with performance change immediate after playing the immersive simulation in AR/VR reality emerged. The AR/VR platforms and systems raise excitement and motivation with providing visual expressions different from traditional training settings, the extent of actual learning and human performance, in particular, is more difficult to decipher. A meta-analysis study conducted by the University of Colorado Denver Business School in 2010 reported that workers gained a greater skill level and higher retention of relevant information with online simulation versus formal classroom or web-based tutorials. The studies also show that the level of interactivity within a learning environment is driving the learning through higher motivation level. The more the learner interacts and collaborates with other learners, the content, and the instructor, the more likely it is that learning with occur [18, p. 21]. According to de Freitas [7], *"learning in the immersive worlds presents us with the ability to remember learning experiences for longer, engaging and motivating us as learners"*. AR/VR solutions simulate the physical presence of the user in a virtual environment, which is categorized as sensory-motoric, cognitive, and emotional experiences [45].

Integrating human body and including human personal spaces to improve their experiences can extend to sensory, cognitive and motor functions [14] and influence to
the both, motivation and memory.

As Jones et al. [15] argue, CRs should aim to provide higher-level learning experience for offensive and defensive behavior. The philosophical and psychological
cognitive foundations can bring value to the research agenda and practical solutions
through visualization, immersion and cognition in cybersecurity learning platforms.
Experiential learning approaches and motor skill development are relevant when
target is in acquisition of knowledge, skills and competences. More critical the
role of learning styles and methods becomes if the training aims to augmenting
(virtual) reality with desired learning outcome to human performance in reality
context. Human cognitive behavior is rich, dynamic and combines deliberative and
autonomous information processing [15]. Cyber domain itself is taking place in
virtual and cyber domain and therefore the competence development learning with
reflection in both realities is even more complex. Dewey, Lewin and Piaget were the
founders of experiential learning approach, which led to commonly used conceptual
tradition to adult education and training. Kolb's [8] dichotomy between real or actual
doing and reflection has strongly influenced to adult education and training [39]. The
acquisition of knowledge and skills is related to sensorimotor learning. In movement
skill, inputs from sensory and cognitive processes are important in order to choose
correctly and organizing and adjusting movement. Task requirements, environmental
conditions and personal characteristics inflict to constraints to determine what person
must do. Most of the motor skill schemes are based on experiences and descriptive
and distinguished with regard to skill complexity, structure and the level of difficulty and familiarity. This possess that the relation between skills acquisition and
experiential learning approach is strong.

"*Learning is a holistic process of adaptation to the world. Not just the result of
cognition, learning involves the integrated functioning of the total person—thinking,
feeling, perceiving, and behaving*" [30; p. 194]. Kolb's (1984) experiential learning
style theory is a four-stage learning cycle of (1) Concrete experience (new experience
or situation is encountered, or a reinterpretation of existing experience (feeling)),
(2) Reflective observation of the new experience—of particular importance are
any inconsistencies between experience and understanding (watching), (3) Abstract
conceptualization—reflection gives rise to a new idea, or a modification of existing
abstract concepts (thinking), (4) Active experimentation—the learner applies the
idea(s) to the world around them to see what happens (doing). The experiences and
adoptions may take place in daily real-life setting and informal form. Persons developing their performance in a training with a purpose to work with the performance in
different environment or reality require comprehensive learning setting and awareness of learning methods, not only among facilitators, trainers, teachers or mentors
but also by the learners [8]. Modelling and mimicing human cognitive behaviour
to cognitive agents as an element of active opposition in virtual environment can
increase the strategic cognitive learning goals as well [15].

3   Results

To understand current CR interactions, capabilities and learning features, qualitative research methods were implemented to collect empirical data part of ECHO (European Network of Cybersecurity Centres and Competence Hub for Innovation and Operations)—project funded by the European Commission. Survey questions were prepared for both, the cyber-range providers and potential users or customers. The questions were framed based on CR capacities and capabilities, and with combinations of open-ended and multiple-choice questions. Multiple choices questions were framed to potential responses or with open ended "something else" opportunity. The survey questionnaire respondents were selected and questionnaire was sent to specified organizational group to ensure CR expertise and experience of the respondents. The pre-selection was decided to ensure proper responses from the organizational representatives who either use the CRs or provide cyber range services.

The survey was conducted among the sample of respondents (n = 49). Data validation ensured that the questionnaire was fully completed and presented the consistent data. This study focuses on presenting the analysis and findings of qualitative responses which focused on understanding CR capabilities by the providers. Quantitative questions of the survey addressed mainly cybersecurity competences by the personnel in the organisations and budgetary constraints. The background data included organization related information and position in the organization. The data was analysed with use of qualitative method content analysis. This empirical research identified the current CR interactions, capabilities and learning based on the survey findings and observations by the researcher during study process. In terms of learning practices (scenarios, simulations, tasks, outcomes), the observation of open access CR descriptions were studied.

3.1   CR Capabilities

Respondents identified that their CR is mainly deployed on the premises. One described the CR in cloud, and another in both, cloud and on the premises. When asking about the technology used for managing and orchestrating the virtualized system, the respondents identified these; Custom-made tools and automations, virtualization platform vendor's tools, VirtualBox—vagrant—Ansible, Vmware, OpenStack, Ansible Packer, COTS hypervisor technology. In terms of predefined scenarios, the maximum was up to 50 and one answered they didn't have predefined scenarios. Respondents identified they had SCADA, IoT and Mobile devices as specific scenarios in their cyber-range. Respondents detailed scenarios of different business sectors for financial, critical infrastructure, ISP, Cloud Service Provider, Supply Chain, Software development and Healthcare. When analysing the technical capabilities within CR scenarios, all respondents answered they had attack-response simulations. Beyond that defence simulation and learning platform were

Deployed	Technicalities	Scenarios	Capabilities	Evaluation
Premises Cloud Premises and Cloud	Custom-made tools and automa- tions, virtualiza- tion platform ven- dor's tools, VirtualBox - va- grant – ansible, VMware, Open- Stack, Ansible Packer, COTS hypervisor tech- nology	SCADA, IoT and Mobile de- vices, Sectors: financial, criti- cal infrastruc- ture, ISP, cloud service pro- vider, supply chain, software development and Healthcare	Attack simula- tion, defence simulation, learning plat- form, perfor- mance evalua- tion, real-time monitoring of the running, traffic simula- tions	hot wash up as debriefing, provide of feedback based on col- lected data during train- ing, debriefing by company per- sonnel

Fig. 1 Summary of results: CR capabilities

commonly found from the CRs by different providers. When asking about "*How training debriefing/performance evaluation is performed in your cyber-range?*", the majority of the respondents addressed that they have evaluation activities such as hot wash up as debriefing, a possibility to provide the feedback based on collected data during training, or debriefing by company personnel. Only half of the respondents identified the possibility to include physical devices part of CRs. Summary of the CR capabilities is presented in Fig. 1.

From the development point of view, respondents would like to further develop their sector or domain specific capabilities (e.g. healthcare, space, transportation, energy), provide better automation for quicker development and deployment and add automatic performance evaluation. The domain-specific approach was seen necessity and one respondent described "we always draw domain-specific expertise when create a cyber exercise". SCADA capabilities were seen the most relevant by the respondents.

3.2   CR Interactions and AR/VR Continuum

CR raises the new level to AR/VR continuum with modelling cognitive and behavioural interact successfully with the task-environment. Cognitive models of network users, defenders, and attackers that can interact with the same software that humans interact with generate offline predictions and adding simulated participants in training sessions. Cognitive modelling aims to integrate with memory dynamics and explaining human awareness to predict system on cyber-attack. Modelling efforts involve scenario building and predict of human behaviour.

CR simulations and game exercises implement several levels of interactions in system or platform; red and blue teams' interaction (defined as attackers and defence), human-network interactions (defined as exploitation) and network-human interactions (defined as defence). Scenarios include specific models of behaviours and interactions. Cyber-attack activities are often modelled from real world descriptions and use cases. CR platforms embed the interactions of human cognition and behaviour in the context of network and computer systems.

3.3 Learning Features

All CR studied provided training and education services for capacity building for any public or private organisation and their personnel. Therefore, the users of CRs also were mainly teachers, students, IT-personnel and teams of companies, SMEs and big enterprises. The number or participants attending to single training session varied from five to fifty participants. The red and blue team exercise format was commonly used and the learners play a role of members in security teams under the scenario or simulation. Scenarios commonly followed phases of reconnaissance, exploitation, escalation and completion of the mission (e.g. "shutdown" or "kill"). Tasks included securing network and services as well as collaboration with different actors.

Observations of five different CRs revealed the general lifecycle of cyber exercises or simulations. In general, the exercise lifecycle could be outlined as follows; a set of learning outcomes or objectives, included scenario tasks or injects, scoring design (gamified features), repetition of the exercise with scenario and simulation tweaks, familiarisation by the learners, execution of the exercise, hot wash up and possible internal lessons learned processes. The survey respondents addressed the performance evaluation as the most important to be further developed, even some of the CRs in the survey already implemented hot wash up or debriefing after the exercise or simulation. From the learning and human performance view point, the evaluation and reflection in real world among other professionals enhances to achieve the learning outcomes.

4 Discussion

The lack of experience by humans in Cyber-Physical Systems and dealing with cyber-security issues and the recent increase in adoption of digital equipment have raised the interest to empower cybersecurity training and education in different sectors. Due to the large turnover and the number of different employees involved in organisations, the comprehensiveness of cybersecurity often stays limited and several vulnerabilities may occur. CR can be defined as a safe environment for cyber-attack scenario simulation and test. It offers the capabilities to create realistic cyber simulations useful for cyber training and exercise, to equip cyber analysts and operators with advanced cyber skills, cyber research and development, to prototype development in realistic cyber scenarios, test and evaluate, adapt test framework for certification testing. CR capabilities include strategies, skills and competences of red and blue team, tools and techniques selected in CR and vulnerabilities. Competences of attackers of defenders are set in a form of nodes in the network to exploit the capabilities to finally perform on the target. The iteration of logics of CR game theories or simulations sets possible actions towards achieving the goals. Simulations are often run for multiple network configurations and demonstrating different behaviours on a network.

Knapp [18, 20] argued that gamification is a time-consuming and difficult process which requires development efforts for interactive learning experiences, multiple interactions and a careful melding of content. Emerging development of CR capabilities (techniques, tools procedures) shows the difference of reality to traditional e-Learning, AR, VR, and gamification discussions. CR capabilities are emerging fast and provides several opportunities for different organizational purposes. The upsurge interest in learning platforms and serious games has not led to convincing evidence when it comes to learning and human improvement. In general, the interest towards games and AR/VR techniques has been speculative, discussing mainly of the potential of games to provide new methods for supporting learning [1, 178].

While traditional learning platforms and games are expected to promote educational and competence improvement goals, they also have further claims of supporting attitudinal and behavioural change by virtue of playing between reality and virtuality, cyber and physical. This can pose additional questions and challenges as to the accurate measurement and evaluation of such changes in attitude, behaviour or competences, especially since it is anticipated that some kind of learning will naturally occur by participating. The subjectivity of all learning approaches also afflicts AR/VR simulation and scenario methods. CR exercises and simulations following rather traditional exercise life-cycle could be improved with human performance and experiential learning features allowing AR/VR continuum. The link with CR activities with their relevance to human performance and learning perspectives enhances the cybersecurity resilience in the organisations. Practicing hands-on practical cyber reality skills in CR platform is not only enough to improve human performance. The improved human performance contains learning with reflection, critical thinking and enhancement of problem-solving competences to cope and manage during real cyber-attacks.

Studies argue that there is an emerging need for cognitive-level synthetic cyber offense and defence in order to ensure realistic cyber simulation and training. The purpose to analyse human cognitive behaviour (which is already seen rich, subtle, and combines autonomous and deliberative information processing) often focuses to capture the dynamic and cognitive-level characteristics of cyber warfare [15]. While the potential for cybersecurity learning platforms in CRs must be acknowledged, over-emphasizing the results in cases where substantial and consistent research is lacking and purposes of human performance and cognitive processes varies. The augmentation of expertise and actual cases to CRs.

5 Conclusions

The organizational representatives value the cybersecurity learning platforms in CRs high and there is emerging need to expand the use of CRs among entire personnel. Currently, the cyber ranges are lacking evaluation and wash-up capacities. Consequently, it is posited that the cyber range or learning platform is best utilised as a complementary training and education resource but not adequate and sufficiently nuanced to act as a stand-alone training tool, nor as a replacement for real-life experiential learning. Often, cyber range learning platforms aim to provide experiences with the combination of virtual-physical touchpoints in learning environment and improvement of human performance. The perceptions may take place in AR forms, tools, simulations or scenarios. Cognitive learning processes may finally lead to competence development and learning. In practice, simulation is one mean to create meaningful learning environment, where learner can apply experience-based constructivist-learning approach.

This study revealed that number of cyber range learning capabilities mainly lack the use of evaluation tools for reflection. Humans have a unique capacity to adapt their skills and competences to different natural and cultural environment and this adaptation requires the building of suitable mental representations, which help in learning and creating new concepts, skills, preferences, motivations, and emotional tendencies on the individual, social and cultural levels. Variety of factors must be considered in the effective design of cyber range learning platforms in use of training and education purposes. Empirical difficulties continue to persist and one of the most significant questions relates to cyber range learning platform expectations and overestimation of potential uses.

The conclusion to be drawn from this study is that there is a role for CR learning platforms within AR/VR and gamification research agenda and in the discussion of cognitive behavior. The CR capabilities are aiming at reaching higher level and is interesting are for future research from different academic disciplines. However, that role should not be over-stated, and this study demonstrated that given the fluid, dynamic nature of professionals working in cyber environments, the nuances and sector-specified challenges cannot be always an exact replicable in AR or VR platforms. The augmentation of human cognitive behavior by the experts for strategies and tactical capabilities for example to cognitive agents should critically assess the different design models with ethical, societal and legal perspectives. There is an inherent danger that over-reliance on behaviourist learning through CRs at the expense of traditional learning approaches could leave individuals ill-prepared, under-informed and under-skilled for their professional roles and work environments. The continuum of AR/VR from learning and performance view point should be further studied in use of CRs.

References

1. Boyle, E., et al.: An update to the systematic literature review of empirical evidence of the impacts and outcomes of computer games and serious games. Comput. Educ. **94**, 178–192 (2016)
2. Collins, J., Futter, A.: Reassessing the Revolution in Military Affairs: Transformation, Evolution and Lessons Learnt. Palgrave, London (2015)
3. Couretas, J.: An Introduction to Cyber Modeling and Simulation. Wiley, New York (2019)
4. Dankbaar, M., et al.: Comparative Effectiveness of a Serious Game and an E-Module to Support Patient Safety Knowledge and Awareness. BMC Medical Education (2017). https://bmcmed educ.biomedcentral.com/articles/10.1186/s12909-016-0836-5
5. Davis, J., Magrath, S.: A Survey of Cyber Ranges and Testbeds. Cyber Electronic Warfare Division. Commonwealth of Australia 2013, Oct 2013
6. Aaltola, K., Taitto, P.: Utilising experiential and organizational learning theories to improve human performance in cyber training. Inf. Secur. Int. J. **43**(2), 123–133 (2019). https://doi.org/10.11610/isij.4311
7. de Freitas, S.: Education in Computer Generated Environments. Routledge, New York and London (2014)
8. Kolb, D.: Experiential learning: Experience as the source of learning and development. 1st Edition. Prentice-Hall, Englewood Cliffs, NJ (1984)
9. Dix, A.: Human–computer interaction, foundations and new paradigms. J. Vis. Lang. Comput. **42**(2017), 122–134 (2017)
10. Flavián, C., Ibáñez-Sánchez, S., Orús, C.: The impact of virtual, augmented and mixed reality technologies on the customer experience. J. Bus. Res. **100**, 547–560 (2019)
11. Futter, A.: Cyber Threats and Nuclear Weapons—New Questions for Command and Control, Security and Strategy. RUSI, London (2016)
12. Grunewald, D., Lützenberger, M., Chinnow, J.: Agent-based network security simulation. In: Proceedings of the 10th International Conference on Autonomous Agents and Multiagent Systems, Taipei, Taiwan, 2–6 May 2011, pp. 1325–1326. International Foundation for Autonomous Agents and Multiagent Systems, Richland, SC (2011)
13. Happa, J., Glencross, M., Steed, A.: Cyber security threats and challenges in collaborative mixed-reality. Front. ICT **6**, 5 (2019). https://doi.org/10.3389/fict.2019.00005
14. Ihde, D.: Technology and the Lifeworld: From Garden to Earth. Indiana University Press, Bloomington (1990)
15. Jones, R., O'Crady, R., Nicholson, D., Hoffman, R., Brunch, L., Bradshaw, J., Bolton, A.: Modeling and integrating cognitive agents within the emerging cyber domain. In: Interservice/Industry Training, Simulation, and Education Conference (I/ITSEC) 2015, Paper No. #15232 (2015)
16. Kabil, A., Duval, T., Cuppens, N., Le Comte, G., Halgand, Y., Ponchel, C.: Why should we use 3d collaborative virtual environments for cyber security? In: IEEE Fourth VR International Workshop on Collaborative Virtual Environments (IEEEVR 2018), pp. 1–8. IEEE, Reutlingen (2018)
17. Kaplan, F.: Dark Territory: The Secret History of Cyber War. Simon & Schuster, London (2016)
18. Knapp, K.: The Gamification of Learning and Instruction: Game-based Methods and Strategies for Training and Education. Pfeiffer and Company Publishers (2012)
19. Linkov, I., et al.: Comparative, collaborative, and integrative risk governance for emerging technologies. In: Environment Systems and Decisions. Springer, Berlin, 4 May 2018. https://doi.org/10.1007/s10669-018-9686-5
20. McGonigal, J.: Reality is Broken: How Games Make Us Better and How they Can Change the World. Penguin Press, USA (2011)
21. Milgram, P., Kishino, F.: A taxonomy of mixed reality visual displays. IEICE Trans. Inform. Syst. E77-D, 1321–1329 (1994)

22. Moskal, S., Yang, S., Kuhl, M.: Cyber threat assessment via attack scenario simulation using an integrated adversary and network modeling approach. J. Defense Model. Simul. **15**(1), 13–29 (2018). https://doi.org/10.1177/1548512917725408
23. Ostrom, A., Parasuraman, A., Bowen, D., Patricio, L., Voss, A.: Service research priorities in a rapidly changing context. J. Serv. Res. **18**(2), 127–159 (2015)
24. Patrício, L., Fisk, R., Falcão e Cunha, J., Constantine, L.: Designing multi-interface service experiences: the service experience blueprint. J. Serv. Res. **10**(4), 318–334 (2011)
25. Piskozub, M., Creese, S., Happa, J.: Dynamic re-planning for cyberphysical situational awareness. In: CPS Conference on Computational Science and Computational Intelligence, Las Vegas, NV, pp. 1–6 (2017)
26. Priyadarshini, I.: Features and Architecture of the Modern Cyber Range: A Qualitative Analysis and Survey. University of Delaware (2018). https://udspace.udel.edu/handle/19716/23789 © 2018 Ishaani Priyadarshini
27. Raybourn, E., Kunz, M., Fritz, D., Urias, V.: A zero-entry cyber range environment for future learning ecosystems. In: Koç, Ç. (eds.) Cyber-Physical Systems Security. Springer, Cham. https://doi.org/10.1007/978-3-319-98935-8_5
28. Tabansky, L.: Basic Concepts in Cyber Warfare. Mil. Strateg. Stud. **3**(1), 75–92 (2011)
29. Tussyadiah, I., Jung, T., Tom Dieck, M.: Embodiment of wearable augmented reality technology in tourism experiences. J. Travel Res. **57**(5), 597–611 (2017)
30. Kolb, A., Kolb, D.: Learning styles and learning spaces: Enhancing experiential learning in higher education. Acad. Manag. Learn. Educ. **4**(2), 193–212 (2005)
31. Wang, B., Cai, J., Zhang, S.: A network security assessment model based on attack-defense game theory. In: Proceedings of the IEEE 2010 International Conference on Computer Application and System Modeling (ICCASM), Taiyuan, China, 22–24 Oct 2010, pp. V3-639–V3-643. IEEE, Piscataway, NJ
32. Wiener, N.: Cybernetics or Control and Communication in the Animal and the Machine. The MIT Press, Cambridge, MA (1948)
33. Kraiger, K., Passmore, J., Rebelo dos Santos, N., Malvezzi, S.: The Wiley blackwell handbook of the psychology of training, development, and performance Improvement. Wiley (2014)
34. Tagarev, T., Pappalardo, S., Stoianov, N.: A logical model for multi-sector cyber risk management & security. Inf. Secur. Int. J. **47**(1), 13–26 (2020)
35. Guttentag, D.: Virtual reality: Applications and implications for tourism. Tourism Manag. **31**(5), 637–651 (2010)
36. Szilárd, S., Benedek, A., Ionel-Cioca, L.: Soft skills development needs and methods in micro-companies of ICT Sector. Soc. Behav. Sci. **238**, 94–103. Elsevier (2018)
37. Burton, J., Lain, C.: Desecuritising cybersecurity: towards a societal approach. J Cyber Policy (2020). https://doi.org/10.1080/23738871.2020.1856903
38. Lehto, M., Limnéll, J.: Strategic leadership in cyber security, case Finland. Inf. Secur. J. Global Perspect. (2020). https://doi.org/10.1080/19393555.2020.1813851
39. Wang, R., Newton, S.: A review of Kolb's learning styles in the context of emerging interactive learning environments. In: AUBEA 2012: Proceedings of the 37th International Conference of the Australasian Universities Building Educators Association (pp. 191–199). University of New South Wales (2012)
40. Beitzel, S., Dykstra, J., Toliver, P., Youzwak, J.: Exploring 3d cybersecurity visualization with the microsoft hololens. In: International Conference on Applied Human Factors and Ergonomics (pp. 197–207). Springer, Cham (2017)
41. Dalsgaard, C.: Social software: E-learning beyond learning management systems. Eur. J. Open, Distance E-learning **9**(2) (2006)
42. Means, B., Toyama, Y., Murphy, R., Bakia, M., Jones, K.: Evaluation of evidence-based practices in online learning: A meta-analysis and review of online learning studies. Project Report. Centre for Learning Technology (2009)
43. Siemens, G.: Connectivism: a learning theory for the digital age. Instruct. Technol. Distance Learning **2**(1) (2005)

44. Friedrich, P.: Web-based co-design. Social media tools to enhance user-centered design and innovation process, University of Tampere (2013)
45. Björk, S., Holopainen, J.: Games and design patterns. The game design reader, pp. 410–437 (2006)
46. Voelcker-Rehage, C.: Motor-skill learning in older adults—a review of studies on age-related differences. Eur. Rev. Aging Phys. Act. **5**, 5–16 (2008)
47. Walker-Roberts, S., Hammoudeh, M., Aldabbas, O., et al.: Threats on the horizon: understanding security threats in the era of cyber-physical systems. J. Supercomput. **76**, 2643–2664 (2020). https://doi.org/10.1007/s11227-019-03028-9

Research on the Overall Attitude Towards Mobile Learning in Social Media: Emotions Mining Approach

Radka Nacheva and Snezhana Sulova

Abstract In this paper, we address the importance of classification and social media mining of human emotions. We compared different theories about basic emotions and the application of emotion theory in practice. Based on Plutchik's classification, we suggest creating a specialized lexicon with terms and phrases to identify emotions for research of general attitudes towards mobile learning in social media. The approach can also be applied to other areas of scientific knowledge that aim to explore the emotional attitudes of users in social media. It is based on the Natural Language Processing and more specifically uses text mining classification algorithms. For test purposes, we've retrieved a number of tweets on users' attitudes towards mobile learning.

Keywords Emotions mining · Text mining · Mobile learning · Social media · Higher education

1 Introduction

The popularity of mobile technologies and their widespread use in everyday life are constantly growing. Globally, around 8.9 billion mobile connections have been reported, of which 5.1 billion have been unique users [1]. Only in Bulgaria there have been 9.8 million [2] with about 7.1 million people population.[1] The penetration of mobile technologies into the users' lives is growing. In 2018, it was reported that between 66% [4] and 68% [12] of the world's population has a mobile phone. According to GSM Association [4], for Europe the share is higher—85% and is

---

[1] According to data published on the official website of the World Bank [3].

R. Nacheva (✉) · S. Sulova
University of Economics—Varna, Knyaz Boris I Blvd. 77, Varna, Bulgaria
e-mail: r.nacheva@ue-varna.bg

S. Sulova
e-mail: ssulova@ue-varna.bg

© The Author(s), under exclusive license to Springer Nature Switzerland AG 2021
T. Tagarev et al. (eds.), *Digital Transformation, Cyber Security and Resilience of Modern Societies*, Studies in Big Data 84,
https://doi.org/10.1007/978-3-030-65722-2_27

rising. It is estimated that the end of 2025, 73% of the planet's population will have a mobile connection [4].

92% of the Millennium generation (Generation Y)[2] own a smartphone compared to 85% of Generation X [7]. Generation Y spend time on their mobile phone which is about 223 min per day [8]. Only Google Play has approximately 2.1 million apps. There are 1.8 million apps in the Apple's App Store [9]. Among the millions of applications most popular during the first quarter of 2019 have been social media ones—more than 135 million downloads only in Google Play [10] and approximately 95% reach global Android mobile users [11].

Quite naturally, social media is gaining more and more popularity in the everyday lives of consumers all over the world. In 2018, it was reported that social media users have reached about 3196 billion [12], with an increase of over 400 million compared to 2017 [13]. With a world population of 7.55 billion [14], the social media penetration was 42% in early 2018 [12], 15] compared to 37% in 2017 [13]. Facebook users are the most active users with 2.23 billion by the beginning of 2018, followed by YouTube with 1.5 billion, Instagram with 813 million and more [16]. Another social network, which is gaining more users, is Twitter with 336 million for the first quarter of 2018, marking a 6 million increase over the last quarter of 2017 [17]. There are over 450,000 published tweets per minute [18]. Millions of Twitter posts and social media as a whole are a rich source of sentiment analysis (SA), opinion mining (OM), and emotion mining (EM). They are, in turn, the basis for analyzes and forecasts in the area of marketing [19, 20], politics [21], e-commerce [22, 23], education [24, 25], cyber security and resilience [26, 27, 28] and many others. In some cases, they are combined with artificial intelligence methods to discover new dependencies [29]. Regardless of the area, one of the major difficulties in retrieving data from this vast mass of information stems from the style of publications, which is mostly everyday use. Users express their opinions, moods and emotions in a variety of ways, depending on the specifics of the language in which they write.

In this regard, **the purpose of the current paper** is to develop and approbate a lexicon-based emotions mining approach, aimed at extracting knowledge from social media. To test it, Twitter social networking data are used to study attitudes towards mobile learning. The approach can also be applied in areas such as cyber security, politics, commerce, etc. It takes into account the specifics of the language in which the posts are written.

---

[2]We do not have a single opinion in the literature on the clear boundaries between different generations. For example, Pew Research Center [5] points out that those born between 1965 and 1980 are representatives of Generation X, between 1981 and 1996—the Generation Y, and those after 1997—the Generation Z. According to Goldman Sachs [6] Millennials were born between 1980 and 2000.

2 Importance of Emotions: Theories and Applications

Studying the emotional states of human beings is not a new direction for the scientific sphere. The nature of emotions has been explored and described by philosophers such as Aristotle, Thomas Aquinas, Descartes, Spinoza and others, and Charles Darwin is the first scientist who has tried to give a scientific explanation and has laid the foundations for the bio-evolutionary theory of basic emotions [30]. Hebb [31] has conducted an analysis of the intuitive processes of recognition by examining emotion in man and animal. He defined the term "emotion" as "distinctive mental state or conscious content, but sometimes it refers to more vaguely conceived states of excitation without any definite implications about consciousness". According to Frijda [32] in his *"The Laws of Emotion"*, the meaning structures of given situations give rise to emotions; different emotions arise in response to different meaning structures.

Emotions play an important role in people's lives. They are associated with temporary body reactions that have been obtained in response to external stimuli or changes in the internal state (thought-driven, for example). Emotions are short term states which duration is strictly individual and dependent on their type—positive or negative, as well as external stimuli [33, 34]. Their significance for man from an evolutionary point of view is related to his survival. They are encoded in the human genes. Through them people study the world around and make decisions that help them survive in this dynamically changing world.

Throughout the twentieth century, psycho-biologists, psychologists and neuroscientists have carried out classifications of underlying emotions. Table 1 summarizes a small part of them.

Table 1 Assumptions about basic emotions

Researcher	Positive emotions	Negative emotions
Watson [35]	Love	Fear, rage
Izard [36]	Interest, joy, surprise, shame	Anger, contempt, disgust, distress, fear, guilt
Plutchik [37]	Anticipation, acceptance, joy, surprise	Anger, disgust, fear, sadness
Ekman [38]	Joy, surprise	Anger, disgust, fear, sadness
Panksepp [39]	Expectancy	Fear, rage, panic
Tomkins [40]	Interest, joy, shame, surprise	Anger, contempt, disgust, distress, fear
Frijda [32]	Desire, happiness, interest, surprise, wonder	Sorrow
Kemper [41]	Satisfaction	Fear, anger, depression
Oatley and Johnson-Laird [42]	Happiness	Anger, disgust, anxiety, sadness
Piryova [43]	Happiness, surprise	Sorrow, anger, fear, disgust

Scientists do not have a single opinion on the number of underlying emotions—the summaries in Table 1 vary between 3 and 10. They also use different names of the same phenomena, which further creates ambiguity when adopting their theories. We see that for the most part the negative emotions prevail, which [44] explains with "recent evolutionary theories that suggest that the affective repertoire of our species has been largely shaped by processes of natural selection".

We can argue with assumptions that include three or four basic emotions because they do not show the multi-faceted perception of the surrounding world by individuals. In our opinion, the advantage of Plutchik's theory over others are colors associations to each emotion, including their consideration of opposite pairs. Plutchik's wheel of emotions also gives insight into the more sophisticated human nature through the so-called secondary emotions, which are more complex psychic phenomena combining basic emotions. But his theory does not fully express the cultural differences between individuals that should be considered even in today's globalized environment. For example, if for some people the red color is an indicator of caution (associated with anger), it is associated as sign of prosperity within Eastern cultures.

Based on these facts, we can point out that there can be no single theory that gives knowledge simultaneously about the cultural differences between people and takes into account their personality peculiarities.

Studies of emotional experiences are the basis for making the so-called extraction of emotions from text. That is called "emotions mining". Many research studies have been carried out in the field of computer science and, in particular, human–computer interaction. There are frequently used social media posts. For example, in the context of cyber security emotions mining could detect irregularities in the behavior of social media players. It also helps analyzing the risk of committing terrorist acts or at least the intentions of terrorism.

Emotions mining is also used in the context of education area to study attitudes to applying a given type of teaching approach, for example. Often the most accurate results are derived from an environment in which learners feel they can express themselves, even on a subconscious level, such as social media. Extracted users' opinions can help create a learner-centered teaching process to increase their motivation and interest in the learning environment.

Emotions mining can also be applied in areas such as the financial sphere to explore consumer attitudes to raising the rate on loans or the drop-in shares of large companies.

There are many examples of how theories about emotional experiences of individuals could be applied in practice. That is why in this paper we have narrowed the scope of the research to the overall attitude towards mobile learning in social media. Our intentions for this are well grounded in the introductory part of the paper through the provided statistics.

3   Extracting Emotions from Text Data

Social networks are platforms that have enormous potential and provide opportunities for interaction and information sharing in different spheres. That leads to them being identified as the largest sources of unstructured data.

Automated processing of unstructured data is a difficult task, and there is no single algorithm for it. In the recent years, it has been a subject of great research interest.

The discovery of knowledge in unstructured text data is known in literature as Text Mining (TM). In order to present the process of retrieving text from sources located on social networks, an adapted model of Cross-industry Standard Process for Data Mining (CRISP-DM) can be used (Fig. 1). In it, in order to retrieve and structure the data prior to the process of (Data mining DM) the following has been done:

- Information retrieval (IR)—Searching social networks for posts, sharing and downloading of the posts;
- Natural Language Processing (NLP)—preprocessing and transforming the text in a suitable for processing by computer programs way;
- Text Conversion—transforming the messages and presenting them as a set of word vectors.

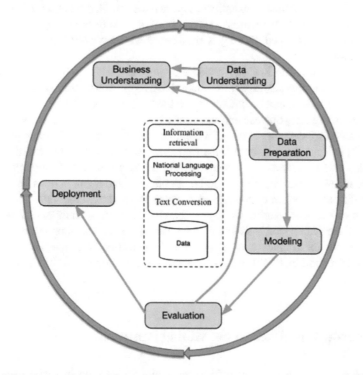

Fig. 1 Retrieving text from social networks

The approaches to handling unstructured content are differentiated according to the specific research tasks and the type of resources that are being used. There are researches for: automatic classification of web pages [45] grouping of documents [46], detection of similarity of a text document [47, 48], extraction of emotions from text [49].

Sharing and accumulating multiple social networking publications creates prerequisites for the search of ways to retrieve attitudes, moods, emotions expressed in them. In literature, the interrelated concepts of Opinion mining (OM) and Sentiment analysis (SA) appear. They deal with analysis of the opinions and attitudes of people towards certain events, products, services. For a more detailed analysis of emotions, when the purpose is to determine not only whether it's positive or negative, but to also mine a particular emotion—affection, joy, sadness, anger, etc.

Khan [50] describes the sentiment analysis of social media as a categorization process dividing social media text to positive, negative or neutral. In his study, Bing Liu classifies the approaches to analyzing feelings from text according to the analyzed components. He examines 3 types of approaches, at the level of a whole document, of phrases and of words [51]. In the document level analysis, the general opinion presented in the document is classified as positive, negative or neutral. It helps determine the overall polarity of a text, but it is difficult to assess the different emotions connected with individual aspects of the text. In the level of sentence analysis, the polarity of each sentence is examined, and it is determined whether it is positive, negative, or neutral. In the words' analysis level, the words are grouped into dictionaries, and an automated grading is performed with keyword retrieval and topic identification.

Medhat, Hassan and Korashy divide the approaches into two main groups: the Machine learning approach and the Lexicon-based approach and one additional group of approaches that cannot be classified into the main groups [52].

Many authors carry out research and offer approaches to analyzing user views from social networks based on the use of a combination of methods. For example, Khan et al. [53] offer the use of both Enhanced Emoticon Classifier, Improved Polarity Classifier, and SentiWordNet based classifier at the same time. Kouloumpis et al. [54] use n-gram, lexicon, microblogging emoticons, and abbreviations features. Li and Xu [55] also uses an emoticon-based method to identify emotions.

Based on the many studies we have researched, as well as on those conducted by us, we believe that lexicon and dictionaries-based approaches are appropriate in cases where the text doesn't have a large meaningful connection between its sentences are to be used for short texts without a complex sentence structure, just like social media publications.

4 Lexicon-Based Emotions Mining Approach

For research of the general attitude towards mobile learning in social media, we suggest creating a specialized lexicon with terms and phrases to identify emotions.

It is necessary to note that there are some lexicons based on the English lexical database WordNet: WordNet-Affect, which works with words grouped in sets of synonyms [56] and SentiWordNet, where words are as-signed a marker showing whether they have a subjective/objective, positive/negative meaning [57]. However, for the dictionary to be universal—applicable not only to English but also to other languages and to include a collection of words appropriate to social media, which is also corroborated with the moral and cultural differences between people, it is necessary to develop and approbate an approach for its creation.

As a base for the dictionary, basic phrases and words are used in the respective language for the 8 emotions from Plutchik's wheel of emotions. To these basic words and phrases, we have added specific ones derived from social networks. As Todoranova and Todoranova [58] point out, the vocabulary and phraseology of one language must be taken into account, and the words, sentences, phrases that are commonly used.

For test purposes in our research, we have used the Twitter social network as one of the most popular among young people today. We believe that these words will help significantly when identifying emotions because they are part of the everyday speech. It is important to note that the texts in social networks are specific, users often use informal and nonstructured language for communication.

The approach we propose can be seen on Fig. 2. It includes the following main

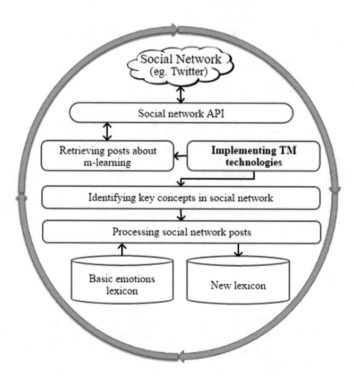

Fig. 2 Proposed emotions mining approach

Table 2 Retrieved additional words for the lexicon

Acceptance	Anticipation	Anger	Disgust	Joy	Fear	Sadness	Surprise
Approval, recognition, acknowledge	Expectance, outlook	Fury, hostility, mad, rage	Distaste, disdain, repulsion, loathe	Bliss, cheerfulness, delighted, euphoria	Trouble, worry, afraid, dread	Grief, misery, gloomy, unhappy	Shock, startled, awe, wonder

stages:

1. Retrieving posts from the Twitter social network, reflecting the opinions and attitudes of users towards mobile learning. At this stage, specially written programming solutions that work with the Twitter API or one of the many software add-ons to the data analysis software can be used.
2. Implementing text mining technologies to identify key concepts in social networking sites.
3. The processing of twitter posts should include the following steps:

    - dividing the text into words, the process is known as tokenization;
    - transforming the text into lower case letters;
    - removing stopwords, which are words that usually have no analytic value, such as "a", "and", "the" etc.;
    - removing digits and punctuations;
    - removing @user and links, the symbol @ is used when a user mentions another one;
    - transforming the words into their base form, known as stemming.

4. Adding the extracted words and phrases to the original basic vocabulary of emotions.

Approximately 5285 Tweets for the period 1.06-10.06.2019 have been derived for the approbation of the approach. The keywords used in the extraction are "mobile learning" and "m-learning". Text extraction and processing are done with the RapidMiner software product. The words we have identified are given in Table 2.

The proposed approach can be periodically approbated and the new words identified should be added to the lexicon.

5 Conclusion

Emotions are an important and fundamental aspect of our lives. By communicating on social networks, users usually express their emotions. This has led to searching for ways to extract and analyze the moods and attitudes inherent in communication to specific events and subject areas.

The proposed lexicon-based emotions mining approach for the analysis of attitudes to mobile learning is based on knowledge from several interdisciplinary areas. The approach has an universal application and can be used for another social network, not only for education but also for other areas, for example security. The main advantages of the approach are that it is based on a lexicon in words that conform to the way of communication in the social networks and the cultural differences of the people. The lexicon can be improved by including phrases showing attitudes and emotions.

References

1. GSMA Intelligence: Definitive data and analysis for the mobile industry (2019). https://www.gsmaintelligence.com/. Accessed 18 June 2019
2. GSMA Intelligence: Definitive data and analysis for the mobile industry, Bulgaria (2019). https://www.gsmaintelligence.com/markets/362/dashboard/. Accessed 18 June 2019
3. WorldBank: Population, total (2017). https://data.worldbank.org/indicator/SP.POP.TOTL?locations=BG. Accessed 18 June 2019
4. GSM Association: The Mobile Economy 2018 (2018). Available via https://www.gsma.com/mobileeconomy/wp-content/uploads/2018/02/The-Mobile-Economy-Global-2018.pdf. Accessed 18 May 2019
5. Pew Research Center: Millennials (2019). https://www.pewresearch.org/topics/millennials/. Accessed 29 May 2019
6. Goldman Sachs: Millennials Coming of Age (2019). https://www.goldmansachs.com/insights/archive/millennials/. Accessed 18 June 2019
7. Jiang, J.: Millennials stand out for their technology use, but older generations also embrace digital life (2018). Retrieved from http://www.pewresearch.org/fact-tank/2018/05/02/millennials-stand-out-for-their-technology-use-but-older-generations-also-embrace-digital-life/. Accessed 18 June 2019
8. Statista.com: Daily time spent on mobile by Millennial internet users worldwide from 2012 to 2017 (in minutes) (2019). https://www.statista.com/statistics/283138/millennials-daily-mobile-usage/. Accessed 18 June 2019
9. Statista.com: Number of apps available in leading app stores as of 1st quarter 2019 (2019). https://www.statista.com/statistics/276623/number-of-apps-available-in-leading-app-stores/. Accessed 18 June 2019
10. Statista.com: Leading non-gaming app publishers in the Google Play Store worldwide in March 2019, by number of downloads (in millions) (2019). https://www.statista.com/statistics/697012/leading-non-gaming-android-app-publishers-worldwide-by-downloads/. Accessed 18 June 2019
11. Statista.com: Market reach of the most popular Android app categories worldwide as of June 2018 (2019). https://www.statista.com/statistics/200855/favourite-smartphone-app-categories-by-share-of-smartphone-users/. Accessed 16 June 2019
12. Kemp, S.: Digital in 2018: world's internet users pass the 4 billion mark. In: We Are Social (2018). Available via https://wearesocial.com/blog/2018/01/global-digital-report-2018. Accessed 21 May 2019
13. Kemp, S.: Digital in 2017: global overview. In: We Are Social (2017). Available via https://wearesocial.com/special-reports/digital-in-2017-global-overview. Accessed 21 May 2019
14. United Nations: World Population Prospects The 2017 Revision: Key Findings and Advance Tables. Available via Department of Economic and Social Affairs, Population Division. https://esa.un.org/unpd/wpp/publications/Files/WPP2017_KeyFindings.pdf. Accessed 21 May 2019
15. Statista.com: Global social network penetration rate as of January 2018, by region (2018). https://www.statista.com/statistics/269615/social-network-penetration-by-region/. Accessed 21 May 2019

16. Statista.com: Most popular social networks worldwide as of April 2018, ranked by number of active users (in millions) (2018). https://www.statista.com/statistics/272014/global-social-net works-ranked-by-number-of-users. Accessed 21 May 2019
17. Statista.com: Number of monthly active Twitter users worldwide from 1st quarter 2010 to 1st quarter 2018 (in millions) (2018). https://www.statista.com/statistics/282087/number-of-mon thly-active-twitter-users/. Accessed 21 May 2019
18. Statista.com: Media usage in an internet minute as of July 2017 (2017). https://www.statista. com/statistics/195140/new-user-generated-content-uploaded-by-users-per-minute/. Accessed 21 May 2019
19. Howells, K., Ertugan, A.: Applying fuzzy logic for sentiment analysis of social media network data in marketing. Procedia Comput. Sci. **120**, 664–670 (2017)
20. Moro, S., et al.: A text mining and topic modelling perspective of ethnic marketing research. J. Bus. Res. (2019). https://doi.org/10.1016/j.jbusres.2019.01.053
21. Bankov, B.: An approach for clustering social media text messages retrieved from continuous data streams. Sci. Bus. Soc. **3**(1), 6–9 (2018)
22. Chang, W., Wang, J.: Mine is yours? Using sentiment analysis to explore the degree of risk in the sharing economy. Electron. Commer. Res. Appl. **28**, 141–158 (2018)
23. Vinodhini, G., Chandrasekaran, R.: Inf. Process. Manage. **53**(1), 223–236 (2017)
24. Rana, R., Kohle, V.: Analysis of students emotion for Twitter data using Naïve Bayes and non-linear support vector machine approaches. Int. J. Recent Innov. Trends Comput. Commun. **3**(5), 3211–3217 (2015)
25. Ortigosa, A., et al.: Sentiment analysis in Facebook and its application to e-learning. Comput. Hum. Behav. **31**, 527–541 (2014)
26. Mansour, S.: Social media analysis of user's responses to terrorism using sentiment analysis and text mining. Procedia Comput. Sci. **140**, 95–103 (2018)
27. Penchev, B.: Enhancing the security of mobile banking through the implementation of automated checks on the mobile device. Comput. Sci. Commun. J. **5**(1), 3–8 (BSU, Burgas)
28. Al-Rowaily, K., et al.: BiSAL—a bilingual sentiment analysis lexicon to analyze Dark Web forums for cyber security. Digit. Invest. **14**, 53–62 (2015)
29. Vasilev, T., Stoyanova, M., Stancheva, E.: Application of business intelligence methods for analyzing a loan dataset. Informatyka Ekonomiczna **1**(47), 97–106 (2018)
30. Minchev,B.: Psychology: Evolutionary Phenomenological Approach (2013) (in Bulgarian)
31. Hebb, D.: Emotion in man and animal: an analysis of the intuitive processes of recognition. Psychol. Rev. **53**, 88–106 (1946)
32. Frijda, N.H.: The laws of emotion. Am. Psychol. **43**(5), 349–358 (1988)
33. Mata, A., et al.: Forecasting the duration of emotions: a motivational account and self-other differences. Emotion **19**(3), 503–519 (2018) (Advance online publication). https://doi.org/10. 1037/emo0000455
34. Verduyn, P., et al.: Determinants of emotion duration and underlying psychological and neural mechanisms. Emot. Rev. **7**(4), 330–335 (2015). https://doi.org/10.1177/1754073915590618
35. Watson, J.B.: Behaviorism. University of Chicago Press, Chicago (1930)
36. Izard, C.E.: The Face of Emotion. Appleton-Century Crofts, New York (1971)
37. Plutchik, R.: A general psychoevolutionary theory of emotion. In: Plutchik, R., Kellerman, H. (eds.) Emotion: Theory, Research, and Experience, vol. 1, pp. 3–31. Academic Press, Cambridge (1980)
38. Ekman, P.: Emotion in the Human Face, 2nd edn. Cambridge University Press, New York (1982)
39. Panksepp, J.: Toward a general psychobiological theory of emotions. Behav. Brain Sci. **5**, 407–467 (1982)
40. Tomkins, S.S.: Affect theory. In: Scherer, K.R., Ekman, P. (eds.) Approaches to Emotion, pp. 163–195. Erlbaum, Hillsdale, NJ (1984)
41. Kemper, T.: How many emotions are there? Wedding the social and the autonomic components. Am. J. Sociol. **93**, 263–289 (1987)

52. Oatley, K., Johnson-Laird, P.: Towards a cognitive theory of emotions. Cogn. Emot. **1**, 29–50 (1987)
43. Piryova, B.: Neurobiological Bases of Human Behavior. The Publishing House of the New Bulgarian University, Sofia (2011) (in Bulgarian)
44. Forgas, J.P.: Can sadness be good for you? On the cognitive, motivational and interpersonal benefits of negative affect. In: The Positive Side of Negative Emotions, edn. 1, Guilford Press, New York, pp. 3–13 (2014). http://dx.doi.org/10.1111/ap.12232
45. Tsukada, M., Washio, T., Motoda, H.: Automatic web-page classification by using machine learning methods. In: Web Intelligence: Research and Development. Lecture Notes in Computer Science, vol 2198, pp. 303–313. Springer, Berlin (2001)
46. Markov, Z., Larosed, D.: Data Mining the Web Uncovering Patterns in Web Content, Structure, and Usage. Wiley, New Jersey (2007)
47. Lakashimi, S., et al.: Analysis of similarity measures for text clustering. Int. J. Eng. Sci. Res. **3**(8), 4627–5463 (2013)
48. Huang, A.: Similarity measures for text document clustering. In: Proceedings of the Sixth New Zealand Computer Science Research Student Conference (NZCSRSC2008), pp. 49–56 (2008)
49. Shelke, N., et al.: Approach for emotion extraction from text. In: Proceedings of the 5th International Conference on Frontiers in Intelligent Computing: Theory and Applications. Advances in Intelligent Systems and Computing, vol. 516. Springer, Singapore (2017)
50. Khan, G.F.: Seven Layers of Social Media Analytics. CreateSpace Independent Publ. Platform, New York (2015)
51. Liu, B.: Sentiment Analysis and Opinion Mining. Morgan & Claypool Publishers (2012)
52. Medhat, W., Hassan, A., Korashy, H.: Sentiment analysis algorithms and applications: a survey. Ain Shams Eng. J. **5**, 1093–1113 (2014)
53. Khan, F., et al.: TOM: twitter opinion mining framework using hybrid classification scheme. Decis. Support Syst. **57**, 245–257 (2014). https://doi.org/10.1016/j.dss.2013.09.004
54. Kouloumpis,E., et al.: Twitter sentiment analysis: the good the bad and the OMG! In: Proceedings of the Fifth International Conference on Weblogs and Social Media, Barcelona, Catalonia, Spain, July, pp. 538–541 (2011)
55. Li, W., Xu, H.: Text-based emotion classification using emotion cause extraction. Expert Syst. Appl. Int. J. **41**, 1742–1749 (2013)
56. Strapparava, C., Valitutti, A.: WordNet-affect: an affective extension of WordNet. In: Proceedings of 4th International Conference on Language Resources and Evaluation (LREC'04), pp. 1083–1086, Portugal (2004)
57. Esuli, A., Sebastiani, F.: SentiWordNet: a publicly available lexical resource for opinion mining In: Proceedings of the 5 the International Conference on Language Resources and Evaluation, pp. 417–422, Italy (2006)
58. Todoranova, L., Todoranova, A.: Incorrectly transmitted computer terminology and strategies to overcome non-equivalence (in Bulgarian). Izvestia J. Union of Sci. Varna. Econ. Sci. Ser **2**, 209–216 (2017)

Chief Assist. Prof. Radka Nacheva, PhD was born on December 26, 1984 in Lovech, Bulgaria. She is a lecturer in the Computer Science Department, University of Economics—Varna. She holds a PhD in Informatics from the same University. Her research interests are in Human-Computer Interaction, Cognitive Science and Software Accessibility. Dr. Nacheva teaches in Web technologies, Software Design and Operating Systems. She is currently a head of a project "Contemporary Approaches to the Integration of Mobile Technologies in Higher Education".

Assoc. Prof. Snezhana Sulova, PhD was born on August 21, 1973 in Nova Zagora, Bulgaria. She received a Ph.D. degree in 2005. In 2006 she joined the Computer Science Department of University of Economics—Varna as an assistant professor. During the period 2008–2012 she was a Chief Assistant Professor, and since 2012 she is an Associate Professor in the same Department. Her research interests are in the field of Data Mining, Text Mining, Machine Learning, Artificial Intelligence, E-commerce. She has more than 60 scientific publications and has participated in 12 scientific projects, in 5 of which she was the scientific team's leader.

Perspectives for Mobile Learning in Higher Education in Bulgaria

Latinka Todoranova and Bonimir Penchev

Abstract The use of mobile technologies in education enables the access to and the implementation of modern teaching methods and tools. Nowadays, higher education faces a number of different challenges. Not only the technologies are developing at an extremely fast pace, but also the generations are changing—children are growing up in an environment where they are surrounded by different technological gadgets. These gadgets influence children's communication, their access to information and also their learning habits. Upon entering the university, the learners continue to use their mobile phones for information exchange and communication. In response, higher education is trying to change in that direction. But it seems that the steps the higher education is doing are not fast enough to meet the expectations of the modern young people. The purpose of this chapter is to define the problems and to outline the perspectives for mobile learning in higher education in Bulgaria.

Keywords Mobile technology · Mobile learning · E-learning · Higher education · Mobile applications

1 Introduction

The research connected with the application of mobile technologies in education began actively 20 years ago. During this period, mobile technologies have undergone a number of different changes like significant reduction of the device size, expanding of the functionality, getting access to high-speed internet, and many more. In the current context of the education, mobile devices are used for online access to learning materials that are usually placed on e-learning platforms. This process leads to the emergence of mobile learning or m-learning as a natural development of e-learning.

L. Todoranova (✉) · B. Penchev
University of Economics, Varna, Bulgaria
e-mail: todoranova@ue-varna.bg

B. Penchev
e-mail: b.penchev@ue-varna.bg

© The Author(s), under exclusive license to Springer Nature Switzerland AG 2021   441
T. Tagarev et al. (eds.), *Digital Transformation, Cyber Security and Resilience of Modern Societies*, Studies in Big Data 84,
https://doi.org/10.1007/978-3-030-65722-2_28

At the dawn of e-learning, lecturers develop electronic learning materials (presentations, text documents, videos) that record on CDs or DVDs. The next stage is the active development of e-learning platforms, where the electronic learning units are published, tests are organized and the communication between the learners and the lecturers is provided in the form of chat, forum and private messages. Nowadays, the access to these platforms is mainly accomplished with the use of mobile phones. In Bulgaria, according to the NSI (National Statistical Institute) data for 2018, 55.9% of the internet users rely on their mobile phone [1].

2 Literature Review

The impact of the information and communication technologies on the education is a research subject to a significant number of authors. They usually study the influence of modern technologies in primary and secondary education as well as in higher education. The penetration of these technologies into the education process is the actual start of the e-learning.

In their study, Parusheva and Alexandrova found out that 23 out of 24 of the examined Bulgarian universities use platforms for electronic and distance learning [2]. However, e-learning platforms do not replace the traditional forms of education, but they rather complement and enrich them.

Through the learning management systems (LMS), learners have access to additional learning resources that can be used in a convenient for them time and place. The access to these platforms is primarily through students' mobile devices, because of their advantages like the lower price of the devices, the availability of mobile internet at a good speed and the fact that mobile phones are an integral part of everyday life of the young people. As a result of the widespread usage of the mobile technologies, the term mobile education or learning is quickly introduced in all forms of education. Doneva, Kasakliev and Totkov define that mobile learning suggests the use of mobile and portable devices applied in the field of information technologies, as well as the use of wireless network and communication technologies for teaching, learning and for support of the learning process [3]. According to Kraeva and Emilova, m-learning aims to create comfortable conditions for teachers and learners in the two-sided process of teaching and learning, allowing the participants not to be attached to the physical location of the learning process [4]. This is an advantage that is of great importance in today's dynamic world. Mobile learning also aims to solve another increasing problem—the working students. According to the Center for Monitoring and Evaluation of Education Quality, in 2009 the share of working students in Bulgaria was 35% [5]. Today, 10 years later, this percentage is likely to be significantly higher. One of the goals defined in the strategic framework for European cooperation in education and training (ET2020 framework) is at least 40% of people aged 30–34 to have completed some form of higher education [6]. Most of the people in this group are employed and thanks to the mobile learning they will have easier access to the learning materials.

In the literature, the different authors are actively studying and analyzing the advantages and disadvantages of applying m-learning and also the different forms of education (when presence of the learners is obligatory and when it is not). But despite the rich features of LMSs, and their good functionalities for communication between educators and learners, they cannot yet completely replace face-to-face communication and education.

However, in order this education to be more effective, it should be paid more attention to the usability of the LMSs, which is a quality criterion seen in a specified context of use [7].

This is a topic that needs to be improved seriously in order to achieve greater efficiency in mobile learning, which (as has been pointed out) is not intended to replace traditional forms of learning but rather to complement and enrich them. And this in turn will allow the learners to be better prepared and to easily combine work and education.

3 Methods

In order to outline the guidelines for future development of the m-learning, this paper have made an analysis both to the LMS capabilities, which are used in the higher education institutions in Bulgaria and to the provided electronic units, published in them. The stages of the research process are shown on Fig. 1.

The identification of the e-learning platforms of higher education institutions in Bulgaria is based on the links to these platforms found on the official web sites of the institutions. The Register of the accredited higher education institutions is maintained by the Ministry of Education and Science [8].

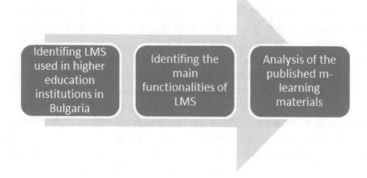

Fig. 1 Stages of the research process of the study

4 Results and Discussion

The results show that the platforms Moodle and Blackboard are the most widely used in Bulgarian higher education. There are also universities using their own custom developed LMS.

Regarding the main features of the LMS, there are a number of publications which present in details their basic functions and also compare them by different criteria. The choice of which platform to be used is made by the higher education institution. Currently there are no established official requirements in Bulgaria for the use of the e-learning platforms. In addition to the selection of a platform, the universities also provide themselves the necessary technical support for the seamless operation of the LMSs.

The research of the LMSs used in Bulgaria indicates that a significant part of the universities take into account the increasingly active use of mobile devices to access platforms and adapt their design according to these trends.

The analysis of the electronic learning units shows that the largest share of the documents published in the platforms are in the pdf and pptx file formats. However, this is an indicator that the modern functionalities (geo location, face recognition, video streaming, etc.) of the mobile devices are not being used. Despite the shortcomings of these devices (small screen size, limited entry capabilities, various operating systems, etc.), they also have many advantages (instant notification, easy way to exchange data and information available anywhere and anytime) that should not be overlooked.

Moreover, the application of mobile technology in education will be more successful, if it is based on a number of pedagogical principles, such as those derived from Burden et al. [9] in early 2019. They must be at the heart of the policies led by the organizations.

The future of m-learning in higher and other education institutions in Bulgaria, depends not only on the measures taken by the institutions themselves, but also on the government policy pursued in this field. In addition, the experience of the leading education institutions, which make significant efforts to achieve effective and quality m-learning, should be taken into account.

In 2016 UNESCO launched 5-year project "Best Practices in Mobile Learning" [10]. The purpose of the project is to present good practices for the planning and implementation of mobile learning that will be of high quality, effective and relevant to the modern digital world. Two approaches have been used in the implementation of the project: top-down and bottom-up. The first approach outlines good practices undertaken by governments and ministries. The second approach reveals initiatives taken by the education institutions themselves. Based on the results presented so far, effective strategies can be developed in planning and implementing mobile learning programs in various contexts and in responding to different needs and challenges.

5 Conclusions

Sulova and Nacheva have researched that the successful development of digital services and the software product as a whole is directly connected to the target audience expectations [11]. These are the main guidelines in which the research should continue—development of new mobile learning applications with the active involvement of learners and trainers, constant research of their attitudes and expectations, and also research of the devices they use to access the learning materials.

From the conducted analysis of the perspectives for mobile learning in higher education it is clear that there are also a number of other problems, such as the need for further training of the teaching staff to prepare training materials suitable for the new environment and also defining the means to compensate the teachers for the additional efforts and time. It cannot be expected that lecturers will develop mobile applications themselves, so it is necessary to hire specialists for the development and maintenance of m-learning applications. Moreover, these specialists should use innovative approaches based on the modern scientific knowledge in order to achieve quality results [12].

References

1. Individuals Using Mobile Devices to Access the Internet (Away from Home or Work). https://www.nsi.bg/en/content/6107/individuals-using-mobile-devices-access-internet. 10.06.2019
2. Parusheva, S., Aleksandrova, Y., Hadzhikolev, A.: Use of social media in higher education institutions—an empirical study based on bulgarian learning experience. TEM J. **7**(1), 171–181 (2018)
3. Doneva, R., Kasakliev, N., Totkov, G.: Towards Mobile University Campuses. CompSysTech'2006. https://ecet.ecs.uni-ruse.bg/cst06/Docs/cp/sIV/IV.3.pdf. 10.06.2019
4. Emilova, P., Kraeva, V.: Mobile training—function and challenges. In: Conference: Higher Education and Business in the Context of the Europe 2020 Strategy (2014)
5. Results from the participation of Bulgaria in Eurostudent III. https://copuo.bg/upload/docs/2013-01/EUROSTUDENT3_report.pdf. 16.06.2019
6. European Policy Cooperation (ET2020 framework). https://ec.europa.eu/education/policies/european-policy-cooperation/et2020-framework_en. 10.06.2019
7. Nacheva, R.: Architecture of web-based system for usability evaluation of mobile applications. Izv. J. Varna Univ. Econ. Varna **61**(2), 187–201 (2017)
8. Register of Higher Education Institutions. https://rvu.mon.bg/. 15.06.2019
9. Burden, K., Schuck, S., Kearney, M.: Is the use of mobile devices in schools really innovative? What does the evidence say? https://impact.chartered.college/article/mobile-devices-schools-really-innovative-what-does-evidence-say/. 1.09.2019
10. Best practices in mobile learning. https://en.unesco.org/themes/ict-education/mobile-learning/fazheng/case-studies. 1.09.2019
11. Nacheva, R., Sulova, S.: Approach to exploring users' expectations of digital services' functionality. Ekonomiczne Problemy Usług, Szczecin: Wydawnictwo Naukowe Uniwersytetu Szczecinskiego **2**(131), 137–145 (2018)
12. Sulova, S.: Association rule mining for improvement of IT project management. TEM J. **7**(4), 717–722. ISSN 2217-8309. https://doi.org/10.18421/TEM74-03, Nov 2018. https://www.temjournal.com/content/74/TemJournalNovember2018_717_722.pdf. 12.06.2019

Hybrid Influence and the Role of Social Networks

Understanding Hybrid Influence: Emerging Analysis Frameworks

Todor Tagarev

Abstract Over centuries countries and alliances in conflict have used all available means at their disposal, in addition to military forces, to gain a competitive edge. 'Hybrid threat' or 'hybrid war' is a new term designating the coordinated use of military and non-military means, that got traction in the analysis of the 2006 Israel-Hezbollah War and became widely used beyond the professional communities after the 2014 annexation of Crimea by the Russian Federation. One specific aspect of such conflicts is the lack of a clear distinction between 'war' and 'peace,' when a skilful player would apply available means to subjugate the opponent's will, and thus achieve his political objectives, without fighting. Economic leverage, energy dependencies, ideology, propaganda and disinformation, cyberattacks and corruption are just a few types of means for such diverse—or 'hybrid'—influence. While using different channels, it is their combined impact that can bring desired effects. Finding an appropriate response to such hybrid influence requires a good understanding of own vulnerabilities, the exposure to and the actual or potential impact over the public institutions, the economy and the society. This paper looks into emerging frameworks allowing to estimate the impact of hybrid influence and the extent to which they may be used to reflect the actual interdependencies and complexity of modern societies. While combining in various ways empirical data and expert assessments, all these frameworks facilitate the application of risk-aware allocation of limited resources to counter hybrid influence.

Keywords Hybrid threats · Hybrid war · Interdependence · Whole-of-society approach · Resilience · Exploratory framework · Disinformation · Propaganda · Kill chain · Cognitive threat · Unvirtuous cycle · Five rings modes · Critical infrastructures

T. Tagarev (✉)
Institute of Information and Communication Technologies, Bulgarian Academy of Sciences, "Acad. G. Bonchev" Str., Bl. 2, 1113 Sofia, Bulgaria
e-mail: tagarev@bas.bg

1 Introduction

A variety of strategies, tactics, and means have been used over the centuries to achieve an advantage over an opponent. The interest in studying such application of multiple diverse instruments grew significantly with the swift annexation of the Crimean Peninsula by the Russian Federation in the Spring of 2014. In that case, Russia used propaganda, disinformation, electronic warfare, local proxies, agents within the Ukrainian security sector and the armed forces and, last but not least, the "polite/little green men" to win decisively an undeclared war without fighting. The combined use of such military and non-military means has been widely designated ever since as "hybrid war" or "hybrid warfare."

However, hybrid instruments are used not only in war or in the period preceding it. Nor do they necessarily include a military component. A set of diverse instruments can, and is, used continuously by both state and non-state actors to influence perceptions and behaviour, create and maintain an environment favouring the achievement of own policy goals and, when possible, to achieve directly desired political outcomes.

Therefore, the author prefers to use the term "hybrid influence," instead of "hybrid war/warfare." The hybrid toolbox includes economic influence, e.g. by ownership in key sectors of the economy, the banking system, sectors of the critical infrastructure and the media, leverage on the politics and state institutions, nurturing patronage networks, corruption, propaganda, disinformation and, possibly, the threat of using military force (but not the actual use of force).

During the Cold war, the combined use of some of these tools had been designated as "active measures" [1]. The end of the Cold war accelerated the processes of globalisation, accompanied with increase of free travel, foreign direct investments and ownership of media and economic assets by foreign nationals, cheaper and ubiquitous communications and access to Internet. While economically, socially and culturally beneficial, these developments increased the opportunities to exercise hybrid influence. In parallel, the importance of the problem has been recognised, along with the need to find ways to address it.

A number of analysis frameworks appeared to examine both the problem of understanding the mechanisms and the dynamics of hybrid influence and its impact, and the problem of synthesis, i.e. how to devise strategies, measures and procedures to counter such influence. This paper is focused on the former.

First, the next section provides selected examples of advanced influence methods and operations, introducing the complexity of the problem of assessing hybrid influence. It is followed by presentation of nine emerging frameworks—some are newly developed, while others have been used for different purposes for approximately two decades and have been subject of interest for the analysis of hybrid influence. The final section summarises the findings, mapping the emerging frameworks on a two-dimensional chart, respectively along the targeted domains and ways of assessing the influence. The paper concludes by summarising the findings, listing major implementation challenges and outlining a way ahead for research and policy-making.

2 Advanced Influence Methods and Operations

The tools and techniques used for hybrid influence are becoming increasingly sophisticated. This section provides just a few illustrative examples and outlines the challenge of identifying and understanding hybrid influence.

In March 2019, Special Counsel Robert S. Mueller published his report on the investigation of Russian interference in the 2016 U.S. presidential elections [2]. In his findings, Mueller outlines a complex Russian campaign consisting of three main strands of activity:

1. a social media influence and infiltration operation led by the Internet Research Agency, a Kremlin's proxy located in Sankt Petersburg, and known as the Troll's factory from Olgino;
2. a cyber hacking operation carried out by GRU, the Russian military intelligence service; and
3. an infiltration operation of the Trump campaign [3].

This campaign was preceded by other operations employing massive use of bots in social networks. According to a 2018 RAND report, Kremlin was the pioneer in using fake social media accounts that are fully or partially automated, alongside accounts operated by humans, and these Russian trolls and bots serve as force multipliers for Kremlin's disinformation operations. Furthermore, the troll types are increasingly diversified, which helps networks evade detection [4]. For example, during a period in the summer of 2014, this troll army reportedly flooded the website of *Guardian* with 40,000 comments a day [4], thus shaping the perceptions of readers in a way favourable to Kremlin.

Another novelty is the use of *deepfakes* (combination of *deep* learning, i.e. a type of artificial intelligence (AI) algorithms, and *fake*). Among the deepfakes are hyperrealistic video or audio recordings, created with the use of AI, of someone appearing to do or say things they actually did not [5]. One of the well known, although not really sophisticated examples is the fake video of Nancy Pelosi, speaker of the U.S. House of Representatives. Deepfakes raise the opportunities to shape perceptions at a new level, and can with time turn into a serious threat to peace and security [6].

These and other, more traditional means of influence, e.g. control over important economic assets in the target country, provide ample opportunities to design and perform influence operations, raising considerably the level of uncertainty for the target. The attacker may use a variety of attack vectors, aim to exploit vulnerabilities in a single domain, or sector, as well as interdependencies between domains and, as a rule, denies his actions and keeps them below a certain threshold, e.g. for triggering

NATO's Article V. Often, the target is unaware or does not recognise the fact that it is subject of hybrid influence.

Therefore, systematic and comprehensive efforts are needed in order to identify attempts at hybrid influence, our vulnerabilities and exposure, and assessment of the impact. The next section presents several frameworks allowing, to an extent, to perform the respective analysis.

3 Emerging Frameworks

The interest in hybrid influence generated a wide spectrum of studies, ranging from utilising information warfare concept and theories of the 1990s [7], through the use of religion to advance geopolitical interests [8], to the role of corruption in hybrid warfare [9]. The reports of some of these studies provide detail on the methods used and implementation processes, and may be examined as emerging analysis frameworks.

This section presents nine such frameworks, starting with an official European document that explains disinformation and prescribes ways to identify disinformation attempts, assess their current and potential impact, and find effective countermeasures.

3.1 Tackling Online Disinformation: A European Approach

In April 2018, the European Commission communicated to the European Parliament, the Council and other key stakeholders its approach and the measures it is taking to deal with the online spread of disinformation [10]. In the preamble of the document the authors emphasise that "new technologies can be used, notably through social media, to disseminate disinformation on a scale and with speed and precision of targeting that is unprecedented, creating personalised information spheres and becoming powerful echo chambers for disinformation campaigns."

Using Internet to spread disinformation, including misleading or outright false information, on a large scale has the potential to:

- hamper the ability of citizens to take informed decisions;
- support radical and extremist ideas and activities;
- impair freedom of expression;
- sow distrust and create societal tensions;
- affect policy-making processes and decisions by skewing public opinion; etc.

The Communication prescribes the following three-step model of studying the process of manipulating social networking technologies to spread disinformation:

1. *Creation* of content—messages consisting mostly of written text, but increasingly including also audio-visual content which may be authentic or false, i.e. to include the so-called deepfakes;
2. *Amplification* through social and other online media, using one or more of the following techniques:

   (a) Algorithm-driven amplification, using the platform's business model to prioritise messages, as well as personalised sharing of sensational content in groups of like-minded users;
   (b) Using the advertising model of the platform to generate clicks and thus further speed up dissemination;
   (c) Using automated services such as bots and fake accounts; this technique may be used on a massive scale by the so-called "troll factories" which, as in the case of Russia's "Internet Research Agency", may be quite sophisticated [11];

3. *Dissemination* by the users, often without verification and consideration of the potential consequences.

The application of this model is expected to help policy makers and practitioners understand better how online disinformation campaigns used by a range of foreign and domestic actors sow distrust and create societal tensions, and thus impact our security.

The Communication then outlines measures to counter online disinformation, but this subject is beyond the scope of the current paper.

3.2 Counter-Disinformation Toolkit RESIST

Also in the month of April 2019, the UK Government launched a toolkit intended to prepare the public sector to counter disinformation [12]. This counter-disinformation toolkit, designated as *RESIST*, is intended to help organisations build their resilience to the threat of disinformation and is divided in six components: **R**ecognise disinformation, **E**arly warning, **S**ituational insight, **I**mpact analysis, **S**trategic communication, and **T**rack outcomes [13]. The first four of these components assist the understanding of an attempt at malign influence and assessment of its impact.

Recognising disinformation is the first step towards its consequent tracking and countering. At this step, analysts try to find the answer of the following questions [13, p. 17]:

- What are the objectives?
- What disinformation techniques are used?
- How these techniques are combined to achieve an impact?

The influencing actors may want to prove their ability, seek an economic gain, undermine the credibility, trust and reputation of an individual or organisation, polarise society by aggravating existing tensions, or undermine national prosperity and security. The toolbox lists a number of respective techniques and possible intended effects to assist analysts in monitoring online sources and identifying attempts at disinformation.

The early warning component provides minimum and recommended standards for monitoring digital media by defining priority policy areas and objectives, key influencers and key audiences, devising an appropriate monitoring toolbox (e.g. platform and usage analytics) and identification of specialised organisational support by UK governmental units, and consequent assessment of potential threats and vulnerabilities.

The knowledge gained in the first two steps needs to be operationalised by achieving situational insight. Towards that purpose, observations are turned into actionable data. That includes definition of a baseline, identifying trends, understanding how disinformation reaches key audiences, and generating hypotheses.

In the impact analysis phase, structured analysis techniques are applied to respond to the following questions [13, p. 35]:

- What is the likely goal of the disinformation?
- What is its likely impact?
- What is the likely reach of the disinformation?
- How to prioritise disinformation monitoring and counter-disinformation tasks?

Thus, the toolbox is expected to provide coherent and consistent elicitation and use of expert opinion for understanding and tackling disinformation across the UK government.

3.3 Information Operations Kill Chain

Another framework assisting the analysis of disinformation campaigns builds on a method, developed in the 1980s to understand the Russian "Operation Infektion." This was an operation to spread the rumour that it is the United States who developed the HIV virus in the search of effective biological weapons [14]. Bruce Schneier updated this "Information Operations Kill Chain" method and adapted it to the Internet era [15]. In his model, the "influencer" performs the following eight steps [15]:

Step 1: *Find the cracks in the fabric of society* exploring disagreements in democratic societies and identifying social, ethnic, demographic, economic and any other existing divisions;

Step 2: *Seed distortion* by creating alternative narratives, contradictory alternative truths that distort the political debate and exploit the propensity of people to share stories that demonstrates their core beliefs without checking whether they are true or not;

Step 3: *Wrap those narratives in kernels of truth* to make exposing untruths and distortions more complicated and exploit the psychologically proven fact that an unprofessional attempt at debunking a fake story amplifies the effect of disinformation;

Step 4: *Build audiences* by establishing direct control of a platform, e.g. RT (formerly, Russia Today), and using social networks to nurture relationships with people receptive to your narratives;

Step 5: *Conceal your hand* by using alternative "original sources" of the story and thus hindering attribution;

Step 6: *Cultivate "useful idiots"* willing to amplify your message and make it even more extreme;

Step 7: *Deny involvement* even when it is obvious;

Step 8: *Play the long game* and try to achieve long-term impact rather than just immediate effects.

Unlike the cybersecurity kill chain [16, 17], in this model steps can be performed in varying order and overlap, while spanning multiple news channels and social media platforms [15], which further defies the ability to understand hybrid influence.

3.4 Study of Russian Propaganda in Eastern Europe

A RAND study on the Russian propaganda in Eastern Europe utilises a set of methods to understand the way in which Russia propagates its influence in Eastern Europe using social networks, identify pro- and anti-Russia players on one of the social networks, Twitter, identify how Russia synchronizes its varied media outlets, and estimate the effects of propaganda, i.e. to what degree Russian-speaking populations in several former Soviet states embrace Russian messages [4]. Then, on that basis, the authors of the report formulate challenges and recommendations for policymakers.

The method of *community lexical analysis* is used to identify a major battle of ideas associated with the conflict in Ukraine. Based on the analysis of Russian language content on Twitter, the study identifies a community of pro-Russia propagandists of approximately 40,000 users and that of anti-Russia pro-Ukraine activists of a similar size [4].

Further, *resonance analysis* has been employed to assess the spread and potential impact of the pro-Russia propaganda. That includes longitudinal studies of Twitter users from Estonia, Latvia, Lithuania, Moldova, Belarus, and Ukraine to create a linguistic "fingerprint" of the pro-Russia propaganda.

More-technical analyses are applied to understand how Russia conducts its media-based influence campaigns, including the use of trolls and complementing automated social media accounts, or bots.

Thus, the RAND study team has applied various analytical tools to understand the aims and themes of Russian information operations and estimate the impact of propaganda.

3.5 Cognitive Threats

Disinformation and propaganda would not succeed if every individual was assessing the received information rationally, setting it in its appropriate context. Obviously, the latter is not the case, and the human mind is vulnerable to exploitation by the so-called "cognitive threats" [18].

In her dissertation, Lora Pitman is framing the analysis of cognitive threats based on "three pillars: (1) psychological studies proving that human cognition is vulnerable; (2) political psychological research that underlines how internal experiences of humans are exploited for political gains; (3) historical overview of cognitive threat showing that different conditions matter for the magnitude and the success of these security threats" [18, p. 7].

One way to describe the cognitive threat is as "any type of information, … in the cyber and physical space, purposefully conveyed to a recipient(s) with the intent to provoke a certain reaction that will cause harm to them and another party" [18, p. 15].

While cognitive vulnerabilities have been exploited throughout history, with the advance of communications and, more generally, the contemporary political, technological and social environment, the cognitive threats are turning into a more significant problem when the influencing actor achieves political (or other) benefits by exploiting psychologically predictable reactions from the victim [18, p. 7].

A number of methods are used to study cognitive threats and their impact [18, p. 11], seeking answers to the following questions:

- How an individual selects, interprets and reacts to certain information?
- How people make inferences and what factors affect them?
- How important is the aggregation of information and the control over that aggregation?
- How social judgments and stereotypes are formed based on the selected information and its interpretation?
- How cognitive processes, emotions, and habits impact inference and the reaction of an individual or a group?

Further, to understand potential consequences, analysts use methods from the fields of political science, international relations, and political psychology, including experimentation, simulations, interviews, content analysis, observer ratings, case studies, etc.

3.6 Hostile Measures Short of War

Another RAND study, published in 2019, adds to propaganda other influence activities of the Russian Federation aiming to achieve its objectives in Europe. The study uses the term "hostile measures" to encompass "a wide range of political, economic, diplomatic, intelligence, and military activities that could be perceived as harmful or hostile" [19].

In addition, these measures can be *directed*, with a specific purpose or goal and, as a rule, employed as part of a plan, or *routine*, conducted by all arms of the Russian state as a matter of course, and fall *short of war* [19, pp. ix–x]. Based on scarce evidence, analysts need to focus on intent, since Russia does not want its sources and methods to be publicly known and regularly denies its involvement.

Inevitably, therefore, the exploratory framework of this study is broader and can be structured around the analysis of:

- *Motives*, stemming from Russia's general goals and objectives;
- *Opportunity*, stemming from the understanding of vulnerabilities of the target country shaped by general political, social, and economic factors; and
- *Means*, i.e. Russia's policies, organizations, and past behaviour in comparable contexts.

Correspondingly, this analysis framework suggests to seek evidence, even circumstantial, on [19, pp. 3–4]:

- Intent and influence by an actor in the Russian state or proxies within Russia;
- Proxies in the target country and how their activity benefits Russia;
- How the activities will contribute to the achievement of Russia's foreign policy objectives.

Applying this framework on variety of sources on several European countries, the RAND analysts conclude that in the pursuit of its objectives across Europe, "Russia appears to be adopting a "soft strategy" in which it seeks to achieve its overall objectives by applying a wide and flexible array of hostile measures across instruments of national power to generate possibilities and shape conditions" [19, pp. 145–146].

3.7 Unvirtuous Cycles

Corruption is increasingly recognised as an important tool in the hybrid toolbox and a "core instrument of national strategy, leveraged to gain specific policy outcomes and to condition the wider political environment in targeted countries" [20]. Treating corruption as a governance challenge or a law enforcement problem is not sufficient anymore, since "weaponized corruption has become an important form of political warfare" [20].

In 2016, the Washington-based Center for Strategic & International Studies and the Centre for the Study of Democracy based in Sofia published a report of a joint study on Russia's economic influence in seven Central and Eastern European countries [21]. The authors used qualitative analysis, as well as quantitative data on the Russian economic influence in these seven countries, in particular in the energy, financial, media, telecommunications, and the infrastructure sectors for the period 2004–2014 and found out that Russia's economic footprint averages 12% of the GDP of the seven countries.

The worrying aspect of this economic influence is that it goes hand-in-hand with opaque patronage networks directly influencing the decisions of local politicians, with corruption as "the lubricant on which this system operates, concentrating on the exploitation of state resources to further Russia's networks of influence" [21, p. x].

The authors framed their analysis using a network-flow model explaining how Kremlin uses intermediaries and local affiliates to penetrate a country economically or politically, and then gain an increasing influence. In an unvirtuous cycle, economic influence is used to establish local partnerships, that raise affiliates to prominence, gaining thus political influence used then to advance Russian interests and projects and expand patronage networks, while corruption is used on all steps of the cycle. In some instances, this cycle leads to a state capture, where Russia builds its "influence over (if not control of) critical state institutions, bodies and the economy and uses this influence to shape national policies and decisions" [21, p. x].

3.8   The "Five Rings" Model

War is one of the most complex human activities, involving practically any societal and economic sector either as a tool to advance own objectives and support the warfighting effort or as a target.

Unless one of the warring parties has an overwhelming advantage in material capacity and morale, e.g. in operation Desert Storm in the Gulf War of 1991, the analysis of the potential effects of an operation during war is complicated and its findings are uncertain for a number of a reasons, including:

- quantitative models can define tactical effects fairly precisely, but the link to consequent strategic outcomes remains rather uncertain;
- the famous Clausewitzian "fog and fiction" of war;
- complex interaction in the opponent's system;
- the will and creativity of the enemy.

Understanding these limitations and still willing to utilise the power of systems analysis, Colonel John Warden, US Air Force, proposed the "Five Rings" model [22]. Warden examines the enemy as a system, consisting of five concentric rings, starting from the centre outwards as follows:

- Leadership;
- Organic essentials, e.g. fuel, electrical power, food;
- Infrastructure;
- Population;
- Fielded military forces.

Each ring then can be represented as a number of sub-systems with a similar "five rings" structure.

Warden suggests that military planners need to analyse the enemy's system in its entirety (and not to overemphasise the opposing armed forces) and to seek, if possible, ways to attack the system or its sub-systems closer to the centre, since in this case both the direct effect and the indirect consequences will be higher. Thus, strategic paralysis of the enemy will be achieved most efficiently. When attacking the centre is not possible, Warden advises to seek other targets, which, when destroyed or degraded, will have the strongest indirect effect on the leadership or other rings closer to the centre.

And while planners need to use available data, when appropriate, their analysis should employ the lessons from history, experience from exercises, and examples from the study of similarly complex systems.

Recently, Nikolic suggested to build the experience accumulated in employing Warden's model to study hybrid warfare [23]. In his analysis, the five rings model provides a good framework for examining systematically numerous modes of hybrid influence and provide tools for identifying attempts of hybrid influence and prioritisation of countermeasures.

3.9 Critical Infrastructure Interdependencies

The final framework, examined in this paper, comes from the field of critical infrastructure protection and, in particular, the work of Steven Rinaldi with focus on dependencies and interdependencies among infrastructures [24, 25]. Rinaldi and co-authors examine how one sector of critical infrastructure depends on another one and provide details on the dependencies of the electric power infrastructure. Then they suggest a structure for describing interdependencies, present a flow model and, for a case study, develop a model explaining first-, second-, ..., and n-th order effects of a disruption of electrical power supply on other critical infrastructures [24].

Interdependencies are structured along the following dimensions:

- Types of Interdependencies—physical, cyber, logical, geographic;
- Infrastructure Environment—economic, legal/regulatory, technical, social/ political, business, public policy, security, health and safety;
- Coupling and Response Behaviour—loose/tight, linear/complex, inflexible, adaptive;
- Infrastructure Characteristics—spatial, temporal, operational, organisational;
- Types of Failures—common cause, cascading, escalating.

In a follow-on study, Rinaldi examines the methods that can be used to model and simulate the propagation of a disruption though critical infrastructures and outlines the application of six groups of methods [25]:

- Forecasting and modelling aggregate supply and demand for infrastructure services;
- Dynamic simulations;

- Agent-based models;
- Physics-based models;
- Population mobility models; and
- Leontief input–output models of economic flows, extended to include nonlineari-
  ties and used to study the spreading of risk among interdependent infrastructures.

The application of methods from these groups requires good understanding of the underlying principles and laws and comprehensive description of the interdependencies, while they may differ in terms of requirements to the available data, resolution, and fidelity.

4   Mapping Analysis Frameworks

The study presented in this paper did not aim to provide comprehensive examination of all existing analysis frameworks, not the least because the subject area—hybrid influence—is not yet well defined and that hinders the application of rigorous methods for identification of relevant frameworks.

Further, some advanced methods and tools to identify hybrid influence, e.g. in applying artificial intelligence and deep learning [26], were not included in the examination, since they usually treat well-defined problems. Even though such problems may be scientifically challenging, at this stage they address a fairly narrow spectre of issues in the analysis of hybrid influence and do not allow, by themselves, to evaluate its impact on the target and the political objectives of the influencer.

Nevertheless, the analysis of the nine frameworks presented above allows to compare approaches and outline a trend in their evolution.

First, more mature frameworks are tailored for analysis of a single domain and implement more specialised methods and techniques. Of highest interest in the last few years has been the use of propaganda, disinformation and fake news targeting individual and group perceptions and behaviour. Not surprisingly, five of the frameworks presented here are in that group. Other frameworks account for interdependencies among two or more domains. Both groups have their merits—the frameworks in the first usually build on a consistent theoretical foundation, while those in the second benefit from a systemic view on hybrid influence and its impact.

Second, frameworks differ in terms of the expected analytical rigour in their implementation. Some rely primarily on social science methods tailored for application by practitioners in public organisations. Others envision meticulous collection and verification of data and evidence, design and tuning of quantitative models and, often, continuous scientific support.

These distinctions between frameworks is visualised in Fig. 1.

Third, it is safe to predict that, with the growing understanding of the phenomenon of hybrid influence and its security implications, and the accumulation of evidence and knowledge, in the coming years we will witness a trend towards multi-domain

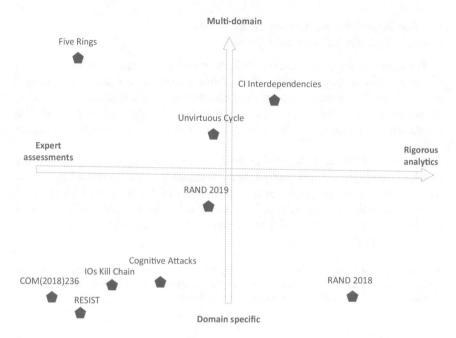

Fig. 1 Mapping frameworks for analysis of hybrid influence

frameworks employing rigorous analytics, and thus more heavily populated upper right quadrant of Fig. 1.

5 Conclusion and Way Ahead

The goal of the study, presented in this paper, was to promote risk-aware decision making on investing in organizational, procedural, technological and educational measures to counter hybrid influence. Towards that goal, there is a clear need for systematic efforts to identify attempts at hybrid influence, our vulnerabilities and exposure, and assessment of the impact.

The paper provided a snapshot on currently existing or emerging frameworks. Two of these come from public organisations, while the remaining seven are proposed by researchers. Five focus on propaganda and disinformation, and the other four account for interdependencies between multiple domains.

The frameworks examined here illustrate the trend towards application of more rigorous analytics. Methods and frameworks for analysis will evolve further towards:

- enhancing and validating domain specific methods;
- studying interdependencies;
- capturing the dependencies among multiple domains;

- collecting and sharing data and evidence; and
- advancing analytical methods and tools.

The further development and the implementation of analytical frameworks faces a number of challenges: need for multidisciplinary competences, scarce data, design of reliable models, even for a single domain, and their verification, and still weak understanding of interdependencies. Of paramount importance, therefore, will be multidisciplinary studies and interagency and international coordination of efforts [27].

As a result of the annexation of Crimea in 2014, the ongoing war in South-Eastern Ukraine and attempts to influence political processes in countries around the world, the focus of the studies presented here is on Russia's activities, with majority of sources on the Russian influence in Europe. Nevertheless, the applicability is much broader and can assist the application of risk-aware allocation of limited resources to counter hybrid influence by any actor.

Acknowledgements This work was supported by the project "Digital Transformation and Security of Complex Adaptive Systems" from the working programme of the Institute of Information and Communication Technologies, Bulgarian Academy of Sciences.

References

1. Abrams, S.: Beyond propaganda: Soviet active measures in Putin's Russia. Connections Q. J.**15**(1), 5–31 (2016). https://doi.org/10.11610/Connections.15.1.01
2. Mueller, R.S.: Report on the Investigation into Russian Interference in the 2016 Presidential Election, vol. I. U.S. Department of Justice, Washington, D.C. (2019). https://www.justice.gov/storage/report.pdf
3. Polyakova, A.: What the Mueller report tells us about Russian influence operations. Order from Chaos, Brookings, 18 Apr 2019. https://www.brookings.edu/blog/order-from-chaos/2019/04/18/what-the-mueller-report-tells-us-about-russian-influence-operations/
4. Helmus, T.C., Bodine-Baron, E., Radin, A., Magnuson, M., Mendelsohn, J., Marcellino, W., Bega, A., Winkelman, Z.: Russian Social Media Influence: Understanding Russian Propaganda in Eastern Europe. RAND Corporation, Santa Monica, CA (2018)
5. Rini, R.: Deepfakes are coming. We can no longer believe what we see. New York Times, 10 June 2019. https://www.nytimes.com/2019/06/10/opinion/deepfake-pelosi-video.html
6. Bressan, S.: Can the EU prevent deepfakes from threatening peace? Judy Dempsy's Strategic Europe, 19 Sept 2019. https://carnegieeurope.eu/strategiceurope/79877
7. Hammond-Errey, M.: Understanding and assessing information influence and foreign interference. J. Inf. Warfare **18**(1), 1–22 (2019)
8. Ivanov, I.: Russia's Orthodox Geopolitics. St. Kilment Ohridksi University Press, Sofia (2019). (in Bulgarian)
9. Allen, D.: A Deadlier Peril': the role of corruption in hybrid warfare. Multinational Capability Development Campaign (2019). https://cids.no/wp-content/uploads/2019/06/20190318-MCDC_CHW_Info_note_7.pdf
10. Tackling Online Disinformation: A European Approach: Communication from the Commission to the European Parliament, the Council, the European Economic and Social Committee and the Committee of the Regions, COM (2018) 236 final, Brussels, 26 Apr 2018. https://eur-lex.europa.eu/legal-content/EN/TXT/?uri=CELEX%3A52018DC0236

11. Linvill, D.L., Warren, P.L.: Troll factories: manufacturing specialized disinformation on Twitter. Polit. Commun. **37**(4), 447–467 (2020). https://doi.org/10.1080/10584609.2020.1718257
12. Owen, J.: Exclusive: government to train public sector comms troops for battle in escalating disinformation war. PR Week, 10 Apr 2019. https://www.prweek.com/article/1581558/exclusive-government-train-public-sector-comms-troops-battle-escalating-disinformation-war
13. Government Communication Service: RESIST: Counter-Disinformation Toolkit, London, Apr 2019. https://gcs.civilservice.gov.uk/wp-content/uploads/2019/03/RESIST_Toolkit.pdf
14. Ellick, A.B., Westbrook, A., Kessel, J.M.: Meet the KGB spies who invented fake news. New York Times, 12 Nov 2018
15. Schneier, B.: Toward an information operations kill chain. Lawfare, 24 Apr 2019. https://www.lawfareblog.com/toward-information-operations-kill-chain
16. Hutchins, E., Clopperty, M., Amin, R.: Intelligence-driven computer network defense informed by analysis of adversary campaigns and intrusion kill chains (2010). https://www.lockheedmartin.com/content/dam/lockheed/data/corporate/documents/LM-White-Paper-Intel-Driven-Defense.pdf
17. Siukonen, V.: Human factors of cyber operations: decision making behind advanced persistence threat operations. In: 18th European Conference on Cyber Warfare and Security, ECCWS 2019, Coimbra, Portugal, 4–5 July 2019, pp. 790–800
18. Pitman, L.: The Trojan Horse in your head: cognitive threats and how to counter them. Ph.D. dissertation, Old Dominion University, Norfolk, VA (2019)
19. Cohen, R.S., Radin, A.: Russia's Hostile Measures in Europe: Understanding the Threat. RAND Corporation, Santa Monica, CA (2019)
20. Zelikow, Ph., Edelman, E., Harrison, K., Gventer, C.W.: The rise of strategic corruption: how states weaponize graft. Foreign Affairs, July/Aug 2020. https://www.foreignaffairs.com/articles/united-states/2020-06-09/rise-strategic-corruption
21. Conley, H.A., Mina, J., Stefanov, R., Vladimirov, M.: The Kremlin Playbook: Understanding Russian Influence in Central and Eastern Europe. Center for Strategic and International Studies and Rowman & Littlefield, Washington, D.C. (2016). https://www.csis.org/analysis/kremlin-playbook
22. Warden, J.A.: The enemy as a system. Airpower J. **9**(1), 40–55 (1995)
23. Nikolic, N.: Connecting conflict concepts: hybrid warfare and warden's rings. Inf. Secur. Int. J. **41**, 21–34. https://doi.org/10.11610/isij.4102
24. Rinaldi, S.M., Peerenboom, J.P., Kelly, T.K.: Identifying, understanding, and analyzing critical infrastructure interdependencies. IEEE Control Syst. Mag. **6**(1), 11–25 (2001)
25. Rinaldi, S.M.: Modeling and simulating critical infrastructures and their interdependencies. In: Proceedings of the 37th Hawaii International Conference on System Sciences (2004). https://doi.org/10.1109/HICSS.2004.1265180
26. Hanot, C., Bontcheva, K., Alaphilippe, A., Gizikis, A.: Automated tackling of disinformation: major challenges ahead. Eur. Parliamentary Res. Serv. (2019). https://doi.org/10.2861/368879
27. Yanakiev, Y.: Promoting interagency and international cooperation in countering hybrid threats. Inf. Secur. Int. J. **39**(1), 7–10. https://doi.org/10.11610/isij.3900

Usage of Decision Support Systems for Modelling of Conflicts During Recognition of Information Operations

Oleh Andriichuk, Vitaliy Tsyganok, Dmitry Lande, Oleg Chertov, and Yaroslava Porplenko

Abstract This chapter describes an application of decision support systems to conflict modelling in information operations recognition. An information operation is considered as a complex weakly structured system. We suggest a model of conflict between two subjects, based on the second-order reflexive model. We describe a method for construction of the design pattern for knowledge bases of decision support systems. We also suggest a methodology allowing to use decision support systems for modelling of conflicts during information operations recognition. The methodology is based on the use of expert assessments and content monitoring data.

Keywords Information operation recognition · Decision support system · Knowledge base · Content-monitoring system · Reflexive model · Conflict modelling

O. Andriichuk (✉) · D. Lande · O. Chertov
National Technical University of Ukraine "Igor Sikorsky Kyiv Polytechnic Institute", Kyiv, Ukraine
e-mail: andriichuk@ipri.kiev.ua

D. Lande
e-mail: dwlande@gmail.com

O. Chertov
e-mail: chertov@i.ua

O. Andriichuk · V. Tsyganok · D. Lande · Y. Porplenko
Institute for Information Recording of National Academy of Sciences of Ukraine, Kyiv, Ukraine
e-mail: tsyganok@ipri.kiev.ua

Y. Porplenko
e-mail: daliss@ukr.net

1   Introduction

Data sources make a significant impact upon the public. During recent years, it has become evident that mass media can be efficiently used for spreading of disinformation [1]. In addition, social experiments show that many people believe unconfirmed news and continue to spread it. For example, in [2] we can find a review of dissemination of deliberately false notions and misinformation in the American society. In [3] the social experiments are described, aimed to study the susceptibility of people to political rumours about the health care reform (approved by the USA Congress in 2010). Depending on the way the information was presented, 17–20% of the experiment participants believed in the rumours, 24–35% of participants did not have a definite opinion, and 45–58% of the respondents rejected them.

Let us define an information operation (IO) [4, 5] as a complex of information events (news posts in Internet and articles in newspapers, news on the radio and TV, comments and "likes" in social networks, forums, etc.), intended to modify public opinion about a certain subject (person, organization, institution, country, etc.). IO can involve a system of information attacks and take a long time. For example, dissemination of rumours about some problems in a bank can make its investors withdraw their deposits, and this, in turn, can lead to bankruptcy. In general, IO, usually, involve disinformation activities.

In [6] it is shown that IO are related to the so-called weakly structured subject domains, since they have some features, specific to these domains, such as uniqueness, impossibility of efficiency function formalization, and, consequently, impossibility of construction of an analytical model, dynamics, incomplete description, presence of the human factor, absence of benchmarks. Such subject domains are formally described using expert decision support systems (DSS) [7].

By a conflict we mean an opposition of several subjects competing for a limited resource in the media. In the case of IO, the conflict is the effect of information on the audience.

2   Approaches to Conflict Modelling

There are several approaches and models in game theory, which are applicable to conflict modeling. Existing mathematical models of a conflict help its participants build their respective optimum strategies. The approaches are related to game theory, suggested by von Neumann and Morgenstern [8]. This theory was developed in many papers in which different aspects of the opposition process were considered [9, 10]. We should also mention Nash equilibrium [11] as a method of resolving non-cooperative games. In addition, in some of the papers published on the subject the opposition process is considered not as a measure of supremacy of one player over another, but rather as a tool to determine the ways of interaction between parties [12]. Shelling was one of the first authors to apply the theory of games to international relations considering the armament race in [12]. In his paper, he considered the long-term conflicts and concluded that establishment of continuous friendly relations between the parties could result in higher profit (even with higher losses, incurred during this period, taken into account), than short-term relations.

In classical works on game theory, the player's profit is determined by a constant predetermined payment matrix, which, in many cases, is rather difficult to obtain. At present, the method of target dynamic estimation of alternatives [13], based on the application of a hierarchic approach, is widely used for decision-making in complex target programs. This method allows us to calculate the effectiveness of each alternative (in the present case—a possible player's turn/move). In most cases it is problematic (or totally impossible) to describe the winning strategy of a player, so the use of a hierarchy of goals becomes a handy tool for describing the players' preferences. It is reasonable to use this model for conflict modelling, taking into account conflict subjects' reflexion, reject the approach based on a certain sequence of steps, and make a transition to a scenario, where each of the subjects performs a complex of actions in dynamics. Reflexive models [14, 15] also allow us to consider the subjectivity of the opposing parties and the presence of compromises on some items.

3   A Reflexive Model of Subjects in a Conflict

The complete model describing the readiness of a subject with reflexion to accomplish some action (model of reflexion subject choice based on the second-order reflexive model [14]) is illustrated by the following function (1):

$$A = (a_3 \rightarrow a_2) \rightarrow a_1 \tag{1}$$

where A is the subject's readiness to make a choice; a_1 is the influence of the environment on the subject; a_2 is the psychological setting of the subject (the influence of the environment expected by the subject); a_3—the subject's intensions.

The equation describing the self-appraisal of the subject in the situation is as follows (2):

$$A_1 = a_3 \rightarrow a_2 \tag{2}$$

where A_1 is the self-appraisal of the subject in the situation.

The self-appraisal of the subject in the situation means the appearance of "self-image" in "self-image", when the subject estimates his own image of the situation, his perception of himself and his intentions.

Let us consider the general model of interaction of two subjects A and B in conflict conditions. The subjects are being influenced by the external environment [14, 15]. Subject A supposes that his counterpart B also possess the reflexion, i.e. has his own notions of the environment effect, his plans and objectives in the situation. In this case subject A (in some manner) is interpreting its own relations with subject B and his ideas about these relations. Then subject A's choice readiness is described by the following function (3):

$$A = (a_3 \ \& \ b_3 \rightarrow a_2) \lor (a_4 \ \& \ b_4 \rightarrow b_2) \rightarrow a_1 \tag{3}$$

where A is the subject A's choice readiness; a_1 is the influence of the environment on both subjects; a_2 is the influence of the environment, expected by subject A; b_2 is the influence of the environment, expected by subject B, from the viewpoint of subject A; a_3 are intentions of subject A; b_3 are intentions of subject B from the viewpoint of subject A; a_4 is the impression of subject A of how subject B imagines the intentions of subject A; b_4 is the impression of subject A of how subject B imagines his own intentions.

The expression which describes the self-appraisal of subject A in the conflict with subject B is as follows (4):

$$A_1 = (a_3 \ \& \ b_3 \rightarrow a_2) \lor (a_4 \ \& \ b_4 \rightarrow b_2) \tag{4}$$

where A_1 is the self-appraisal of subject A in the conflict with subject B.

The subject B choice readiness is described by the following function (5):

$$B = (c_3 \ \& \ d_3 \rightarrow c_2) \lor (c_4 \ \& \ d_4 \rightarrow d_2) \rightarrow a_1 \tag{5}$$

where B is the subject B's choice readiness; a_1 is the influence of the environment on both subjects; c_2 is the influence of the environment, expected by subject B; d_2 is the influence of the environment, expected by subject A, from the viewpoint of B; c_3 are the intensions of subject B; d_3 are the intensions of subject A from the viewpoint of subject B; c_4 is the impression of subject B of how subject A sees the intensions of subject B; d_4 is the impression of subject B of how subject A sees his own intentions.

The expression, which describes the self-appraisal of subject B in the conflict with subject A, is as follows (6):

$$B_1 = (c_3 \ \& \ d_3 \rightarrow c_2) \lor (c_4 \ \& \ d_4 \rightarrow d_2) \tag{6}$$

where B_1 is the self-appraisal of subject B in the conflict with subject A.

Let us show, how the described model of reflexive behaviour of subjects in a conflict can be used for construction of a knowledge base (KB) for a DSS.

4 Reflexive Model of a Conflict in Knowledge Bases of Decision Support Systems

The result of a conflict of subjects will depend on the degree of goal achievement for each subject, participating in the conflict. The winner will be the subject with higher goal achievement degree. Thus, in order to be able to calculate the respective values of goal achievement degrees, we should construct the model of the subject domain of the conflict in DSS KB.

Based on the features of hierarchic decomposition process and of the method of target dynamic estimation of alternatives [13], we can associate certain logical operations with DSS KB objects and their relations. By means of a DSS, in which the above-mentioned method is implemented, it is possible to model logical "or" ($\vee$) as sub-goals of the same goal, logical negation as negative influence of the respective goal, and "XOR" as groups of goal compatibility.

Having analysed the above-mentioned reflexive model of the conflict between two subjects, we can suggest the following design pattern for DSS KB (Fig. 1). Black solid arrows indicate positive influence of goals, while the dashed red arrows indicate negative influence. Titles of the goals correspond to designations in the above-listed equations.

During construction of this design pattern for DSS KB (Fig. 1) the above-mentioned features of modelling of logical operations have been taken into account. In order to associate logical equations with objects and relations in DSS KB, we had

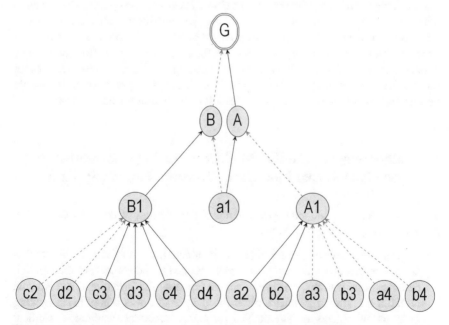

Fig. 1 Decision support system's knowledge base design pattern for a conflict between two subjects

to transform the functions describing the choice readiness of subjects A and B in the following way (see (7) and (8)):

$$A = \neg A_1 \vee a_1 \tag{7}$$

$$B = \neg B_1 \vee a_1 \tag{8}$$

Respectively, the functions describing the self-appraisals of the subjects in the conflict situation were transformed as follows in (9) and (10):

$$A_1 = a_2 \vee b_2 \vee \neg a_3 \vee \neg b_3 \vee \neg a_4 \vee \neg b_4 \tag{9}$$

$$B_1 = c_2 \vee d_2 \vee \neg c_3 \vee \neg d_3 \vee \neg c_4 \vee \neg d_4 \tag{10}$$

As the winner will be the subject with higher value of the goal achievement degree, let us assume that the main goal of the hierarchy G (Fig. 1) is affected positively by the goal of subject A, and negatively by the goal of subject B. Then if the achievement degree of the main goal G is above zero, then the subject A wins in the conflict, while if this value is less than zero, then subject B is the winner (and in the case of zero value there is a tie). The DSS also allows us to calculate the goal achievement degrees for subject A and subject B separately.

The DSS KB design pattern, obtained in such a way, should be complemented to the full-fledged KB using the methods of DSS KB construction during identification of information operations [16, 17]. Extension of KB with the use of expert knowledge only (as described in [17]) requires the group of experts. The work of experts is rather expensive and time-consuming. Besides that, for small expert groups the competency of experts should be taken into account [18], leading to additional time expenses for the search and processing of extra information. Therefore, expert information should be used moderately in the process of DSS KB construction for IO recognition.

5   Methodology of Conflict Modelling During Information Operations Recognition Using Decision Support Systems

Thus, the essence of the suggested method for modelling of conflicts using DSS during IO recognition is as follows:

1. Preliminary study of the IO object is conducted, the subjects of the conflict are determined, together with their goals, related to the subjects of the conflict, persons, organizations, companies. In the process of informational and analytic research a common problem, often faced by analysts is compiling a ranking of a set of objects or alternatives (products, electoral candidates, political parties etc.) according to some criteria. For this task we propose to apply

special approaches for considering importance of information sources during aggregation of alternative rankings [19].

2. The respective design pattern for DSS KB is selected and modified, if necessary. During modification of the design, one should take into account the features of modelling of logical operations in DSS KB.
3. The selected design pattern for DSS KB is extended to a full-scale KB. The group expert examination for determining and decomposition of IO goals is performed. Thus, the decomposition of IO as a complex weakly structured system is taking place. For this purpose, the system for distributed acquisition and processing of expert information (SDAPEI) is used.
4. The respective KB is complemented (extended) using DSS tools, and taking into account the results of the group expert examination, carried out by means of SDAPEI, as well as available objective information. For clarification of queries to content monitoring systems (CMS) and for complementation of DSS KB with missing objects and links, the keyword network of the subject domain [16] of respective IO is used.
5. Using CMS tools, the analysis of dynamics of the thematic data stream is performed. DSS KB is complemented with partial influence coefficients.
6. Using DSS tools, based on the constructed KB, the recommendations are calculated.

Recommendations (in the form of dynamic estimates of efficiency of topics related to the IO object and values of the goals achievement degrees of subjects of the conflict), obtained using the above-mentioned technique, are used for estimation of the IO-related damage [17], for organization of the information counter-actions (with impacts of information sources taken into account), and for prediction of the conflict's outcome.

The suggested method of modelling of opposition between two subjects can be used for IO recognition during modelling of confrontation of lobbyists and their opponents, for example, in such events as Brexit [20].

6   Conclusions

In the paper, the advantages of DSS usage for modelling of conflicts during information operation recognition are substantiated. An information operation is considered as a complex weakly structured system.

A model of a conflict between two subjects, based on the second-order reflexive model, is presented.

A method for construction of DSS knowledge bases based on the model of the conflict between two subjects, is described.

A method for application of DSS to conflict modeling during information operation recognition is suggested. The range of applicability of this method is described.

Acknowledgements This study is funded by the NATO SPS Project CyRADARS (Cyber Rapid Analysis for Defence Awareness of Real-time Situation), Project SPS G5286.

References

1. Chertov, O., Rudnyk, T., Palchenko, O.: Search of phony accounts on Facebook: Ukrainian case. In: International Conference on Military Communications and Information Systems, ICMCIS, 22–23 May 2018
2. Lewandowsky, S., Ecker, U.K.H., Seifert, C.M., Schwarz, N., Cook, J.: Misinformation and its correction continued influence and successful debiasing. Psychol. Sci. Public Interest **13**(3), 106–131 (2012). (Department of Human Development, Cornell University, USA)
3. Berinsky, A.J.: Rumors and health care reform: experiments in political misinformation. Br. J. Polit. Sci. **47**(2), 241–262 (2017)
4. Information operations roadmap. DoD US. GPO, Washington, D.C. (2003)
5. Military Information Support Operations. Joint Publication 3-13.2. 07 Jan 2010, Incorporating Change 1 20 Dec 2011, p. 10. https://fas.org/irp/doddir/dod/jp3-13-2.pdf
6. Kadenko, S.V.: Prospects and potential of expert decision-making support techniques implementation in information security area. In: CEUR Workshop Proceedings (ceur-ws.org), vol. 1813 urn:nbn:de:0074-1813-0. Selected Papers of the XVI International Scientific and Practical Conference "Information Technologies and Security" (ITS 2016) Kyiv, Ukraine, 1 Dec 2016, pp. 8–14. https://ceur-ws.org/Vol-1813/paper2.pdf
7. Lee, D.T.: Expert decision-support systems for decision-making. J. Inf. Technol. **3**(2), 85–94 (1988)
8. von Neumann, J., Morgenstern, O.: Theory of Games and Economic Behavior: 60th Anniversary Commemorative Edition, Princeton Classic edn, 776 p. Princeton University Press (2007)
9. Takehiro, I.: Relational Nash equilibrium and interrelationships among relational and rational equilibrium concepts. Appl. Math. Comput. **199**(2), 704–715 (2008)
10. Colin, R.: Non-linear strategies in a linear quadratic differential game. J Econ. Dyn. Control **31**(10), 3179–3202 (2007)
11. Martin, J.: Osborne, Ariel Rubinstein, "A Course in Game Theory," p. 14. MIT, Cambridge, MA (1994)
12. Schelling, T.: The Strategy of Conflict. Harvard University Press, Cambridge, MA, (1960, 1963) 1980, 309 p
13. Totsenko, V.G.: One approach to the decision making support in R&D planning. Part 2. The method of goal dynamic estimating of alternatives. J. Autom. Inf. Sci. **33**(4), 82–90 (2001)
14. Lefebvre, V.A.: Algebra of Conscience, 358 p Springer, The Netherlands (2001)
15. Novikov, D., Korepanov, V., Chkhartishvili, A.: Reflexion in mathematical models of decision-making. Int. J. Parallel Emergent Distrib. Syst. **33**(3), 319–335 (2018)
16. Lande, D.V., Andriichuk, O.V., Hraivoronska, A.M., Guliakina, N.A.: Application of decision-making support, nonlinear dynamics, and computational linguistics methods during detection of information operations. In: CEUR Workshop Proceedings (ceur-ws.org), vol. 2067 urn:nbn:de:0074-2067-8. Selected Papers of the XVII International Scientific and Practical Conference on Information Technologies and Security (ITS 2017), Kyiv, Ukraine, 30 Nov 2017, pp. 76–85
17. Andriichuk, O., Lande, D., Hraivoronska, A.: Usage of decision support systems in information operations recognition. Recent developments in data science and intelligent analysis of information/advances in intelligent systems and computing (ISSN: 2194-5357). In: Proceedings of the XVIII International Conference on Data Science and Intelligent Analysis of Information, 4–7 June 2018, Kyiv, Ukraine. Springer Nature, Berlin, vol. 836, pp. 227–237 (2018)

18. Tsyganok, V.V., Kadenko, S.V., Andriichuk, O.V.: Simulation of expert judgements for testing the methods of information processing in decision-making support systems. J. Autom. Inf. Sci. **43**(12), 21–32 (2011)
19. Tsyganok, V.V., Kadenko, S.V., Andriichuk, O.V.: Considering importance of information sources during aggregation of alternative rankings. In: CEUR Workshop Proceedings (ceur-ws.org), vol. 2067 urn:nbn:de:0074-2067-8. Selected Papers of the XVII International Scientific and Practical Conference on Information Technologies and Security (ITS 2017), Kyiv, Ukraine, 30 Nov 2017, pp. 132–141
20. Bachmann, V., Sidaway, J.D.: Brexit geopolitics. Geoforum **77**, 47–50 (2016)

A System for Determining the Overall Estimate of the Level of Tension in Society

Maksym Shchoholiev and Violeta Tretynyk

Abstract This chapter presents a system developed for state structures responsible for ensuring security and stability in the society. The main task of the system is to determine the impact of information stored on various Internet resources on society. The system determines the level of emotionality of comments on certain posts and also the overall average tension provoked by news. The methods TF-IDF and word2vec are used for keywords determination in the text and their transformation into a numerical form. The level of emotionality is defined by an artificial neural network. The overall estimate is obtained by calculating the geometric mean of the coefficients indicating how many times the average number of comments with a negative emotional connotation exceeds the average number of comments for each of the valuations of emotionality of these comments.

Keywords Level of social tension · Keyword · TF-IDF · word2vec representation · Context · Data aggregation · Social networks

1 Introduction

Today the world is going through a phase of aggravation of information wars. Due to the high level of accessibility of information, millions of people, regardless of their moral and socio-political views, are the target of information attacks. The basis for a successful information attack is conflicts and divergences among people and authorities or among certain groups of people. The purpose of many attacks is to increase tensions between these groups. It is especially advisable to single out information attacks aimed at cultivating a sense of threat among people. These threats include

M. Shchoholiev (✉) · V. Tretynyk
National Technical University of Ukraine "Igor Sikorsky Kyiv Polytechnic Institute", Kyiv, Ukraine
e-mail: shchoholiev.maksym@gmail.com

V. Tretynyk
e-mail: viola.tret@gmail.com

threats to life of an individual or lives of family members, property loss threats, and the so on.

The system is designed to detect real-time information attacks targeting Ukraine, in particular by Russia, and to determine the impact of these attacks on Ukrainian society.

The purpose of work is to develop a system for the dynamic determination of the level of social, informational, political or other tension in the society according to the data from social networks. The principle of work of the system is to determine the tension provoked by publication, using comments to it, with the subsequent scaling of this tension across all readers of the information resource under study.

The first section of the article presents the architecture of the proposed system. The basic structural blocks, their mathematical interpretation and interconnections are described. The second section describes software implementation, the structure of training and test data, and test results. The final section presents conclusions and plans for future research.

2 System Architecture

The schematic representation of the developed system is shown in Fig. 1.

Evaluation of the Emotionality of Comments on News by Keywords

The level of tension is determined on the basis of analysis of comments on news in social networks. The emotionality of comments is determined by their keywords. The authors of the article conducted studies that prove that keywords most often contain emotional vocabulary.

In order to use word groups to classify text corpuses, the words are presented as numeric vectors. This representation is created using the word2vec method with the CBOW (continuous bag-of-words) learning algorithm [1]. In this way a dictionary

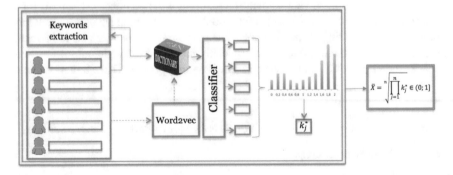

Fig. 1 The scheme of the system of operative determination of the level of tension in society

is formed consisting of ordered pairs $(w_s, l_s) \in (\mathbb{N}^n \times \mathbb{R}^m)$; $s = \overline{1, \ldots, q}$; q—current number of words in the dictionary; $w_s \in \mathbb{N}^n$—words in Unicode encoding; $l_s \in \mathbb{R}^m$—numerical presentation of these words in context, having a fixed length.

For keywords extraction the TF-IDF (term frequency-inverse document frequency) method is used [2]. From each comment on news M keywords are extracted which theoretically are the most important in the context of this comment. These M keywords $\widetilde{w}_j \in \widetilde{W} \subseteq \mathbb{N}^n$; $j = \overline{1, \ldots, M}$ get into the dictionary. There each word is assigned a numeric vector $\widetilde{l}_j \in \mathbb{R}^m$; $j = \overline{1, \ldots, M}$, using Word Embedding technique. Therefore, each comment is represented as a vector $\tilde{w} = (w_{1k}, w_{2k}, \ldots, w_{Mk})$; k—sequence number of the explored comment. At the output of the dictionary a vector $\tilde{l} = (l_{1k}, l_{2k}, \ldots, l_{Mk})$ is obtained.

This vector is submitted to the input of the classifier, which evaluates the degree of emotionality of the comment associated with this vector. As a classifier, the method of artificial neural networks is used. The classifier can be presented as a function $F : \mathbb{R}^n \to \mathbb{R}$:

$$F\left(\tilde{l}\right) = r_k, \tag{1}$$

$r_k \in \mathbb{R}$—valuation of comment k; $r_k \in [0, 2]$, where the most emotionally positive comments have value 0, the most emotionally negative have value 2.

It is known that comments can have different nesting levels. For example, comments at level 0 are reviews for news or opinions expressed by readers about the described events. Comments at level 1 are comments on the comments of the previous level, etc. Obviously, the valuations of the comments of higher levels should have a greater impact on the overall value of the level of tension. Thus, it will be considered that with each subsequent level of nesting the valuation of the comment is reduced by half. Then the estimates of emotionality of the comments r_k^*, taking into account the levels of their nesting, can be calculated by the formula:

$$r_k^* = (r_k - 1)\left(\frac{1}{2^n}\right)_k + 1, \tag{2}$$

where n is the level of nesting.

The received estimates of emotionality of the comments can be presented in the form of a bar graph, each column of which shows the number of comments with a certain level of emotionality to the publication under consideration.

Data Aggregation

Different news agencies cover the same news in different ways. More popular and reputable agencies often cover events on their own. The level of emotionality of their articles is largely determined by the policy of the agency itself. Some news may be published with links to other sources. However, generally the interpretation of the same news on the pages of different news resources is different.

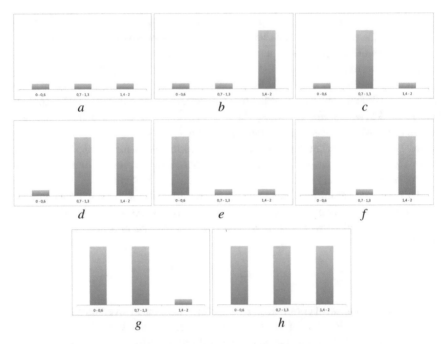

Fig. 2 Simplified schemes of comments distribution by emotionality

To get an objective picture of people's reaction to a particular news, comments from the pages of different news resources need to be explored. The emotional distribution of the comments will be different for each of them. To show all variants of this distribution schematically, it is advisable to group the comments with the emotionality estimates 0–0.6, 0.7–1.3 and 1.4–2. Then any distribution can be attributed to one of the 8 schemes shown in Fig. 2.

After analysing the distribution of comments by their emotionality, only those publications that have a significant number of comments with negative emotional connotation will be considered, since they indicate that the level of tension in some groups is increasing. Schemes b, d, f and h in Fig. 2 have such type of distribution. All other schemes are either non-informative (a) or neutral/positive emotional connotation prevails in them (c, e, g).

It is a well known fact that the number of comments on different news is generally different. Therefore, it is inappropriate to compare absolute values. So, it was decided to compare the ratios between the average number of comments with negative emotional connotation which have estimates 1.4–2 and the total average number of comments for each estimate.

Table 1 Coefficient distribution for the comments with emotion estimates 1.4–2

Estimates of comments emotionality	1.4	1.5	1.6	1.7	1.8	1.9	2
Weighting coefficients	0.7	0.8	0.9	1	1.1	1.2	1.3

The average number of comments with estimates 1.4–2 will be calculated as a weighted arithmetic mean:

$$x_1 = \frac{\sum_{i=1}^{n_1} N_i w_i}{n_1},$$ (3)

$N_i \in \mathbb{N}$—number of comments with a certain estimate of emotionality, $n_1 = 7$—number of possible estimates of emotionality in the interval 1.4–2, w_i—weights of estimates. To reduce the deviation of the weighted arithmetic mean from the usual arithmetic mean, it was decided to set the weighting coefficient 1 to the central estimate from the interval 1.4–2. Let the coefficients of the emotion estimates, located adjacent, differ by 0.1. Then the coefficient distribution shown in Table 1 is obtained.

The overall average number of comments will be calculated as the usual arithmetic mean:

$$x_2 = \frac{\sum_{i=1}^{n_2} N_i}{n_2},$$ (4)

$N_i \in \mathbb{N}$—number of comments with a certain estimate of emotionality, $n_2 = 21$—overall number of possible estimates of emotionality.

It is then necessary to calculate a coefficient showing the ratio between the total average number of comments and the average number of comments with negative emotional connotation:

$$k = \frac{x_2}{x_1}.$$ (5)

If $k > 1$, it means that the number of comments on a particular publication with a neutral and/or positive emotional connotation is greater than the number of comments with a negative emotional connotation. Such publications will not be considered. The publication estimate will be taken into account in the overall estimate of tension only if $k \leq 1$ for this publication.

To determine the level of social tensions, one must first scale the tensions provoked by each publication individually across all subscribers of the information resource under study, and then calculate the mean value of tension, taking into account data from all information resources. Suppose that publications from pages of different news sources are independent. At the time of the study, only the articles that have been published within a short period of time, such as 24 or 12 h, will be considered. This will give an opportunity to determine the current level of tension and speed up the system.

To take into account the popularity of a news resource, the value of k^* is calculated for each k by the formula:

$$k_j^* = 1 - a_j(1 - k_j),\qquad(6)$$

a_j—number of subscribers, normalized by all news resources; $a_j \in [0, 1]$; $j = [1, n]$, n—number of publications under study. The graph of values k_j^* for different a_j and k_j is shown in Fig. 3.

It can be seen that with increasing a_j and decreasing k_j the value of k_j^* decreases. It should also be noted that the number of publications may be greater than the number of news resources, that is, a variant is allowed, in which more than one article about the same news is published on one news resource.

Now the geometric mean formula can be applied to calculate the overall estimate of the level of tension:

$$\overline{X} = \sqrt[n]{\prod_{j=1}^{n} k_j^*},\qquad(7)$$

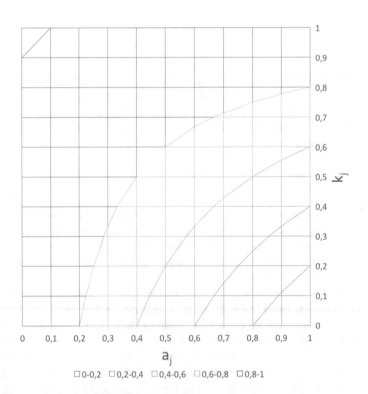

$\square$0-0,2 $\square$0,2-0,4 $\square$0,4-0,6 $\square$0,6-0,8 $\square$0,8-1

Fig. 3 The graph of values k_j^* for different a_j and k_j

$\overline{X} \in [0, 1]$.

3 Program Realization

The program realization was developed using MATLAB programming language and concerns the subsystem for evaluating emotionality of comments on news. The neural network was created with the help of the framework Neural Network Toolbox.

For quality check of the program 783 comments were collected from the social network Facebook pages of the 8 most popular news resources in Ukraine. On each of the resources the reaction of people on one news was explored. Comments to each of the news were saved in text files. The structure of all files is the same. Each row keeps data about one comment and consists of 4 elements, separated by a "|" character. The first element is the order number of the comment within one news resource. The second element is the level of comment's nesting (0 or 1). The third element is the level of emotionality of the comment (on the interval [0, 2]). The fourth element is the text of the comment. The level of emotionality of each comment was determined manually by the experts.

All comments were divided into two groups: 683 comments were prepared for training the classifier, 100 comments—for testing. Accuracy was determined by the mean-square deviations of the results obtained at the output of the neural network from the actual results. For the word2vec method the window with the length 5 was chosen. It's the longest possible window because there are comments which have only 5 words. The length of the vectors which correspond to each word in the dictionary and the optimal number of neurons on the hidden layer of neural network were chosen experimentally. The feedforward neural network with one hidden layer was used. There were conducted numerous experiments for different combinations of numbers of elements in the word vectors and numbers of neurons on the hidden layer. For each combination 10 experiments were performed and the arithmetic mean of their results was counted. The results of these experiments are presented in Fig. 4.

Thus, we have concluded that the increasing number of neurons and the length of word vectors lead to decreasing of accuracy of the results.

4 Conclusions and Future Work

The architecture and the mathematical model of the system of operative determination of the level of tension in the society were described in detail. Also, the main schemes, graphs and formulas that allow to understand the principles of the system work were presented. Besides, the brief description of program realization and its results were presented.

Future work will focus primarily on several areas.

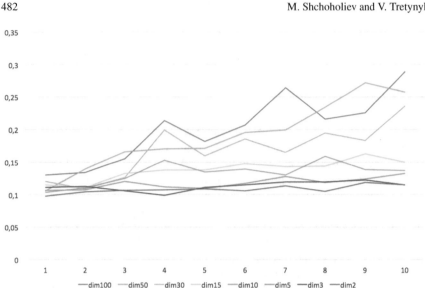

Fig. 4 Graphs of the dependencies of mean-squared deviations from the number of neurons on the hidden layer of neural network for the word vectors of different length

First, research will be conducted to develop methods for determining an author's value position. He may support, condemn, or express a neutral attitude to an event described in the news. The emotional connotation of the comments will accordingly depend on the position of the author of the post. In this way it will be possible to determine not only the fact of the existence or absence of tension as such, but also the attitude of people to the event described in the article.

Secondly, subsystems that will take into account other ancillary characteristics such as the number of reposts, views, likes, etc. in the overall estimate of tension will be developed. Also the subsystem for searching bots and fake pages will be applied. If suspicious authors are found among commentators, their posts will not be evaluated and taken into account when calculating the overall tension.

Third, the system and its software implementation will be improved. Tests will be conducted on the system with more comments. The work will also be continued to improve the quality of the system by more careful parameter selection.

Fourth, a data visualization system will be developed. It will display a map of Ukraine where each region is painted in colour on a continuous palette from green to red. The green colour will show a low level of tension, and red will show a high level of tension. This system will also determine what kind of news is affecting the level of tension in the region and to what extent.

Acknowledgements This study is funded by the NATO SPS Project CyRADARS (Cyber Rapid Analysis for Defense Awareness of Real-time Situation), Project SPS G5286.

References

1. Mikolov, T., Le, Q.V., Sutskever, I.: Exploiting similarities among languages for machine translation. In: Computation and Language, 10 p (2013)
2. Manning, D.C., Raghavan, P., Schütze, H.: An Introduction to Information Retrieval, 544 p (2009)

Main Avenues for Hybrid Influence Over the National Security System

Todor Tagarev

Abstract The main purpose of the national security system is to reveal, prevent and protect against attempts to undermine the national sovereignty and independence. However, in itself it is not immune to hybrid influence. This chapter suggests a framework, representing a simplified version of John Warden's 'Five Rings Model,' for arranging and analysing publicly available information on attempts at degrading the national security sector. It then describes the respective venues of influence and presents a number of cases, likely related to the interests of the Russian Federation to undermine Bulgaria's national security. The suggested framework, along with innovative models of cooperation between public authorities, academia and private actors, can be used to collect and verify information on influence attempts, to analyse their actual or potential impact, and find effective remedy.

Keywords Hybrid threats · Security sector · Framework · Strategic paralysis · State capture · Bulgaria · Russian Federation

1 Introduction

Throughout the ages, states and non-state actors have used variety of ways and means to influence a potential opponent and gain advantages, that could complement their military and political leverage when necessary. During the Cold war, for example, the Soviet security apparatus pursued a rather comprehensive strategy abroad, involving "a broad range of influence activities, including: covert media placement, forgery, agents of influence, 'friendship' societies, front organizations," etc., designated as "active measures" [1]. Since then, and while contributing to trade, welfare and cultural exchange, the processes of globalization, free movement of

T. Tagarev (✉)
Institute of Information and Communication Technologies, Bulgarian Academy of Sciences, "Acad. G. Bonchev" Str., Bl. 2, Sofia 1113, Bulgaria
e-mail: tagarev@bas.bg

people and capital, and the advancements in information and communications technologies and internet in particular, increasing connectivity and the speed of disseminating information, provided new venues of hybrid influence. These opportunities have been readily used by assertive actors, including non-state and hybrid actors, as well as states.

Since the 2006 Lebanon War, also called the 2006 Israel-Hezbollah War, the use of military and non-military means in a conflict has been designated as 'hybrid warfare' [2]. The interest in the concept grew significantly after the illegal annexation of the Crimean Peninsula in the Spring of 2014 and the follow-up support by the Russian Federation to separatist forces in South-Eastern Ukraine [3, 4].

These early works focused on the use of combinations of military and non-military tools and tactics to achieve political and military objectives. Soon, research and analytical studies went beyond the examination of conflict per se and addressed the variety of ways through which a state actor attempts to achieve its political objectives directly, i.e. without use of military force, or to gain a leverage allowing it to weaken the will and/or the capacity of a potential opponent to resist coercive actions. An example of the first case is the interference in democratic elections or direct organization of a coup [5, 6], designated in some cases as 'political warfare' [7]. The studies on the latter grew exponentially in recent years. Among the examples are the analyses of Russia's economic and political influence in selected countries of Eastern [8] and Western [9] Europe, attempts to use ethnicity concepts, such as 'pan-Slavism' [10], propaganda, disinformation [11] and, more generally, cognitive attacks [12], the use of cyber space [13], etc.

Bulgaria—a member of NATO for the past 15 years—is often seen as one of the most vulnerable targets of Russian influence operations. According the influential 'Kremlin Playbook' study, in Bulgaria "Russia's economic presence averaged over 22% of the GDP between 2005 and 2014 [and] there are clear signs of both political and economic capture, suggesting that the country is at high risk of Russian influenced state capture" [8]. Kremlin is using Russia's traditional influence in Bulgaria, "enhancing it with hybrid means to place the cohesion and future" of NATO and the European Union under question, to shape the public debate and de facto dictate what defence capabilities and equipment Bulgaria may have [14]. In a recent study, Naydenov enumerated ways in which the Russian Federation manages to interfere into the development of Bulgaria's defence system [15]. There are indications, that Russia's influence goes beyond the issues of defence modernization and NATO membership and encompasses other security institutions, as the former Prime Minister of Bulgaria, Ivan Kostov, points out [16].

This chapter makes probably the first attempt to examine systematically the foreign, in this case—exclusively Russian, influence over Bulgaria's system for national security, i.e. on the deterrence and defence capacity of the country and the 'immune system' of the state. The next section provides a brief examination of methodological issues and suggests a framework—a simplified model of John Warden's 'rings model'—for collecting information on and analysing the influence over the security sector. Section 3 elaborates on the main avenues for hybrid influence over the national security system and provides examples. The paper concludes by a call for systematic collection of information on hybrid influence over the security sector and the search for innovative models for cooperation in this regard among public authorities, the academic sector and civil society.

2   A Framework for Assessing Hybrid Influence Over the National Security System

Focusing efforts on a particular point of the enemy's system is a time-honoured principle of military art. These 'points' are selected so that when they become dysfunctional as a result of a military attack, or 'influence' more generally, one can achieve his or her own objectives. Karl von Clausewitz designated these points as 'centres of gravity' [17]. The concept continues to enjoy considerable interest. A recent analysis underlines that the Clausewitzian centres of gravity are not characteristics, capabilities, or locations; rather, they are "dynamic and powerful physical and moral agents of action or influence with certain qualities and capabilities" [18].

Strategists and operational planners need to understand the relations between the effects one can exercise, the centres of gravity and the opponent's vulnerabilities. As a rule, the information on vulnerabilities, centres of gravity, and effects is incomplete, even scarce. Their relations are often nonlinear, thus making difficult to make 'what if' predictions in a rigorous manner. Therefore, while efforts to systematically study the opponent's system are likely underway, simpler, empirically-based frameworks are subject of interest by both academics and practitioners.

One of these frameworks is the 'Five Rings Model,' elaborated by the US Air Force Colonel John Warden at the end of the 1980s [19] and allegedly used in the planning process for the 'Desert Storm' air campaign in the beginning of 1991—part of the Gulf War, 2 August 1990—28 February 1991. The five concentric rings, from the centre outwards, are:

- Leadership;
- Organic or System Essentials and key production lifelines;
- Infrastructure;
- Population;
- Fielded Military Forces.

Each of the rings, in turn, has nested 'five rings' models. Warden considers the relations among the rings and affirms that, as a rule, attacks closer to the centre of the model have higher utility for paralysing the enemy and achieving own political and strategic objectives. Hence, Warden's underlying concept is that one needs to select targets focusing his efforts to achieve *strategic paralysis* of the enemy.

Nikolic suggested to use the 'Five Rings Model' in the study of hybrid warfare [20]. This idea is developed further in this paper, with a few modifications. Four rings are used (see Fig. 1) to analyse actual cases or the potential of hybrid influence over the security sector:

Fig. 1 A framework for
analysing hybrid influence
over the security sector

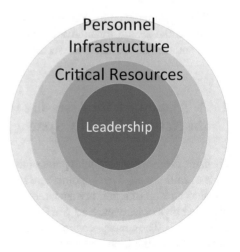

1. *Leadership*—the people in senior positions in the security sector, i.e. director or
   general/admiral in the armed forces, as well as people in positions exercising
   control and oversight over the security sector;
2. *Critical resources*—the provision of knowledge, energy, spare parts, ammunition,
   etc. that are critical for the operation and maintenance of the security sector;
3. *Infrastructure*—the security and defence technological and industrial base developing new technologies, know-how, weapon systems, equipment, ammunition;
   training ranges, etc.; and
4. *Personnel*—the military and civil servants employed by the security sector.

The second key methodological consideration relates to the scarcity of data, in
particular of proven and publicly accessible data and information on the actual influence of foreign interests in the national security system. None of the suspected cases
of influence have stood the full procedure of the Bulgarian justice system, while the
public annual reports on national security, defence and armed forces, and of individual security agencies are notoriously silent on Russia's interferences in Bulgaria's
security affairs.

The next section, therefore, provides examples of *possible* or *likely* hybrid influence in the four rings of the model, presented here, based on inconclusive information
from public sources. The actual understanding of the nature and impact of such influence would require innovative approaches [21, 22] for systematic and meticulous
collection and verification of information of suspected influence over the security
sector.

3   Main Avenues for Hybrid Influence Over the National Security System

3.1   Leadership

Political and media influence in the country can be translated into influence over the security sector, even through the highest political positions. For example, the political analyst Daniel Smilov states that the Bulgarian President has consistently taken positions in line with Kremlin's interests, whereas Bulgaria has a parliamentary form of governance, but the President is Supreme Commander of the Armed Forces with important influence over the country's foreign and security policy [23].

In 2014, for example, the Bulgarian Ministry of Defence published the document "Vision: Bulgaria in NATO and in European Defence 2020" clearly pointing to the Russian 'hybrid warfare model' as a direct threat to the security of the country. According to the then minister of defence, Dr. Velizar Shalamanov, the text of the document has been agreed by all relevant ministries and agencies.

After pressure from the highest political levels, the document has been edited to avoid the reference to Russia as posing a direct threat to Bulgaria's national security. The version, adopted by the Council of Ministers, states:

> Heterogeneous regional and geostrategic interests suggest a continuation of the trend of conflict and confrontation in the Black Sea and Caucasus regions. The illegal annexation of Crimea by the Russian Federation and the conflict in Eastern Ukraine has become one of the most serious threats to peace and security in Europe since the Second World War. This has led to a serious change in the balance of power in the Black Sea region and has direct implications for Bulgaria's security [24].

The change of the position was thoroughly dissected by analysts and the media (e.g. [25]).

Furthermore, officially Bulgaria supports positions and declaration of NATO and the European Union on the illegal annexation of the Crimean Peninsula, the introduction of sanctions against the Russian Federation, conducting military exercises on NATO's Eastern flank and in the Black Sea, etc. However, often senior politicians make public statements in the sense that it is in the best interest of Bulgaria to have "yachts, tourists and pipelines" in the Black Sea, and not a military build-up (notwithstanding Russia's determined investments in naval, anti-access and areal denial (AA/AD) and medium-range strike capabilities in the Black Sea and on the Crimean Peninsula).

Practically the same was the behaviour in the Skripal case, when Bulgaria supported the declaration of the European Council on the Salisbury attack while, in a "balanced and moderate" position, decided not to expulse Russian diplomats until "fully convincing evidence" for the involvement of the Russian Federation is provided [26].

These statements and actions at highest political level create the environment, in which experts in the security sector perform their duties. For example, and possibly as a result of self-censorship, the annual reports on the status of national security, defence and armed forces, and the State Agency "National Security"—public documents sent for approval to Parliament—avoid references to assertive activities of the Russian federation.

Another venue for influencing the security sector is through the former communist security services, that at the time were practically subordinated to and to a significant extent controlled by the Soviet security services—KGB and GRU (the Main Intelligence Directorate of the ministry of defence). At the end of 2006, the People's Assembly (the Bulgarian Parliament) adopted a *Law ... on announcing the association of Bulgarian citizens to the state security and intelligence services of the Bulgarian People's Army.* When the respective commission started announcing the names of people and their positions, it became clear that a significant number of senior military officers, civil servants and political appointees to the security sector have been associated with the coercive apparatus of the former regime.

The scale of this phenomenon is so massive, that it manages to perpetuate itself through the promotion and appointment of likeminded individuals, both military and civilian. Where the actual loyalties of these individuals lie, on occasion becomes visible once they leave active service. There are several publicly known cases when retired flag officers join parties running on anti-NATO/anti-US and/or pro-Russia agenda or become otherwise instrumental in spreading anti-Western and pro-Russian sentiments.

With time, the number of people that have been associated with the former communist services and are still professionally active quickly declines. However, the respective groups found another way to preserve their influence. i.e. through the masonic lodges. From a number of documents leaked online it becomes clear that senior members of the military and the security sector, along with politicians, members of the judiciary, and businessmen have joined one or another masonic lodge, and this seems to be a fashionable trend. Personnel or associates to the communist security services are widely present in these lodges. The question here is whether a member of a lodge currently serving in the security sector would be loyal to his 'brothers' or to the oath taken to adhere to the Constitution and the laws of the Republic of Bulgaria and serve the country honestly, selflessly, objectively without any bias.

3.2 Critical Resources

Bulgaria's security sector seems less vulnerable to Russian influence through the resources it uses in its operations. One—yet critical exception in this regard—is in the maintenance of main combat capabilities of the Bulgarian armed forces and in particular airframes and guided missiles.

The key lacking resource is the know-how in extending the life cycle of airframes and certification of respective overhaul and life extension works. This problem is well known at least since the beginning of the century, when the ministry of defence announced an international tender for upgrading its fleet of MiG-29s. Although the tender attracted a significant international interest, at a late stage of the procedure the MOD introduced the requirement that any modifications of the aircraft need to be certified by the original designer or manufacturer. With this move the defence ministry eliminated all competitors to RSK MiG and solidified the dependence on Russia for any overhaul and life extension works, including the delivery of spare parts and components.

In practice Bulgaria, through the respective defence authorities, is only nominally party to NATO and EU air worthiness policies and mechanisms and relies entirely on the original designer and/or manufacturer of the equipment for assessing air worthiness. Almost exclusively, this is Soviet legacy equipment and this reliance translates directly into dependence on Russian companies, often not taking account of alternative sources of expertise available, in particular sources from Ukraine, other former Soviet Republics or Western companies. This applies to fighter, fighter-attack aircraft and helicopters, as well as to ground-to-air and airborne missiles.

The current defence minister deepened these dependencies by extending the plans to use Soviet made airframes practically till the end of their anticipated life cycle and for all platforms still in service [27]. In addition, by implementing this decision the MOD allocates a significant amount of money to the Russian military-industrial complex, which is under sanctions from major NATO and EU allies, and diverts considerable resources from the badly needed rearmament of the Bulgarian armed forces.

To a lesser extent, the armed forces are vulnerable to channels for supply of aviation fuel and other POL which are under the control of Russia. In an example from 2011, an attempt of the state to install measurement devices in "Lukoil Bulgaria" let to difficulties in the delivery of aviation fuel to the civilian airports and Bulgaria had to release 1800 tons of aviation kerosene from the state reserve to prevent difficulties in air traffic [28].

3.3 Infrastructure

Traditionally, Bulgaria has a strong defence industry. It is mostly export oriented, but in dire times would provide important support to the armed forces and other security services. Although till 1990 Bulgarian security and defence companies were part of the defence industrial base of the Warsaw Pact, over time they increased the cooperation with European and North American partners, introduced new technologies, diversified their products and often compete with Russian companies in third countries.

Since the 1990s, the Russian Federation is raising the issue of license-based defence production in Bulgaria, requesting financial remuneration and the right to sanction deliveries to international markets. These demands have never been recognised by Bulgaria, but the pressure in recent years is growing. According to Boyko Noev, former defence minister and Bulgarian Ambassador to NATO, there are indications that the current government might succumb to the pressure and include for the first time at least some of the Russian claims in the process of bilateral negotiations [29]. A potential renegotiation, under unforeseeable set of hybrid influence on the process, may have a detrimental impact on Bulgaria's defence industrial sector.

In April 2015, an attempt was made to poison the owner of a major Bulgarian defence company, Emil Gebrev, his son and the company's trade director. The investigation into the assassination attempt had been closed, but new information in beginning of 2019 linked the poison used to the Novichok class of weapon grade chemical weapons used against Sergei and Yulia Skripal, as well as the visit of 'Fedotov'— a suspected GRU officer, who had visited England at the time the Skripals were poisoned—to Bulgaria when Gebrev was poisoned [30]. The prosecution and the security services, however, seem hesitant in relaunching the investigation, not to mention linking the assassination attempt to the Salisbury attack and the involvement of GRU officers.

Other attempts to pressure Bulgaria's defence industry involve Russian propaganda channels around 'discoveries' (all proven fake at a later stage) of Bulgarian munitions in the Syrian city of Aleppo supposedly used to attack civilian population, the use of Bulgarian portable ground-to-air missile to shoot down a Russian SU-25 close air support aircraft, etc., as well as indications on involvement of competitors in the series of accidents in defence production facilities or test ranges in recent years [31].

Less attention has attracted the use by defence and security agencies of Russia-owned telecom infrastructure. Practically, there is no related public information. It is however known that the Russian VTB Capital is among the owners of Bulgaria's backbone telecommunications infrastructure.

More conventional Russian capabilities able to deny the use of Bulgarian land, air and port infrastructure go beyond the scope of the current elaboration.

3.4   Personnel

The fourth ring represents the military and civilian experts in the security sector, or its 'foot soldiers.' As any other citizen, they are subject of general propaganda and disinformation campaigns in traditional and electronic media, as well as in social networks.

The environment created by prevailing leadership attitudes certainly has an influence on—even if it may not predetermine—the selection and promotion of personnel in the security sector.

The unreformed system of military education, and particularly in advancing its academic personnel, is a main vehicle for perpetuating legacy dependencies and attitudes. Examples in that regard can be found in the special issue of the journal 'Bulgarian Science,' dedicated to national security [32], which featured department heads, faculty members and doctoral students from the "G.S. Rakovski" National Defence Academy who, with a single exception, are active military officers.

In one of the articles, examining possible triggers for a new world war, the author lists Ukraine as a place potentially causing a world war, pointing out that "until recently it was part of the Russian sphere of influence but, after the dissolution of the Soviet Union, only one hegemonic alliance exists which allowed for invasion of Western influence that led to an enormous political crisis and caused the [post-2014] civil war" [32, p. 13].

Another officer, contributing to the special issue, builds his examination of hybrid threats exclusively on Russian sources and, consequently, claims that the Russian Federation is under hybrid threats through sanctions, imposed by the US, European Union, Australia, Canada, Japan, etc., imposed in 2014 "after the unification of Crimea to Russia" [32, pp. 35–36].

A third contributor builds the examination on soft and colour revolutions exclusively on Russian sources and echoes respectively the Russian view that outside support is key to the success of the colour revolutions, underpinned by effective information operations and leading to a change in the foreign policy orientation of the country [32, pp. 107–108].

The same uncritical reliance on and promotion of the Russian concept of 'colour revolutions' [33] takes place even in the 'Military Journal'—the official theoretical journal of the Ministry of Defence of Bulgaria published since 1888.

A number of specialised publications on aviation, naval and general military issues also actively promote anti-Western and pro-Russian sentiments in their magazines, websites and discussion forums, and through social networks.

The final venue included in this paper is through para-military organisations, possibly sponsored by Russia, which seek cooperation with the active military and call for a coup, for exiting NATO, killing Bulgarian politicians, diplomats, academics, and journalists [34].

4 Conclusion

The purpose of the national security system is to reveal attempts and counter, with military means when that becomes necessary, any action undermining the national sovereignty and independence of Bulgaria. In itself, however, it is not immune from hybrid influence which most likely aims at its strategic paralysis. There are indications, based on publicly available information, that point at attempts to impair Bulgaria's security sector using a variety of hybrid tools and channels.

Finding a proper response to these attempts requires innovative forms of complementing and coordinating the efforts of public authorities, academia and private actors. Examples of such coordination are already available, e.g. in the experience of the Czech Republic [35].

This chapter had a more modest ambition. It suggests a framework that can be used to arrange and analyse cases of influence over the national security system and describes the respective venues and likely cases of hybrid influence. This framework, possibly enriched and detailed, can be used to collect and verify information on influence attempts, to analyse their actual or potential impact, and find effective remedy.

Acknowledgements This work was supported by the project "Digital Transformation and Security of Complex Adaptive Systems" from the working programme of the Institute of Information and Communication Technologies, Bulgarian Academy of Sciences.

References

1. Abrams, S.: Beyond propaganda: Soviet active measures in Putin's Russia. Connections Q. J. **15**(1), 5–31 (2016). https://doi.org/10.11610/Connections.15.1.01
2. Hoffman, F.: Conflict in the 21st Century: The Rise of Hybrid War. Potomac Institute for Policy Studies, Arlington, VA (2007)
3. Iancu, N., Fortuna, A., Barna, C., Teodor, M. (eds.): Countering Hybrid Threats: Lessons Learned from Ukraine. IOS Press, Amsterdam (2016)
4. Veljovski, G., Taneski, N., Dojchinovski, M.: The danger of 'hybrid warfare' from a sophisticated adversary: the Russian 'hybridity' in the Ukrainian conflict. Defense Secur. Anal. **33**(4), 292–307 (2017)
5. Gardasevic, I.: Russia and Montenegro: how and why a centuries' old relationship ruptured. Connections Q. J. **17**(1), 61–75 (2018). https://doi.org/10.11610/Connections.17.1.04
6. Bechev, D.: The 2016 coup attempt in Montenegro: is Russia's Balkans footprint expanding? Russia Foreign Policy Papers. Foreign Policy Research Institute, Philadelphia, PA (2018). https://www.fpri.org/article/2018/04/the-2016-coup-attempt-in-montenegro-is-russias-balkans-footprint-expanding
7. Chivvis, C.S.: Hybrid war: Russian contemporary political warfare. Bull. Atomic Sci. **73**(5), 316–321 (2017)
8. Conley, H.A., Mina, J., Stefanov, R., Vladimirov, M.: The Kremlin playbook: understanding Russian influence in Central and Eastern Europe. Center for Strategic and International Studies and Rowman & Littlefield, Washington, D.C. (2016), https://www.csis.org/analysis/kremlin-playbook.
9. Conley, H.A., Ruy, D., Stefanov, R., Vladimirov, M.: The Kremlin Playbook 2. Center for Strategic and International Studies, Washington, D.C. (2019)
10. DeDominicis, B.E.: Pan-slavism and soft power in post-cold war Southeast European international relations: competitive interference and smart power in the European theatre of the clash of civilizations. Int. J. Interdiscip. Civic Polit. Stud. **12**(3), 1–17 (2017)
11. Richey, M.: Contemporary Russian revisionism: understanding the Kremlin's hybrid warfare and the strategic and tactical deployment of disinformation. Asia Eur. J. **16**(1), 101–113 (2018)
12. Pocheptsov, G.: Cognitive attacks in Russian hybrid warfare. Inf. Secur. Int. J. **41**, 37–43 (2018)
13. Bagge, D.P.: Unmasking Maskirovka: Russia's Cyber Influence Operations. Defense Press, New York (2019)
14. Hadjitodorov, S., Sokolov,M.: Blending new-generation warfare and soft power: hybrid dimensions of Russia-Bulgaria relations. Connections Q. J. **17**(1), 5–20 (2018). https://doi.org/10.11610/Connections.17.1.01
15. Naydenov, M.: The subversion of the bulgarian defence system—the Russian way. Defense Secur. Anal. **34**(1), 93–112 (2018)
16. Kostov: Russia's Influence here is a Fact, News.bg, 17 October 2018, https://news.bg/bulgaria/kostov-ruskoto-vliyanie-u-nas-e-fakt.html
17. Von Clausewitz, C.: On War, edited and translated by Michael Howard and Peter Paret. Princeton University Press, Princeton (1976)

18. Strange, J.L., Iron, R.: Center of Gravity. What Clausewitz Really Meant. Joint Force Q. **35**, 20–27 (2004)
19. Warden, J.: The Air Campaign: Planning for Combat. Pergamon-Brassey's, Washington, D.C. (1989)
20. Nikolic, N.: Connecting conflict concepts: hybrid warfare and warden's rings. Inf. Secur. Int. J. **41**, 21–34 (2018). https://doi.org/10.11610/isij.4102
21. Bell¿ngcat—"the home of online investigations," https://www.bellingcat.com/
22. StopFake—"debunking fakes of Russian propaganda," https://www.stopfake.org/en/news/.
23. Smilov, D.: "Rumen Radev: Composition on a Painting by Reshetnikov. Deutsche Welle, 7 February 2019 (in Bulgarian).
24. Bulgaria in NATO and in European Defence 2020, Sofia, 2 September 2014.
25. Hristov, J.: In the edited version of Vision 2020 Russia is not pointed at as direct threat. OffNews, 2 September 2014 (in Bulgarian), https://offnews.bg/politika/v-redaktiranata-vizia-2020-rusia-ne-e-posochena-kato-priaka-zaplaha-383589.html (in Bulgarian)
26. "The Great Expulsion," Information Centre of the Ministry of Defence, 1 April 2018 (in Bulgarian). https://armymedia.bg/archives/116811
27. The Council of Ministers Approved the Projects for Overhaul of the MiG-29 and Su-25 Airplanes. News.bg, 28 November 2018, https://news.bg/politics/ms-odobri-proektite-za-remont-na-samoletite-mig-29-i-su-25.html (in Bulgarian)
28. "'Lukoil Aviation': The Kerosene from the Reserve is not Sufficient; Import is Difficult. Dnevnik, 1 August 2011, https://www.dnevnik.bg/bulgaria/2011/08/01/1132407_lukoil_avieishun_kerosinut_ot_rezerva_ne_e_dostatuchen/#comments-wrapper (in Bulgarian)
29. Boyko Noev: Russia's Interests to Control Bulgaria's Defence Industry are openly Stated, FrogNews, 13 February 2019. https://frognews.bg/novini/boiko-noev-interesite-rusiia-kontrolira-balgarskata-otbranitelna-industriia-otkprito-zaiaveni.html (in Bulgarian)
30. A GRU Agent has visited Bulgaria at the time Gebrev was poisoned. Capital, 8 February 2019, https://www.capital.bg/politika_i_ikonomika/bulgaria/2019/02/08/3387392_agent_na_gru_e_bil_v_bulgariia_po_vreme_na_otravianeto/ (in Bulgarian)
31. Boyko Noev: Russia's Interests to Control Bulgaria's Defence Industry are openly Stated
32. Bulgarian Science 117, Special National Security Issue (February 2019)
33. Military Journal 122, no. 2 (2015)
34. There is no Place in Bulgaria for Para-military Formations, Assassination Threats, and Russian Media Aggression. Declaration of the Atlantic Council of Bulgaria, 14 November 2017 (in Bulgarian)
35. Daniel, J., Eberle, J.: Hybrid warriors: transforming Czech security through the 'Russian Hybrid Warfare' assemblage. Sociologicky Casopis **54**(6), 907–931 (2018)

Printed in the United States
by Baker & Taylor Publisher Services